FORMAL METHODS IN THE METHODOLOGY OF EMPIRICAL SCIENCES

SYNTHESE LIBRARY

MONOGRAPHS ON EPISTEMOLOGY,

LOGIC, METHODOLOGY, PHILOSOPHY OF SCIENCE,

SOCIOLOGY OF SCIENCE AND OF KNOWLEDGE,

AND ON THE MATHEMATICAL METHODS OF

SOCIAL AND BEHAVIORAL SCIENCES

VOLUME 103

FORMAL METHODS IN THE METHODOLOGY OF EMPIRICAL SCIENCES

PROCEEDINGS OF THE CONFERENCE FOR FORMAL METHODS IN THE METHODOLOGY OF EMPIRICAL SCIENCES, WARSAW, JUNE 17–21, 1974

Edited by

MARIAN PRZEŁĘCKI, KLEMENS SZANIAWSKI
AND RYSZARD WÓJCICKI

(*Associated Editor:* GRZEGORZ MALINOWSKI)

D. REIDEL PUBLISHING COMPANY
DORDRECHT-HOLLAND / BOSTON-U.S.A.

OSSOLINEUM PUBLISHING COMPANY
WROCŁAW-POLAND

Library of Congress Cataloging in Publication Data

Conference for Formal Methods in the Methodology of
Empirical Sciences, Warsaw, 1974.
Formal methods in the methodology of empirical sciences.

(Synthese library; v. 103)
1. Science—Methodology—Congresses. 2. Science—Philosophy—Congresses. I. Przełęcki, Marian. II. Szaniawski, Klemens. III. Wójcicki, Ryszard. IV. Title.
Q174.C66 1974 501'8 76-4586
ISBN 90-277-0698-0

Published by D. Reidel Publishing Company,
P.O. Box 17, Dordrecht, Holland

Sold and distributed in the U.S.A., Canada, and Mexico
by D. Reidel Publishing Company, Inc.
Lincoln Building, 160 Old Derby Street, Hingham,
Mass. 02043, U.S.A.

Sponsored by the Polish Academy of Sciences
and the International Union of History
and Philosophy of Sciences

Printed in Poland

TABLE OF CONTENTS

A

B

(Papers presented by title)

PART A

RYSZARD WÓJCICKI

SOME PROBLEMS OF FORMAL METHODOLOGY OF SCIENCE

O

Most of empirical theories can be applied in different areas, to different phenomena, which, in general, cannot be combined into a single model structure of the sort examined in metamathematics. What we can at best achieve is to represent the different phenomena being possible referents of a theory by different model structures. This observation seems to underlie the well known Suppes (cf. [9], § 12, 2) account of empirical theories (shared by Simon, Sneed and others) according to which an empirical theory should be given in the form of a set-theoretic predicate that defines a class of model structures. As far as I am concerned, I shall stick to the traditional point of view and dealing with a theory I shall assume that it is defined by its components such as its language, theorems (laws), methods of proof, methods of gathering empirical data, and criteria of distinguishing the intended referents of the theory from those phenomena which do not belong to its range.

The variety of types of referents of empirical theories and the complexity of relationships which may occur between referents and the language of an empirical theory make the task of setting up the foundations of semantics of empirical theories extremaly difficult. In this situation two different strategies can be applied. One may try to deal separately with different kinds of empirical theories in hope that certain general ideas will emerge from partial results. But as well, one may proceed in the opposite direction; starting with some general assumptions, one may try to adjust them to peculiarities of different theories.

These two approaches may complement each other and thus both of them can be useful. Selecting the latter, however, we should be aware that our considerations cannot be conducted in fully precise terms. Any attempt to make the syntactical or semantical properties of the theories under discussion definite immediately restricts the area of considerations to a particular sort of theories. Perhaps then, the results of general considerations could be best pronounced in the form of general suggestions which are to determine further considerations. Everyone knows, however, how little heed is taken of all good adwice he can offer. For this reason, those philosophers who, like Montague in his 'Deterministic Theories' [7] or van Fraassen in his papers on semantics of Quantum Mechanics (cf. e.g. [3]), restricted their considerations to a well defined theory or a well defined set of theories chose more sound and more promising approach.

Contrary to what you might expect after these last remarks my considerations will have a general character. What can I say in self-defense? Well, I hope that I will be able to show that the general proposals I am going to discuss are related to some well known and well elaborated solutions to certain more specific problems. But first of all there is a practical reason I have had in mind. Our common aim to exchange points of view, to discuss different approaches and different ideas, searching, in this way, for inspiration for our further work. Being of general character, my talk will perhaps better serve this purpose than a more technical one.

I

Whenever I shall speak about a language $\mathscr{L}$, I shall assume that its semantic framework is defined. By the *semantic framework* of $\mathscr{L}$, I shall understand a singled out and fixed set of interpretations (model structures) of $\mathscr{L}$, call them *admitted*, paired with a function $Ver_{\mathscr{L}}$ which assigns to each admitted interpretation I a set $Ver_{\mathscr{L}}(I)$ of sentences of $\mathscr{L}$. A sentence h of $\mathscr{L}$ will be said to be *true* under an admitted interpretation I if and only if $h \in Ver_{\mathscr{L}}(I)$.

We shall need a few familiar semantic notions. A set of sentences A of $\mathscr{L}$ will be called *consistent*, or else *inconsistent*, when there is at least one admitted interpretation such that all sentences in A are true under this interpretation. A single sentence h of $\mathscr{L}$ will be said to be consistent (inconsistent) if the unit set h is consistent (inconsistent).

Whenever h is true under each admitted interpretation under which all sentences in A are true, we shall say that *A entails h*, $A \models h$. If a sentence h is true under each admitted interpretation I, h will be said to be *analytically true*.

II

Given a theory Θ, I shall denote by $\mathscr{L}_\Theta$ the language of Θ. Any part of physical (empirical) reality, say a phenomenon or a set of phenomena, which under a suitable interpretation of $\mathscr{L}_\Theta$ can be described in $\mathscr{L}_\Theta$, will be called a *possible* (*empirical*) *referent* of Θ. Clearly there is a sort of correspondence between admitted interpretations of $\mathscr{L}_\Theta$ and possible referents of Θ. By no means, however, these two notions coincide. Two languages which are based on entirely different conceptual framework and consequently have no interpretation in common can nevertheless be used to describe the same phenomenon, though considered under different aspects; consequently this phenomenon becomes their common referent.

In what follows I shall discern between a possible referent of Θ and a *possible applications* of Θ. The latter will be defined as a pair (Z, u) which consists of a possible referent Z of Θ and a set of methods u of gathering empirical evidence about Z. We assume then that whenever we use two different sets of methods, say u and v, of collecting data about Z, we deal with two different possible applications (Z, u) and (Z, v) of Θ.

III

In order to carry out further our discussion we need not inquire into the nature of methods which can be involved into possible applications of Θ. They may be reduced to simple observations but on the other hand they may consist of sophisticated techniques based on highly theoretical assumptions and involving non-trivial calculations. We must, however, define some formal properties of the sets of methods which will be considered.

Let (Z, u) be a possible application of Θ. We shall assume that:

(i) We know exactly which hypotheses on Z are decidable by means of the methods in u. Denote the set of all such hypotheses by $E(Z, u)$.

(ii) Given a finite sequence $e_1, \ldots, e_m$ of elements of $E(Z, u)$ we know whether they are codecidable by means of u, i.e. whether we can decide not only *each* hypothesis e_i but also *all* of them. (It may happen that methods in u are such that whenever we apply them to some of the hypotheses in the sequence we lose the possibility of applying them to the others, and vice versa).

(iii) The set $E^+(Z, u)$ is consistent.

Under certain week and obvious assumptions, the last condition implies that the methods in u are 'dispersion free', i.e. for each e in $E(Z, u)$, succesive applications of methods in u in order to decide whether e result in the same outcome; we obtain either constantly 'yes' or constantly 'no'. The set of those hypotheses in $E(Z, u)$ for which the outcome is "yes" will be denoted by $E^+(Z, u)$.

Observe that even if a set of methods is not dispersion free, all the same the data established by means of such methods can furnish us with valuable and coherent results, provided that they underwent an appropriate statistical elaboration. Thus, at least in some cases, we can transform dispersive methods into those being dispersion free by redefining them in a suitable manner. An alternative approach which in a way 'accepts' the phenomenon of dispersion of empirical data has been developed by R. Giles (cf. e.g. [4]).

Given an infinite subset E of $E(Z, u)$ we shall define it to be *codecidable* (*by means of the methods in* u) if and only if for each finite $E_f \subseteq E$, E_f is codecidable. Applying Zorn's lemma one can prove that each codecidable subset E of $E(Z, u)$ can be extended to a maximal codecidable subset E' of $E(Z, u)$.

A possible application (Z, u) of Θ will be said to be *complementary* if and only if $E(Z, u)$ is not codecidable.

IV

In general the sentences in $E(Z, u)$ need not belong to $\mathscr{L}_\Theta$. We shall assume that they belong to an extension $\mathscr{L}_\Theta(Z, u)$ of $\mathscr{L}_\Theta$. Some comments on this matter seem to be desirable.

Suppose (the case which has been discussed a number of times) that $E(Z, u)$ consists of 'observational' sentences while $\mathscr{L}_\Theta$ is 'theoretical'. Observe, that enlarging $\mathscr{L}_\Theta$ to $\mathscr{L}_\Theta(Z, u)$ we must incorporate the terms appearing in sentences in $E(Z, u)$ into $\mathscr{L}_\Theta(Z, u)$ in such a way that some meaning connections will be established between them and terms of $\mathscr{L}_\Theta$. Otherwise, no empirical evidence $E \subseteq E(Z, u)$ will be relevant to theoretic hypotheses stated in terms of $\mathscr{L}_\Theta$. Perhaps the best way in which the problem can be solved is to set up some meaning postulates (correspondence rules) which will relate denotations of 'observational' terms to those of 'theoretical' ones. Quite independently of the dubious doctrine of the observational-theoretical dychotomy, there is a number of empirical theories which are related to empirical data only through an appropriate set of correspondence rules.

Observe that meaning postulates restrict the class of interpretations of $\mathscr{L}_\Theta(Z, u)$. The admitted interpretations of $\mathscr{L}_\Theta(Z, u)$ should be only those under which the postulates are true. Denote by $\mathscr{L}'_\Theta$ the restriction of $\mathscr{L}_\Theta(Z, u)$ to the terms of $\mathscr{L}_\Theta$. The languages $\mathscr{L}_\Theta$ and $\mathscr{L}'_\Theta$ need not be the same. The set of admitted interpretations of $\mathscr{L}'_\Theta$ may happen to be narrower than that of $\mathscr{L}_\Theta$. In this sense the extension from $\mathscr{L}_\Theta$ to $\mathscr{L}_\Theta(Z, u)$ need not be trivial (for some details cf. Przełęcki and Wójcicki [8]).

To give an example of a situation in which the extension from $\mathscr{L}_\Theta$ to $\mathscr{L}_\Theta(Z, u)$ need not involve any correspondence rules, let us assume that Θ deals with quantities $F_1, \ldots, F_n$. Assume that the atomic formulas formed by means of F_i are of the form

(1) $$F_i(t, r_1, \ldots, r_k) = x,$$

(read: F_i measured at the time t on $r_1, \ldots, r_k$ takes the value x, x being a real number).

The case when each admitted interpretation of $\mathscr{L}_\Theta$ is uniquely defined by a sequence

(2) $$\mathbf{F}_1, \ldots, \mathbf{F}_n$$

of denotations of $F_1, \ldots, F_n$, respectively, will be of special interest for us. Thus the remaining symbols of $\mathscr{L}_\Theta$ possess fixed (standard) meanings. A sequence of the form (2) will be said to be a *partial interpretation* of $\mathscr{L}_\Theta$. For an illustration of this notion cf. Montague [7].

Consider a possible application (Z, u) of Θ. If each sentence in $E(Z, u)$ is of the form

(3) $F_i(t, r_1, \ldots, r_k) \in \Delta x,$

where Δx is a (perhaps infinite) real number interval, we shall say that (Z, u) is an *approximative application* of Θ. If all Δx involved in formulas in $E(Z, u)$ are pointwise, i.e. of the form (x, x), the application will be said to be *strict*. Under this definition a strict application is the 'limit case' of an approximative application.

Comparing (3) and (1) we immediately see that if (Z, u) is an approximative application of Θ, the extended language $\mathscr{L}_\Theta(Z, u)$ should result from $\mathscr{L}_\Theta$ by adding some (if any) individual constants which are of the same logical types as certain variables which appear in $\mathscr{L}_\Theta$. Thus the languages $\mathscr{L}_\Theta(Z, u)$ and $\mathscr{L}_\Theta$ do not differ essentially as to their semantic properties. In particular, we may assume that each admitted interpretation of $\mathscr{L}_\Theta$ can be extended to an admitted interpretation of $\mathscr{L}_\Theta(Z, u)$ and conversely, each admitted interpretation of $\mathscr{L}_\Theta(Z, u)$ can be reduced to an admitted interpretation of $\mathscr{L}_\Theta$.

V

In what follows Th_Θ will denote the set of theorems of the theory Θ. We shall assume that $h \in Th_\Theta$ whenever $Th_\Theta \models h$. In order to simplify our further discussion I shall not take into account the fact that, apart from their theorems, theories may involve different methods of proof which incidentally need not be sound i.e. need not provide us with true conclusions whenever they are applied to true (under a fixed interpretation) premisses.

Definition A. A theory Θ is *confirmed* (*finitely confirmed*) by a possible application (Z, u), if and only if there is no codecidable (codecidable and finite) subset $E \subseteq E^+(Z, u)$ such that $Th_\Theta \cup E$ is inconsistent.

Definition B. A theory Θ is *disconfirmed* (*finitely disconfirmed*) by a possible application (Z, u) if and only if Θ is not confirmed (finitely compatible) with (Z, u).

The following general remark is worth while here. If one wants to evaluate Θ on the base that Θ is confirmed by an application (Z, u), or that (Z, u) disconfirms Θ, the application cannot be completely arbitrary. It should satisfy the following two conditions: Z has to be an *intended referent* of Θ, i.e. it has to belong to those possible referents which Θ is expected to describe, and u has to consist of methods which on the ground of Θ are considered to be legitimate.

Denote by $\mathscr{K}_\Theta$ the *range*, i.e. the set of all intended applications of Θ, and by $\mathscr{W}_\Theta$ the set of its legitimate methods of gathering empirical data. If $Z \in \mathscr{K}_\Theta$ and $u \subseteq \mathscr{W}_\Theta$, the possible application (Z, u) will be said to be *proper*. Thus in evaluating empirical theories we have to take into account their proper applications.

VI

The notion of confirmability is a cognate of the notion of truth. This can be especially easily seen when Θ is a noncomplementary theory of the sort which has been described at the end of the Section 4, and (Z, u) is an approximative application of Θ such that it can be represented by a sequence

(4) $\mathbf{F}_1^u, \ldots, \mathbf{F}_n^u,$

of functions of the form

(5) $\mathbf{F}_i^u(t, r_1, \ldots, r_k) = \Delta x,$

defined for exactly those $t, r_1, \ldots, r_k$, for which at least one sentence of the form

(6) $F_i(t, r_1, \ldots, r_k) \in \Delta y$

is in $E^+(Z, u)$ and Δx is the intersection of all Δy for which the sentence (6) is in $E^+(Z, u)$.

Observe that the sequence (4) can be called an approximation of the partial interpretation (2), provided that for each $t, r_1, \ldots, r_k$ for which the corresponding functions are defined we have

(7) $\mathbf{F}_i(t, r_1, \ldots, r_k) \in \mathbf{F}_i^u(t, r_1, \ldots, r_k).$

Clearly the following holds true: Θ is confirmed by (Z, u) if and only if among all partial interpretations approximated by (4) there is at least one such that all theorems of Θ are true under the interpretation it defines. Exactly along these lines the notion of truth for empirical theories was examined by Toraldo di Francia and Dalla Chiara Scabia

[2] and by me [10] (cf. also: Wójcicki [11]). This is rather a purely verbal matter, but at least at present I am inclined to believe that the term 'confirmability' fits the relation defined in Definition A better than the term 'truth' (cf. [12]).

The fact that Θ is compatible with the data available in a particular application and thus it is confirmed by this particular application says us nothing about the explanatory power of Θ. Observe that if all theorems in Th_Θ were analytically true, Θ would be confirmed by any application whatsoever, being at the same time completely useless.

In fact Definition A can be criticized for it says nothing about how strongly (Z, u) confirms Θ. But, in actual science the degrees of confirmation are computed only in some special circumstances and in a special way. The comparative concept of confirmation seems to be fundamental.

VII

In connection with the issues delt with in the previous section let us discuss briefly the following matter. Thus far we have not considered explicitly the possibility that the theory under discussion is statistical. Suppose that the symbol of probability, say P, appears among the symbols of Θ. Does Θ fall under the theories we have discussed?

As known, there is no unique, standard interpretation of probability. Therefore the exact meaning which the symbol P may possess should depend on possible applications in a way similar to that in which the meaning of a quantity F_i depends on them. The central point is to define in an appropriate way the semantical framework of the language $\mathscr{L}_\Theta$ of a statistical theory Θ, and this can be done in the way proposed and discussed by Łoś [6]. If the semantic framework of $\mathscr{L}_\Theta$ is given, then each possible application (Z, u) of Θ distinguishes from among all admitted interpretations of Θ those which are compatible with (Z, u). It is clear, however, that (Z, u) cannot serve this purpose unless it encounters certain requirements. In general, Z cannot be a single phenomenon but rather a set of phenomena, and at least some sentences in $E(Z, u)$ should assign probabilities to certain evenets. All these are very sketchy remarks, but perhaps they give some general idea how the case of a statistical theory can be handled.

VIII

In what follows the expression 'absolute confirmability' will be sometimes used as synonym of 'confirmability'. This should facilitate discerning between this notion and its 'finite' counterpart.

What are the major differences between the concept of finite confirmability on one hand and the concept of absolute confirmability on the other. It is rather easily seen that though confirmability implies finite confirmability, in general, however, if Θ is finitely confirmed by (Z, u) it needs not be confirmed by (Z, u). A proof of the second part of this assertion can be established by constructing a suitable example. Moreover, quite often, an empirical theory cannot be finitely confirmed by any of its proper applications, though it can be absolutely confirmed by at least some of them. For instance, as it was convincingly argued by Bunge (cf. e.g. [1]), in most cases in order to confirm a physical theory Θ on the ground of a set of empirical data concerning a particular application, say (Z, u) of Θ, we have to start with setting up a 'model' of the referent Z, i.e. a set $A(Z)$ of theoretic assumptions about the nature of Z. In this way, however, this is the union $Th_\Theta \cup A(Z)$ which becomes the subject of finite confirmation, not Th_Θ alone.

Whenever we test a theory against experience, we do this by means of a finite set of data. For this reason the notion of finite confirmability seem to serve better the purpose of methodological analyses of the scientific practice than the 'absolute' notion corresponding to it. In fact, the actual state of affairs is still more complicated.

What we can really do is to test the theory Θ we deal with against this, that, or still another finite collection of data; in general, however, we would be able to establish in this way that Θ is finitely confirmed by an application (Z, u) only if we took into account infinitely many of such finite classes of empirical statements. In fact then, neither absolute nor finite concepts of confirmability correspond directly to any effective properties of empirical theories; if we are 'lucky' enough we can establish, by means of effective procedures, that Θ is finitely disconfirmed (and thus disconfirmed) by (Z, u). Their significance lies in that they help us to define certain ends we try to achieve while developping an empirical theory and consequently they help us to define and to examine certain tendencies which determine changes of empirical theories.

BIBLIOGRAPHY

[1] Bunge, M., *Philosophy of Physics*, D. Reidel Publ. Co., Dordrecht, 1973.

[2] Dalla Chiara Scabia, M. L. and Toraldo di Francia, F., 'A Logical Analysis of Physical Theories', *Rivista del Nuovo Cimento*, Serie 2, **3** (1973), 1–20.

[3] Van Fraassen, B., 'Semantic Analysis of Quantum Logic' in C. A. Hooker (ed.), *Contemporary Research in the Foundations of Quantum Theory*, D. Reidel Publ. Co., Dordrecht-Holland, 1973.

[4] Giles, R., A Pragmatic Approach to the Formalization of Empirical Theories, this volume.

[5] Hempel, C. G., *Philosophy of Natural Science*, Prentice-Hall, INC, Englewood Cliffs, N.J., 1963.

[6] Łoś, J., 'Semantic Representation of the Probability of Formulas in Formalized Theories', *Studia Logica*, **14** (1963), 183–194.

[7] Montague, R., 'Deterministic Theories', in *Decisions, Values and Groups* (ed. by C. Washburne), New York, 1957, 325–370.

[8] Przełęcki, M. and Wójcicki, R., 'The Problem of Analyticity', *Synthese* **19** (1969), 374–399.

[9] Suppes, P., *Introduction to Logic*, Van Nostrand, New York, 1957.

[10] Wójcicki, R., 'The Semantic Conception of Truth in the Methodology of Empirical Sciences', *Dialectic and Humanism*, **1** (1974), 103–116. (An English version of the paper which was originally published in Polish in *Studia Filozoficzne* **3** (1969)).

[11] Wójcicki, R., 'Set Theoretic Representations of Empirical Phenomena', *Journal of Philosophical Logic*, **3** (1974), 337–343.

[12] Wójcicki, R., 'The Factual Content of Empirical Theories', in *Rudolf Carnap, Logical Empiricist: Materials and Perspectives* (ed. by J. Hintikka), Dordrecht 1975, 55–82.

RISTO HILPINEN

APPROXIMATE TRUTH AND TRUTHLIKENESS*

I

Our standard logic is two-valued: every meaningful statement is regarded as being either true or false. Thus it may seem pointless or misleading to speak of *degrees* of truth or of *partial* truth. Nevertheless these expressions are commonplace in writings on epistomology and the philosophy of science, and it has been argued that an explication of the concept of partial truth is necessary for an adequate analysis and understanding of scientific method. For instance, in his book *The Myth of Simplicity* Mario Bunge says that "philosophers still owe scientists a clarification of the concept of *relative and partial truth* as employed in factual science".[1] Some philosophers have attempted to meet this challenge by defining quantitave measures of partial truth. Perusal of the literature on partial truth and degrees of truth indicates, however, that these terms are, as the saying goes, 'vague and ambiguous'. Different measures of partial truth correspond to different *explicanda*, and some attempts to explicate this notion seem to involve a confusion between several distinct concepts. In recent philosophical literature the expressions 'degree of truth' and 'partial truth' have been used e.g. in the following contexts (the following list does not pretend to be exhaustive):

(i) A doctrine of 'degrees of truth' was part of the traditional coherence theory of truth. This doctrine can be described, roughly, as the view that all statements are both partly true and partly false.[2]

(ii) According to Hans Reichenbach, two-valued logic is inapplicable to empirical scientific knowledge: "Scientific propositions are not used as two-valued entities but as entities having a weight within a continuous scale".[3] In Reichenbach's 'probability theory of knowledge' the continuous concept of weight replaces (or is a practical surrogate of) the classical concept of truth.[4]

(iii) It has been maintained that the application of two-valued logic to inferences containing inexact or 'fuzzy' concepts leads to paradoxes (for instance, the paradox called *sorites* or *falakros*). These paradoxes can be avoided in a logic with continuous truth-values or degrees of truth.[5]

(iv) It has often been pointed out that many accepted scientific laws are in fact in error, and known to be in error: 'the laws of nature' are only *approximately* true. Approximation or closeness to truth is obviously a matter of degree, and the expression 'degree of truth' can thus be used to refer to the degree of approximate truth.[6]

(v) C. S. Peirce and Karl R. Popper have described the progress of scientific knowledge as an 'approach towards the truth'. According to Popper, a theory T_1 which supersedes another theory T_2 can be said to be closer to the truth, and to possess a higher degree of truth, than T_2.[7]

The doctrine of degrees of truth associated with the coherence theory is notoriously obscure. In his recent book *The Coherence Theory of Truth* Nicholas Rescher has attempted to explicate this doctrine in terms of a quantitative measure of the degree of truth.[8] According to Rescher, the degree of truth of a proposition P depends on the derivability of P from various 'coherent' subsets (maximal consistent subsets) of a given inconsistent set of propositions. Rescher's measure can be regarded as an index of the *plausibility* of a proposition in a situation in which we have to adjust our corpus of beliefs so as to make it consistent with new 'recalcitrant' data. It is a measure of acceptability or plausibility, and it is misleading to call it a measure of (the degree of) *truth*.[9] Probabilistic theories of degrees of truth can be criticized on the same grounds; they are based on a confusion between truth and verification.[10]

Rescher is aware of the ambiguity of the expression 'degree of truth'. He notes that

> The concept of degrees of truth could be construed as referring to adequacy in the use or presentation of information. On this approach, the 'degree of truth' of a body of discourse would derive from its success in conveying information regarding its object of reference... Discourse with a 'high degree of truth' in this sense must, in the familiar legal jargon, not simply 'tell the truth and nothing but the truth', but it must also succeed in a good measure in telling 'the whole truth'. On this proposal, then, the *degree of truth* of a body of discourse is to be assessed by (1) the negative

but absolute requirement of avoidance of explicit falsehood, and (2) the positive but graduated and comparative extent to which this discourse conveys information regarding its object of reference.[11]

This concept of degree of truth is essentially the same as that used in (v) above. According to Popper, the idea of 'a degree of better (or worse) correspondence to truth' is a combination of the ideas of *truth* and of *content* (or information). Following Popper, I shall call this concept of degree of truth *truthlikeness*.[12] However, in view of (iv) it seems reasonable to liberalise Rescher's first criterion of truthlikeness, and replace it by the condition that a statement or a body of discourse can have a high degree of truthlikeness if it is *approximately* true or *close* to the truth in sense (iv). The idea of truthlikeness was also present in the coherence theorists' doctrine of degrees of truth: Insofar as a statement or a discourse represents its object of reference incompletely, it tells only part of the truth. The coherence theorists expressed this valid observation in a somewhat misleading form, viz. by saying that every statement is only 'partly true'.[13]

It is important to distinguish the concept of approximate truth (or closeness to truth) from the concept of truthlikeness: the truthlikeness of a proposition depends partly on its closeness to truth (or its degree of approximate truth in sense (iv)), partly on the amount of information conveyed by the statement regarding its subject-matter. The expression 'approximation to truth' is sometimes used in the sense of truthlikeness; for instance, when Popper speaks of a "better or worse approximation to truth", he seems to mean truthlikeness, not approximate truth in sense (iv).[14]

In *The Myth of Simplicity* Mario Bunge defines a measure of partial truth which is intended as an explication of what Bunge calls the "epistemological concept of truth". According to Bunge,

Scientists use two different concepts of truth: one for propositions regarded as conceptual reconstructions of certain traits of reality, another for propositions regarded as premisses or conclusions of reasoning. In the former case they view propositions as being *approximately true* or *approximately false*; in the latter case they treat those same propositions *as if* they were either altogether true or definitely false. Let us call *epistemological concept of truth* the one involved in the material adequacy of propositions and—for want of a better name—*logical concept of truth* the one underlying the use of ordinary, two-valued logic and mathematics based thereupon.[15]

According to Bunge, the epistemological concept of truth (or the concept of partial truth) can be represented by a function $V(-, -)$ whose range of values (the range of possible truth-values) is the (closed) interval $[-1, +1]$. The truth-value of a proposition P is conditional on a *system* or *theory* $\mathscr{S}$; thus truth in Bunge's 'epistemological' sense is not only *partial*, but also *relative*. I shall ignore this complication here and suppress the second argument of the V-function.

Bunge says that $V(P) = 1$ if and only if P is "completely true"; if $V(P)$ is close to 1, P is "approximately true"; if $V(P)$ is close to -1, P is "almost false", and if $V(P) = -1$, P is said to be "completely false".[16] However, it is easy to see that Bunge's V-function is not an adequate *explicatum* of the concept of approximate truth (as used in (iv)). The very first condition of adequacy for the V-measure accepted by Bunge fails to hold for the concept of approximate truth; this is the condition

(1) $V(\neg P) = -V(P)$.[17]

According to Bunge's interpretation, (1) implies that $\neg P$ has a very low degree of truth whenever P is approximately true. (If P is approximately true, $\neg P$ is "almost false".) However, if P is approximately, but not strictly true, it is false, and consequently $\neg P$ is true *simpliciter*. In this case the degree of (approximate) truth of $\neg P$ should presumably be maximal. According to Bunge's V-measure, various degrees of approximate truth are symmetrical with respect to truth and falsity, but this conception of approximate truth is fundamentally mistaken. This can be seen e.g. from the fact that the expressions 'approximate truth' and 'approximation to truth' are understandable and meaningful, whereas the expression 'approximately false' (which, incidentally, is used by Bunge in the paragraph cited above) is nonsensical. Condition (1) is plausible in the case of measures of acceptability, but not in the case of measures of approximate truth.[18] Bunge's theory of partial truth does not fare any better if it is interpreted as a theory of truthlikeness or as a theory of coherence. It includes theorems which do not hold for any philosophically interesting interpretation of 'degrees of truth', for instance, the theorem

$$\text{If } V(P) \neq -V(Q), \text{ then } V(P \wedge Q) = V(P \vee Q),$$

that is, "the conjunction of two propositions has the same truth-value as their disjunction unless their truth values are equal and opposite".[19] This theorem is extremely implausible, no matter how the expression 'degree of truth' is interpreted.[20] Bunge's distinction between two fundamentally different uses of the expressions 'true' and 'false' is also unwarranted; the analysis of the concepts of approximate truth and truthlikeness does not require any unorthodox conceptions of truth and falsity.

On the basis of the preceding discussion we can distinguish three reasonably clear *explicanda* to which the expression 'degree of truth' has been applied in recent philosophical literature:

(a) the degree of truth of a *vague* proposition (a proposition involving vague or inexact concepts),

(b) the degree of *approximate* truth, and

(c) the degree of *truthlikeness.*

In addition, the expression 'degree of truth' has sometimes been used to refer to the acceptability or plausibility of a proposition, but this usage is philosophically unsound for reasons mentioned earlier. In this paper I shall make some remarks on the logic of the notions (b) and (c). The following discussion is based on the assumption that talk about 'degrees of truth' in sense (b) or (c) does not involve any departure from classical two-valued logic. The question whether this is true of the concept (a) as well is more problematic; I shall touch on this question briefly in the end of section II of this paper. No attempt will be made to define quantitative measures of approximate truth and truthlikeness; this paper discusses only the qualitative and comparative forms of these notions. In the end of the paper I shall make some comments on Popper's analysis of truthlikeness.

II

A statement '*P*' is approximately true if and only if the statement '*P* is approximately true' is true. A simple logic of approximate truth can thus be obtained by adding to the standard propositional logic an intensional operator 'A' which represents the concept of approximate truth. The propositional logic of approximate truth will be termed here '*A*-logic'. The vocabulary of *A*-logic includes propositional letters $P_0, P_1, P_2, \ldots,$ the constants $\top$ (truth) and $\bot$ (falsity), the usual truth-functional connectives $\neg$, $\wedge$, $\vee$, $\rightarrow$, and $\leftrightarrow$, and an intensional (unary) connective A. Below I shall use the letters $P, Q, R, \ldots$ (with or without subscripts) as metalogical symbols which stand for arbitrary formulae of *A*-logic; the expressions of *A*-logic will be used autonymously. The semantics of *A*-logic is based on the standard semantics of intensional logic: an *interpretation function* of *A*-logic is a function $|\ |$ which assigns to every formula P of *A*-logic a subset $|P|$ of a given set of 'situations' or 'possible worlds' U. $|\ |$ is subject to the following standard conditions:

($\bot$)	$\lvert\bot\rvert = \emptyset$;
(T)	$\lvert\top\rvert = U$;
($\neg$)	$\lvert\neg P\rvert = U - \lvert P\rvert$;
($\wedge$)	$\lvert P\wedge Q\rvert = \lvert P\rvert \cap \lvert Q\rvert$;
($\vee$)	$\lvert P\vee Q\rvert = \lvert P\rvert \cup \lvert Q\rvert$; etc,

where '$\emptyset$' denotes the empty set. P is also termed the *truth-set* of P. If $u \in |P|$, we say that P is true at u; this will be abbreviated '$\models_u P$'.

'Approximation to truth' means *closeness* to truth. We shall exploit this intuitive idea freely in the semantics of *A*-logic. A sentence P is true at u if and only if u belongs to the truth-set of P, and we can say that P is approximately true (or almost true) at u if and only if u is *close* to the truth-set of P. Thus we shall assume that the models of *A*-logic contain a function N which assigns to every situation a nonempty subset of U; N_u is the set of those possible worlds close to (or similar to) u.

The members of U differ from each other in various respects; the similarity between possible worlds depends on the basis of comparison. If a sentence P is not strictly, but only approximately true at u, then P is true in some world which is in some respect(s) slightly different, but nevertheless close to u. In other words, if P is approximately true at u, P is true in some world $v \in U$ which is in no respect essentially dissimilar to u. Thus the concept of approximate truth is subject to the following truth-condition:

(2) *P is approximately true at u if and only if P is true in some world similar to (or close to) u.*

In our formal semantics the concept of similarity is conceptualised by the function N. A *model* of A-logic is a triple $\langle U, N, |\ |\rangle$, where U and $|\ |$ are as before and N is a function from U into $\mathscr{P}(U)$. (2) is equivalent to the following condition on $|\ |$:

(A) $\models_u \mathsf{A}P$ *if and only if* $|P| \cap N_u \neq \emptyset$,

in other words,

(A′) $\models_u \mathsf{A}P$ *if and only if* $\models_v P$ *for some* $v \in N_u$.

Every world is obviously similar to itself; thus all models satisfy the following condition:

(R) *For every* $u \in U, u \in N_u$.

(R) also guarantees that every N_u is nonempty. It is immediately obvious from (A') that A is logically similar to the $\Diamond$-operator of modal logic. Moreover, according to (R),

(3) $P \rightarrow \mathsf{A}P$

is true in all models; thus A resembles the alethic $\Diamond$-operator rather than the P-operator of deontic logic. Similarity relations are usually symmetric; in the present case the assumption of symmetry is tantamount to the assumption that all models satisfy the condition

(S) *If* $v \in N_u$, *then* $u \in N_v$.

By virtue of (S), all instances of the schema

(4) $P \rightarrow \neg \mathsf{A} \neg \mathsf{A}P$

are valid (true in all models). (4) is analogous to the 'Brouwerian Axiom' of modal logic. According to (A), (R) and (S), the logic of approximate truth is at least as strong as the Brouwerian system of modal logic. Similarity relations are not always transitive, and it is clear that in the present case the transitivity assumption (If u belongs to N_v and v belongs to N_w, then u belongs to N_w) is unacceptable. 'It is approximately true that P is approximately true' does not imply 'P is approximately true'.

According to the semantics sketched above, the logic of approximate truth turns out to be just another interpretation of modal logic (more specifically, of the Brouwerian system of modal logic). Has the counterpart of $\Box$ a natural interpretation in the logic of approximate truth?

Let us express the dual operator of A by 'C'; thus we define:

(5) $\mathsf{C}P =_{\mathrm{df.}} \neg \mathsf{A} \neg P.$

According to (5),

(C) $\models_u \mathsf{C}P$ *if and only if* $\models_v P$ *for every* $v \in N_u$.

CP is true at u if and only if P is true at every situation similar to u. The C-operator can perhaps be interpreted as a concept of *clear* or *distinct truth*; a sentence which is true in every situation similar to a given situation u is *clearly* true at u. If P is not approximately true at u (i.e., if P is false in every situation similar to u), we can perhaps say that P is *clearly false* at u. On this interpretation, the Brouwerian Axiom (4) says that if P is true *simpliciter*, it is clearly true that P is approximately true.

It may be argued that our modal-logical analysis of approximate truth ignores, or distorts, an important semantic feature of the concept of approximate truth, viz. its *vagueness*. The expressions 'it is approximately true that' and 'it is almost true that' increase the vagueness of the sentences to which they are applied: even if 'P' is a sentence with sharp truth-conditions, 'AP' is a vague sentence.[21] For this reason, it might be argued, the truth-sets of the sentences of A-logic do not always have sharp boundaries, and the truth-conditions cannot be formulated in terms of the classical values True and False.[22] According to this argument, the concept of approximate truth is bound up with degrees of truth also in sense (a).

This objection is based on the assumption that classical logic is inapplicable to vague concepts. In his paper 'Vagueness, Truth, and Logic'[23] Kit Fine has argued that this assumption is mistaken. According to Fine, we can say that a vague sentence is true if and only if it is true for all ways of making it completely precise. In the semantics of approximate truth sketched above, different functions N defined for a given set of possible worlds U correspond to different ways of making the concept of approximate truth completely precise. Consider a set of models S such that every model $\mathfrak{A} \in$ S has the same domain U, and the interpretation function of each model assigns the same subsets of U to various atomic sentences and their truth-functional combinations. (I assume here that the atomic sentences of our language are not vague).

The members of S differ from each other only with respect to the function N. Each such set S corresponds to what may be called a *vague* model. A vague sentence is true in a vague model S if and only if it is true in every 'sharp' model $\mathfrak{A} \in S$; consequently a sentence is true in all vague models if and only if it is true in all 'sharp' models. Thus the argument from vagueness does not refute the logic of approximate truth based on modal logic.

III

As was mentioned above, possible worlds differ from each other in various respects; the degree of similarity between possible worlds (or possible situations) depends on the basis of comparison. Comparisons between possible worlds can be made with respect to different *characteristics* C^i. Thus we can assign to every $u \in U$ a family of nonempty sets N^i_u, one for each characteristic C^i. The function N mentioned in (A) can be defined in terms of the functions N^i by

(6) $$N = \bigcap_i N^i.$$

According to (6), N_u contains all worlds which do not differ essentially from u in *any* respect C^i. Moreover, for each characteristic C^i, we can distinguish different degrees of similarity. Degrees of similarity can be conceptualized in semantics in various ways.[24] We could e.g. introduce a similarity metric on u, and represent the similarity between worlds by a numerical distance function d. However, if we are only interested in comparative judgments of similarity and distance, there is no need to make the assumption that the similarity between worlds can be measured numerically. In the sequel I shall represent degrees of similarily by David Lewis's method of *nested spheres*.[25] We assign to every characteristic C^i and situation $u \in U$ a *nested system of spheres* $\mathcal{N}^i_u$, that is, a family of subsets of U such that

(7) *For every* $K, L \in \mathcal{N}^i_u$, $K \subseteq L$ or $L \subseteq K$,

and

(8) u *belongs to every* $K \in \mathcal{N}^i_u$.[26]

Different sets $K \in \mathcal{N}^i_u$ correspond to different degrees of similarity to u with respect to C^i. It will be assumed that the systems of spheres under consideration are closed under nonempty intersections and unions:

(9) If $\mathscr{M}$ *is a nonempty subset of* $\mathscr{N}_u^i$, $\bigcap \mathscr{M} \in \mathscr{N}_u^i$ *and* $\bigcup \mathscr{M} \in \mathscr{N}_u^i$.[27]

If $\mathscr{N}_u^i$ is closed under nonempty intersections and unions, $\bigcap \mathscr{N}_u^i$ is the smallest and $\bigcup \mathscr{N}_u^i$ the largest sphere around u. In what follows I shall assume that we are speaking about similarity with respect to some unidentified characteristic C^i, and that the basis of comparison will always be the same. Thus the superscript 'i' used in (6)–(9) becomes idle and can be omitted.

Given a nested system of spheres $\mathscr{N}_u$, we can define a comparative concept of similarity (or distance) as follows: Let

(E) $E_u(P) = \{K | K \in \mathscr{N}_u \text{ and } K \cap |P| = \emptyset\}$.

$E_u(P)$ contains all sets $K \in \mathscr{N}_u$ whose intersection with $|P|$ is empty. The size of $E_u(P)$ is a measure of the distance of the proposition P from u: the smaller is the set $E_u(P)$, the closer P (or the truth-set of P) is to u. If we assume that u is our 'actual world' (or the 'true world'), and call it simply 'the truth', we can say that $E_u(P)$ measures the distance of P from the truth. In the following discussion I shall employ this idiom and usually omit the subscript 'u' from 'E_u' and '$\mathscr{N}_u$'. Now we can define a comparative concept of closeness to truth as follows:

(10) *P is at least as close to the truth as Q (or Q is not closer to the truth than P) if and only if* $E(P) \subseteq E(Q)$.

If $E(P)$ is a proper subset of $E(Q)$, P is closer to the truth than Q. A qualitative concept of closeness to truth (or similarity to truth) can be defined in terms of E as follows:

(11) $\models_u \mathsf{A}^*P$ *if and only if* $E_u(P) \subseteq \mathscr{K}$,

where $\mathscr{K}$ is a proper subset of $\mathscr{N}_u$ such that if $L \in \mathscr{K}$, every $K \in \mathscr{N}_u$ included in L belongs to $\mathscr{K}$. By choosing the set $\mathscr{K}$ in different ways we can vary the strength of the notion defined by (11). If $\mathscr{K} = \emptyset$, we get the following special case of (11):

(A″) $\models_u \mathsf{A}_0 P$ *if and only if* $E_u(P) = \emptyset$.

'$A_0 P$' may be read as 'P is almost true' (or 'P is approximately true'). The concept of closeness to truth defined by (A″) is, unlike that defined by (A), thought of as being relative to some characteristic C^i, otherwise the two notions are logically similar.[28]

Definition (E) yields e.g. the following theorems concerning $E(P)$:

(12) $E(\top) = \emptyset$;

(13) $E(\bot) = \mathcal{N}$;

(14) $E(P \vee Q) = E(P) \cap E(Q)$;

(15) $E(P) \subseteq E(P \wedge Q)$; and

(16) $E(P) = \emptyset$ *or* $E(\neg P) = \emptyset$.

Note that the distance of a conjunctive proposition from the truth is not determined by the distance of the conjuncts. If we attempt to represent degrees of approximate truth in terms of continuous truth-values, the degree of truth of a conjunction is not a function of the truth-values of the conjuncts. In 'fuzzy logic', conjunction is not always a truth-functional connective.[29]

Worlds which do not belong to any sphere $K \in \mathcal{N}$ do not bear any resemblance to the truth (or to the actual world). Such worlds may be termed *distant worlds*. If $E(P) = \mathcal{N}$, P is true only in distant worlds or in no worlds at all; in this case we can say that P is *completely false*. (Thus all contradictions are completely false). If $\neg P$ is completely false, P is true in all nondistant worlds (i.e., in all worlds that bear some resemblance to the actual world); in this case P may be termed a *trivial truth*.[30]

IV

The truthlikeness of a proposition depends on two factors: on its closeness to truth, and on the degree to which it conveys information about the truth. The former factor (the *truth-factor*) is expressed in terms of the set $E(P)$ assigned to P. The latter factor (which may be termed the *information-factor*) can be expressed in terms of a system of spheres in a somewhat similar way. It is usually assumed that a proposition is the more informative, the more possibilities (or possible worlds) it *excludes*. The view that the amount of information carried by a proposition is inversely related to its probability is based on this assumption.[31]

However, the number of possible worlds (or a measure of the set of possible worlds) excluded by P is not a good measure of the information conveyed by P about the truth, since different possible worlds are not on equal footing as regards to the truth: some possible worlds are closer to the truth than others. A proposition which excludes worlds bearing little or no resemblance to the truth is more informative than a proposition which excludes worlds close to the truth, other things being equal. The following analysis of information is based an this idea. Let

(17) $I_u(P) = \{K | K \in \mathcal{N}_u \text{ and } |P| \subseteq K\}$.

As before, I shall usually omit the subscript 'u' and speak of the information conveyed by P about the truth, not about information conveyed by P concerning u. $I(P)$ contains all sets $K \in \mathcal{N}$ such that $|P|$ is totally included in K. It is clear that the larger the set $I(P)$ is, the more P excludes worlds bearing little or no resemblance to the truth. Thus we can define a comparative concept of information as follows:

(18) *P is at least as informative about the truth as Q (or Q is not more informative about the truth than P) if and only if $I(Q) \subseteq I(P)$.*

According to (17), $I(\bot) = \mathcal{N}$, since the empty set is a subset of every set. Thus (17) and (18) imply that a contradiction is maximally informative about the truth. Many probabilistic measures of information yield the same result, but this result is highly counter-intuitive in the case of the notion *information concerning the truth.* It is more natural to say that inconsistent sentences tell us nothing about the truth. For this reason I shall accept, instead of (17), the following definition of '$I(P)$':

(I) $I(P) = \{K | K \in \mathcal{N} \text{ and } |P| \text{ is a nonempty subset of } K\}$.

(I) implies e.g. the following theorems on $I(P)$:

(19) $I(\top) = \emptyset$;

(20) $I(\bot) = \emptyset$;

(21) *If* $\Diamond P$ *and* $\Diamond Q$, $I(P \vee Q) = I(P) \cap I(Q)$;

(22) *If* $\Diamond (P \wedge Q)$, $I(P) \subseteq I(P \wedge Q)$; *and*

(23) $I(P) = \emptyset$ *or* $I(\neg P) = \emptyset$.

In (21) and (22), '$\Diamond$' is the (universal) possibility operator of alethic modal logic: $\Diamond P$ is true if and only if P is true at some $u \in U$. In terms of the I-function, we can define a number of methodologically interesting concepts:

(H) *P is an informative hypothesis* (*i.e., informative about the truth*) *if and only if I(P) is nonempty.*

If $I(P) = \emptyset$, P is an *uninformative* proposition. It is clear from the definitions of E and I that if $I(P)$ is nonempty, $\bigcup E(P)$ is a proper subset of $\bigcap I(P)$; consequently $E(P) = \mathcal{N}$ implies that $I(P) = \emptyset$. Thus every completely false hypothesis is uninformative (about the truth). According to (23), the negation of an informative hypothesis is always uninformative. (21) and (H) imply that a disjunction of two logically possible propositions is informative if and only if both disjuncts are informative propositions. The 'if'-part of this theorem may seem objectionable: it might be argued that a disjunction of several highly informative alternative hypotheses can be totally uninformative, even a logical truth. This objection rests on a confusion between two closely related concepts of information: If the disjunction of several alternative hypotheses is not informative, then the hypotheses cannot all be informative *about the truth*; they might be informative about different possible worlds $u \in U$, but not about the 'true world'. The objection is correct, however, if by an 'informative hypothesis' we mean a hypothesis which is informative in an *absolute sense*: this concept can be defined as follows:

(H′) *P is an informative hypothesis* (*in an absolute sense*) *if and only if P is informative about some world* $u \in U$, *i.e., if* $I_u(P) \neq \emptyset$ *for some* $u \in U$.

It is possible that a number of hypotheses P_i are each informative in the absolute sense, but their disjunction is not. Hypotheses which are maximally informative in sense (H′) are termed *sharp hypotheses*:

(24) *P is a sharp hypothesis if and only if* $I_u(P) = \mathcal{N}_u$ *for some* $u \in U$.

A hypothesis P has a higher degree of truthlikeness than Q if and only if P is closer to the truth than Q and more informative about the truth than Q. Thus we can define a comparative concept of truthlikeness (the truthlikeness of Q does not exceed that of P) as follows:

$$(\succcurlyeq) \qquad P \succcurlyeq Q \quad \textit{if and only if} \quad E(P) \subseteq E(Q) \textit{ and } I(Q) \subseteq I(P).$$

If $P \succcurlyeq Q$ but not $Q \succcurlyeq P$, the degree of truthlikeness of P exceeds that of Q:

$$(\succ) \qquad P \succ Q \quad \textit{if and only if} \quad P \succcurlyeq Q \text{ and not } Q \succcurlyeq P.$$

The concept of equality in truthlikeness is defined in an obvious way:

$$(\approx) \qquad P \approx Q \quad \textit{if and only if} \quad P \succcurlyeq Q \text{ and } Q \succcurlyeq P.$$ [32]

The relation $\succcurlyeq$ is reflexive and transitive, but not connected; in some cases two propositions are *incomparable* (with respect to truthlikeness):

$$(||) \qquad P||Q \quad \textit{if and only if} \quad \textit{neither} \quad P \succcurlyeq Q \text{ nor } Q \succcurlyeq P.$$

P and Q are incomparable if and only if

$$(25) \qquad E(P) \subset E(Q) \quad \textit{and} \quad I(P) \subset I(Q), \textit{ or}$$

$$(26) \qquad E(Q) \subset E(P) \quad \textit{and} \quad I(Q) \subset I(P).$$ [33]

In these cases the two factors of truthlikeness favor different hypotheses: in (25), P is closer to the truth than Q but Q is more informative than P, and in (26) Q is closer to the truth but less informative about the truth than P. Since situations of this type of are possible, any *numerical* measure of truthlikeness (e.g., based on a numerical distance function defined on U) must involve some mechanism by which the two main components of truthlikeness can be balanced against each other, for instance, a weighting parameter similar to those employed in some recent theories of acceptance.[34]

V

The expression 'truthlikeness' used here is borrowed from Karl R. Popper, and it may be of interest to compare the present analysis of truthlikeness with Popper's approach. Popper's informal remarks on the concept of truthlikeness agree with the characterization given above in section I. According to Popper, the concept of truthlikeness (or *verisimilitude*) represents the "idea of better (or worse) correspondence to truth or of greater (or less) likeness or similarity to truth", and is a combination of the ideas of truth and of content.[35]

In *Conjectures and Refutations* Popper defines a numerical measure of verisimilitude.[36] However, Popper's measure does not express the degree of verisimilitude of a proposition directly in terms of truth and content, but as a function of the "truth-content" and the "falsity-content" of the proposition: the degree of truthlikeness of P is defined as the difference between the truth-content and the falsity-content of P. The measure of the total content of P is defined by the familiar formula

$$(27) \qquad Ct(P) = 1 - p(P),$$

where $p(P)$ is the probability of P. Let $Ct_T(P)$ be a measure of the truth-content of P and $Ct_F(P)$ a measure of the falsity-content of P. The truth-content of P includes all true consequences of P, and the falsity-content of P includes all false consequences of P. If P is true, $Ct(P) = Ct_T(P)$; otherwise the truth-content of P comprises only a part of the total content of P. Let Tr be a deductive system which includes all true sentences of the language to which P belongs. For simplicity, I shall treat Tr here as if it were a single sentence. Now obviously

$$(28) \qquad P_T = P \vee Tr$$

is the strongest *true* consequence of P; thus

$$(29) \qquad Ct_T(P) = Ct(P_T) = Ct(P \vee Tr)$$

can be taken as the measure of the truth-content of P. If P is true, it is a logical consequence of Tr, and $P \vee Tr$ is equivalent to P. Thus (28) and (29) imply that if P is true, $Ct_T(P) = Ct(P)$, as was required above. $P \vee Tr$ expresses the *common content* of P and Tr, and (29) is a measure of the amount of information that P conveys concerning the subject-matter of Tr—that is, the amount of information that P conveys *concerning the truth*.[37] Thus Popper's concept of truth-content corresponds roughly to what has been termed above the *information-factor* of truthlikeness. An informative false hypothesis may have more truth-content than an uninformative truth: the truth-content of a tautology is always 0, whereas the (measure of the) truth-content of an informative falsehood can be close to 1.

According to Popper, the falsity-content of P can be measured by

$$(30) \qquad Ct_F(P) = Ct(P \mid P_T),$$

where $Ct(P \mid P_T)$ is a *conditional* content-measure, defined in terms of the conditional probability of P:

(31) $$Ct(P|P_T) = 1-p(P|P_T).$$

According to (28), (30) is equal to

(32) $$Ct(P|P_T) = 1-p(P|P \vee Tr)$$

This measure of falsity-content has the following properties:

(33) (i) *If* P *is true,* $Ct_F(P) = 0$, *and*
(ii) *If* P *is false,* $0 < Ct_F \leqslant 1$.

We can say that the measure of the falsity-content of a proposition expresses how much falsity there is in the proposition. The degree of truthlikeness of a proposition should presumably be directly related to its truth-content and inversely related to its falsity-content. Popper takes the degree of truthlikeness of P to be proportional to

(34) $$\begin{aligned} Vs(P) &\sim Ct_T(P) - Ct_F(P) \\ &= p(P|P_T) - p(P_T), \end{aligned}$$

that is,

(35) $$Vs(P) \sim p(P|P \vee Tr) - p(P \vee Tr).$$

By using the normalizing factor $1/(p(P|P_T)+p(P_T))$, Popper arrives at the following measure of truthlikeness:

(36) $$Vs(P) = \frac{p(P|P_T)-p(P_T)}{p(P|P_T)+p(P_T)}.$$

According to (33), the falsity-content of every false hypothesis is greater than the falsity content of any true hypothesis. Thus Popper's concept of falsity-content corresponds partly to the *truth-factor* of truthlikeness. But the falsity-content of a proposition does not depend only on its truth-value, but also on its content. According to (27), (29) and (32), the following theorem holds for Ct, Ct_T and Ct_F:

(37) *If* P *and* Q *are both false,* $Ct_T(P) \gtreqless Ct_T(Q)$ *according as* $Ct(P) \gtreqless Ct(Q)$, *and* $Ct_F(P) \gtreqless Ct_F(Q)$ *according as* $Ct(P) \gtreqless Ct(Q)$.

Thus, in the case of false hypotheses, both the truth-content and the falsity-content of P are always monotonically related to total content of P. If P is false, the right-hand side of (35) is, for a constant $p(Tr)$, a (nonlinear) function of $p(P)$, and consequently of the total content of P.[38] Moreover, if $p(Tr)$ is small in comparison with $p(P)$ (as usually is the case), the right-hand side of is approximately equal to $1-p(P) = Ct(P)$, and the normalized measure of truthlikeness is monotonically related to $Ct(P)$.

In his paper 'A Theorem of Truth-Content' Popper proves a somewhat similar result.[39] Let $Cn(P)$ be the content of P (that is, the consequence class of P, not the content measure Ct), and let $Cn_T(P)$ be the truth-content of P (the set of true consequences of P). Popper shows that

> $Cn_T(P) = Cn_T(Q)$ *if and only if* $Cn(P) = Cn(Q)$, *and* $Cn_T(P)$ *is a proper subsystem of* $Cn_T(Q)$ *if and only if* $Cn(P)$ *is a proper subsystem of* $Cn(Q)$.[40]

From this result Popper draws the conclusion that "any *measure of content* can be used, for many purposes of comparison, as a *measure of the truth-content*; and vice versa". In these cases the formula for verisimilitude can be simplified by replacing '$Ct_T(P)$' in (34) by '$Ct(P)$'; thus we get

$$(38) \qquad Vs^*(P) = \frac{1-p(P_T)}{1+p(P_T)}.$$

Popper notes that in most cases (38) can be simplified further to

$$\frac{1-p(P)}{1+p(P)} = \frac{Ct(P)}{2-Ct(P)}.$$

According to Popper, "this indicates that, for most cases of competing (and consistent) theories, the comparison of their contents will provide a *rough first comparison of their verisimilitude*".[41]

Thus, if we compare a number of false hypotheses between each other (or a number of true hypotheses), the verisimilitude of a hypothesis depends only on its total content. In many interesting cases this result is clearly unacceptable. In *Conjectures and Refutations* Popper says:

> Ultimately, the idea of verisimilitude is most important in cases where we know that we have to work with theories which are *at best* approximations—that is, theories of which we actually know that they cannot be true. (This is often the case in social

sciences). In this case we can still speak of better or worse approximations to the truth (and we therefore do not need to interpret these cases in an instrumentalist sense).[42]

In this paragraph the expression 'approximate truth' is used in sense (b), not in the sense of truthlikeness. According to (38), comparisons of verisimilitude turn on comparisons of total (or 'absolute') content if the hypotheses under consideration are all false, as in the cases described above. However, it should be obvious that hypotheses with the same *Ct*-measure need not be equally close approximations to the truth. According to Popper, the content of a hypothesis is measured by its *a priori improbability*, and we certainly cannot identify the degree of approximate truth of a hypothesis with its *a priori* improbability.

Popper seems to be aware that (35) is not always an adequate measure of truthlikeness. In *Conjectures and Refutations* he illustrates the use of the *Vs*-measure by simple numerical examples of the following kind. Consider a single throw of a die, and let P^i be the proposition that i ($i = 1, 2, \ldots, 6$) will turn up. Let us assume that 4 will in fact turn up; thus $P^4 = Tr$. According to (28) we have, e.g.,

(39) (i) $P^5_T = P^4 \vee P^5$, *and*

(ii) $P^6_T = P^4 \vee P^6$.

If we assume that the *a priori* probability distribution over the outcomes P^i is even, P^5 and P^6 have, according to the definitions given above, equal truth-content, equal falsity-content, and also the same degree of truthlikeness.[43] Now Popper distinguishes two types of examples: (i) "Examples of the type of ordinary dicing. Here if, say, 4 turns up, while our guess was that 5 would turn up, we consider this as no better or worse a guess than, say, the guess that 6 will turn up. (Better or worse are here being used in the sense of nearer to, or further from, the truth.)" (ii) "Examples in which we have a kind of measure of the distance of our guesses from the truth". Popper says that "we can represent this by the assumption that, if in fact 4 turns up, the guess or the proposition that 6 will turn up (or that 2 will turn up) is separated from the truth by the proposition that 5 will turn up (or that 3 will turn up)."[44] Situations in which some false hypotheses can be regarded as better approximations to the truth than others belong clearly to the second type, and in these cases the assumption ihat the versimilitude of a hypothesis depends only on its (total) content ts unacceptable. Popper applies the *Vs*-measure (34) to both kinds of examples, but in case (ii) he revises the definition of P_T. According to Popper, P_T and $Ct_F(P)$ can be defined according to (28) and (32) only in examples of the first kind. If the example described above is regarded as an example of the second kind, and we assume that the distance between P^i and P^j depends on the difference between i and j, we should take as P_T^6 the proposition[45]

$$(40) \qquad P_T^{6*} = P^4 \vee P^5 \vee P^6,$$

not (39.ii). Popper introduces (40) *ad hoc*, without justifying it in terms of a general definition of P_T^{i*}, but the general principle underlying the choice of (40) is fairly clear: P_T^{i*} is defined as the smallest proposition which includes both P^i and Tr, and all outcomes between them, that is, P_T^{i*} is the smallest 'connected' proposition including both P^i and Tr. If P_T^i is chosen in this way, the result (37) concerning Ct, Ct_T and Ct_F does not hold any more, and in the present example P^5 will have a higher degree of verisimilitude than P^6. This analysis improves the analogy between Popper's concept of truth-content and the information-factor of truthlikeness defined in section IV: P_T^{i*} represents the (maximal) distance between P^i and the truth. However, P_T^{i*} is only a rough *ad hoc* substitute for a genuine distance-measure on the space of outcomes; it is clear than the distance of a proposition from the truth cannot always be represented in this way. (In Popper's example this happens to be possible since the possible outcomes can be arranged in a 'natural' linear order). Thus Popper's theory of verisimilitude is not generally applicable to situations of the second type. It is debatable whether it provides reasonable quantitative judgements of truthlikeness even in situations of the first kind. The definition of $Vs(P)$ in terms of the arithmetical difference between $Ct_T(P)$ and $Ct_F(P)$ (or in terms of $Ct(P)$ and $Ct_F(P)$) is an arbitrary decision for which Popper offers no justification. As was suggested earlier, a flexible and realistic measure of truthlikeness should involve some mechanism by which the truth-factor and the information-factor can be weighted and balanced against each other.

According to Popper's formula for *Vs* (see (38) above), the verisimilitude of the conjunction of two false hypotheses will always be higher than the verisimilitude of each conjunct (unless one of the hypotheses is redundant). In other words, the addition of any false conjunct to a (false) hypothesis will always increase the truthlikeness of the hypothesis, which seems absurd. It is hard to see how this result can be avoided in situations of the first kind. This result suggests that the concept of verisimilitude should be defined in terms of the truth-value and the content of a proposition, not in terms of different content-components, since these components tend to be correlated with each other. On the other hand, it also indicates that the concept of verisimilitude (as distinguished from the concept of content) is of interest mainly in situations of the second kind, and thus supports the distance analysis of truthlikeness presented in section IV.[46]

In his informal discussion of verisimilitude Popper explains this notion in terms of the idea of *similarity* to truth.[47] However, he was also anxious to prove that verisimilitude is not the same as probability, and that the verisimilitude of a hypothesis is often inversely related to its (a priori) probability. Somewhat paradoxically, this concern led him to propound a probabilistic theory of verisimilitude which did not do justice to the idea of verisimilitude as truth-*likeness*.

REFERENCES

* The author is indebted to the Academy of Finland and to the University of Queensland for support of research.

[1] Prentice-Hall, Englewood Cliffs, N.J. 1963, p. 116.

[2] See Nicholas Rescher, *The Coherence Theory of Truth*, Clarednon Press: Oxford University Press, London 1973, pp. 198–199, and A. C. Ewing, *Idealism: A Critical Survey*, Methuen Co., London 1934, p. 308.

[3] *Experience and Prediction*, The University of Chicago Press, Chicago 1938, p. 393.

[4] In *The Theory of Probability* (University of California Press, Berkeley 1949, pp. 389–398), Reichenbach also discusses another concept of degree of truth which is similar to the concept of approximate truth mentioned in (iv) below. Reicbenbach calls the logic of this notion "a quantitative logic of individual verifiability" (as opposed to probability logic which is termed the logic of "nonindividual verifiability").

[5] See J. A. Goguen, 'The Logic of Inexact Concepts', *Synthese* **19** (1968–69), pp. 325–373; Lofti Zadeh, 'Quantitative Fuzzy Semantics', *Information Sciences* 3 (1971), pp. 159–176; George Lakoff, 'Hedges: A Study in Meaning Criteria and the Logic of Fuzzy Concepts', *The Journal of Philosophical Logic* **2** (1973), pp. 458–508. The following simple argument is an instance of the *sorites* (heap) paradox: If you add

one stone to a small heap, it remains small. A heap containing one stone is small. Therefore (by induction) every heap is small (Goguen, *op. cit.*, p. 328).

[6] Michael Scriven, 'The Key Property of Physical Laws: Inaccuracy', in *Current Issues in the Philosophy of Science*, ed. by Herbert Feigl and Grover Maxwell, Holt Rinehart and Winston, New York 1961, pp. 91–101; 'Explanations, Predictions, and Laws', in *Minnesota Studies in the Philosophy of Science III: Scientific Explanation, Space and Time*, ed. by Herbert Feigl and Grover Maxwell, University of Minnesota Press, Minneapolis, 1962, pp. 170–230.

[7] See C. S. Peirce, *Collected Works, Vol. V*, ed. by Charles Hartshorne and Paul Weiss, Harvard University Press, Cambridge 1934, paragraph 5.565, and Karl R. Popper, *Conjectures and Refutations*, Routhledge and Kegan Paul, London 1963, pp. 231–232.

[8] Clarendon Press: Oxford University Press, London 1973, pp. 197–200 and 356–360.

[9] See Rescher, *op. cit.*, pp. 356–359. By 'Rescher's measure' I mean here the Δ-measure defined on p. 359. Rescher also defines a probabilistic measure of 'degree of truth' (p. 360), even though he notes elsewhere (p. 197) that "in speaking of 'degrees of truth', no reference whatever to *degrees of probability* is intended".

[10] Cf. Carnap's classic paper 'Truth and Confirmation' in *Readings in Philosophical Analysis*, ed. by Herbert Feigl and Wilfrid Sellars, Appleton-Century-Crofts, New York 1949, pp. 119–127 (especially p. 122).

[11] The *Coherence Theory of Truth*, pp. 197–198.

[12] See *Conjectures and Refutations*, p. 219.

[13] Cf. A. C. Ewing, *op. cit.*, p. 211.

[14] Cf. *Conjectures and Refutations*, p. 232. However, when Popper says that "the idea of verisimilitude is most important in cases where we know that we have to work with theories which are *at best* approximations" (*Ibid.*, p. 235), he is using the expression "approximation to truth" in sense (iv). Popper does not make a clear distinction between the concept of approximate truth and the concept of truthlikeness; his confusion on this point will be discussed further in section V.

[15] *The Myth of Simplicity*, pp. 116–117. Cf. also Mario Bunge, *Scientific Research II: The Search for Truth*, Springer Verlag, Berlin-Heidelberg-New York 1967, pp. 300–301.

[16] *Ibid.*, pp. 117-119.

[17] *Ibid.*, p. 118.

[18] Rescher's Δ-measure of 'degree of truth' satisfies a condition similar to (1); see *The Coherence Theory of Truth*, p. 359.

[19] *The Myth of Simplicity*, p. 121.

[20] Bunge is aware of the shortcomings of his theory of partial truth, but he says that "a quantitative theory of truth which attempts to reconstruct some basic traits of the hazy notion of partial truth can at best be partially true itself" (*op. cit.* p. 134).

[21] George Lakoff has termed expressions of this type "hedges": hedges are words "whose job is to make things fuzzier or less fuzzy" ('Hedges: A Study in Meaning Criteria and the Logic of Fuzzy Concepts', p. 471).

[22] See the papers mentioned in note 5.

[23] *Synthese* **30** (1975), pp. 265–300.

[24] See David Lewis, 'Completeness and Decidability of Three Logics of Counterfactual Conditionals', *Theoria* **37** (1971), pp. 71–85, and *Counterfactuals*, Basil Blackwell, Oxford 1973, pp. 13–19 and 44–64.
[25] See *Counterfactuals*, pp. 13–19.
[26] This condition is termed by Lewis the condition of *weak centering*; see *Counterfactuals*, pp. 29.
[27] Condition (9) is automatically satisfied if $\mathcal{N}_u$ is a finite set but not if $\mathcal{N}_u$ is infinite; see *Counterfactuals*, pp. 14–15. In *Counterfactuals* the condition of closure under unions is not restricted to nonempty subsets of $\mathcal{N}_u$; consequently in Lewis's system the empty set belong to every $\mathcal{N}_u$ (pp. 14–15). In the present context it seems convenient to restrict (9) to nonempty unions. If (9) is not so resticted, some of the definitions given below must ne modified.
[28] If $\mathsf{A}_0 P$ holds, P cannot be distinguished from the truth in terms of the system $\mathcal{N}$. The operator A_0 and its dual operator C_0 correspond to David Lewis's "inner modalities" ; cf. *Counterfactuals*, p. 30. Lewis says that a system of spheres $\mathcal{N}_u$ is *centered* on u if $\{u\} \in \mathcal{N}_u$ (p. 14). The assumption of centering implies that $\mathsf{A}_0 P$ and $\mathsf{C}_0 P$ are logically equivalent to P; here we accept only the weaker assumption of *weak* centering (8) (cf. note 26).
[29] According to George Lakoff (*op. cit.*, p. 464), the truth-values of complex sentences can be defined in fuzzy logic as follows:

(i) $V(\neg P) = 1 - V(P)$

(ii) $V(P \wedge Q) = min(V(P), V(Q))$

(iii) $V(P \vee Q) = max(V(P), V(Q))$.

Neither (i) nor (ii) holds for degrees of approximate truth. In systems of fuzzy logic the extreme values 0 and 1 are sometimes termed 'truth' and 'falsity' (see Goguen, *op. cit.*, p. 333). This suggests that the values $0 < V < 1$ are 'intermediate' between truth and falsity. If we assume that the V-values correspond to degrees of approximate truth, this interpretation is incorrect, and involves an unwarranted departure from classical logic. If P is true (in the classical sense), its degree of approximate truth is maximal ($V(P) = 1$), whereas a classically false sentence P can receive any value $0 \leqslant V(P) < 1$, depending on its 'distance' from the truth. In an important sense truth does not admit of degrees; it corresponds to perfection, but false sentences can fall short of perfection in various degrees. Thus we get instead of (i):

(i′) $V(P) = 1$ *if and only if* $V(\neg P) < 1$.

Contradictions are 'completely false', and should have zero degree of approximate truth. It is possible that two hypotheses have both a very high degree of approximate truth (are both very close to the truth), but their conjunction is nevertheless inconsistent; consequently (ii) does not hold for degrees of approximate truth. However, a conjunction is never closer to the truth than both conjuncts; thus we can replace (ii) by

(ii′) $V(P \wedge Q) \leqslant min(V(P), V(Q))$.

[30] If '*P* is a trivial truth' is expressed briefly '$\mathsf{G}P$',' $\neg\mathsf{G}\neg P$' can be read '*P* bears some resemblance to the truth'. G and its dual operator correspond to David Lewis's outer modalities' $\Diamond$ and $\Box$; cf. *Counterfactuals*, p. 22. A system of spheres is termed *universal* if $\bigcup \mathcal{N} = U$ (*Ibid.*, p. 16). In a universal system of spheres outer necessity and possibility coincide with universal necessity and possibility. Here I assume that the systems under consideration are not universal, and that U contains some distant worlds (for each $u \in U$).

[31] See Yehoshua Bar-Hillel, 'Semantic Information and Its Measures', in *Language and Information* (by Y. Bar-Hillel), Addison-Wesley, Reading, Mass., 1964, pp. 298–312. (See especially p. 299).

[32] ($\succcurlyeq$) and ($\succ$) imply e.g. the following theorems on the concept of truthlikeness:

(i) $\top \succ \bot$;

(ii) $\top \succcurlyeq P$ *if and only if P is uninformative about the truth:*

(iii) $\bot \succcurlyeq P$ *(and* $\bot \approx P$*) if and only if P* is *completely false;*

(iv) $P \succcurlyeq \top$ if and only if $\mathsf{A}_0 P$;

(v) $P \succcurlyeq \bot$;

(vi) If $P \succcurlyeq Q$, $P \succcurlyeq P \vee Q$; *and*

(vii) If $P \wedge Q \succcurlyeq P \wedge \neg Q$, *then* $P \wedge Q \succcurlyeq P$.

[33] '$\subset$' is here used as a sign of proper set inclusion.

[34] See Isaac Levi, *Gambling with Truth*, Alfred, A. Knopf, New York 1967, pp. 86–90.

[35] See *Conjectures and Refutations*, pp. 229–233.

[36] *Ibid.*, pp. 233–234 and 391–397.

[37] Cf. Jaakko Hintikka, 'Varieties of Information and Scientific Explanation', in *Logic, Methodology and Philosophy of Science III*, ed. by B. van Rootselaar and J. F. Staal, North-Holland, Amsterdam 1968, pp. 311–331. (See especially pp. 315–316).

[38] If *P* is false, (35) is equal to

$$\frac{p(P)}{p(P)+p(Tr)} - p(P) - p(Tr).$$

If *P* is true, (35) is equal to $1-p(P)$; in this case the dependence of *Vs* on *Ct* is even more direct.

[39] In *Mind, Matter and Method: Essays in Philosophy and Science in Honor of Herbert Feigl*, ed. by Paul K. Feyerabend and Grover Maxwell, University of Minnesota Press, Minneapolis 1966, pp. 343–353.

[40] Ibid., pp. 350–352. The proof is based on the assumption that the set of true sentences is nonaxiomatizable.

[41] *Ibid.*, pp. 352–353.

[42] *Conjectures and Refutations*, p. 235.

[43] P^5 and P^6 do not, strictly speaking, have the same content, but their content-measures are equal.

[44] *Conjectures and Refutations*, p. 397.

[45] *Ibid.*, p. 398.

[46] Note that this result does not hold for the comparative concept of truthlikeness defined in section IV. According to ($\succcurlyeq$), the degree of truthlikeness of $P \wedge Q$ may be higher or lower than, or equal to, the truthlikeness of P, or $P \wedge Q$ and P may be incomparable. (Cf. note 32, theorems (vi) and (vii).)

[47] See especially Popper's discussion of the history of the concept of verisimilitude in *Conjectures and Refutations*, p. 399–401.

MARIA LUISA DALLA CHIARA

A MULTIPLE SENTENTIAL LOGIC FOR EMPIRICAL THEORIES

INTRODUCTION

The aim of this paper is to develop some logical and metatheoretical results, which arise from the definition of *physical structure* given in [1] (and studied also in [2]).

A physical structure is conceived as a generalization of the standard model-theoretical notion of *structure* and gives rise to an abstract concept of *physical theory* which seems to be a somewhat adequate formal description of 'concrete' physical theories. The concept of *truth* for such theories turns out to determine a multiple system of sentential logics which correspond to different intuitive ideas of logical operations (all coincident in the case of classical Model Theory).

One deals with a general 'logical situation' where it is possible to meet at the same time some requirements that appear to be all 'highly desired' in the case of physical theories:

1) The Semantics is bivalent (thus any physical law is either true or false).

2) The Mathematics is classical and all laws of classical logic are valid.

3) Nevertheless the sentential connectives turn out to be *non truth-functional.*

4) By extending the logical language in order to have at our disposal also truth-functional connectives, we obtain a somewhat 'realistic' logical situation which admits of certain kinds of contradictions and violations of *tertium non datur*. Further this setting gives rise to a form of *implication* which seems to be very close to the intuitive meaning of 'physical implication'.

A modal interpretation of this 'multiple empirical logic' (*E.L.*) permits better clarification of the meanings of the logical operations involved and the connections between E.L. and other logics that have also been called 'empirical' (for instance quantum logic). It turns out that standard quantum logics, in comparison with *E.L*, involve a very strong mathematical idealization, for they are totally determined in the mathematical part of our physical structures. However it is possible to introduce a mixed 'quantum-empirical logic' which seems to describe formally in a more adequate way, some logical procedures of 'concrete reasoning' in quantum physics.

In this paper we shall refer mainly to physical theories. However, a number of considerations can be clearly extended to the case of a generic 'theoretical field' which involves certain abstraction-operations to be performed on a set of empirical data.

I. PHYSICAL STRUCTURES AND PHYSICAL THEORIES

By *physical structure* we mean a structure $\mathbf{M} = \langle \mathbf{M}_0, S, Q_1, \ldots, Q_k, \varrho \rangle$ where:

1) $\mathbf{M}_0$ is the standard model of a mathematical theory (for instance the standard model of Analysis, or an extension of it).

2) S is a non empty set of *physical situations* (physical objects or systems of physical objects in specified states).

3) $Q_1, \ldots, Q_k$ are *physical quantities*. They are supposed to be *operationally defined*[1] and to associate with each physical situation s (for which they are defined) an interval of real numbers (depending on the precisions of the instruments used).[2]
We admit the possibility of carrying out in one and the same physical situation s more than one measurement for a quantity Q_i (for instance, one may measure several different masses, or times, etc.). Let us indicate by $Q_{i_1}(s)$, $Q_{i_2}(s)$, $Q_{i_3}(s)$, ... the intervals representing the different results associated with Q_i by a number of measurements for s. If Q_i is a function of Q_j,[3] we shall indicate by $[Q_i(Q_j)](s)$ the interval containing all possible values of Q_i associated by a measurement for s with the interval resulting from a measurement of Q_j for s.

4) ϱ (called *mathematical realization* of the quantities) is a function which associates with each Q_i a mathematical entity $\varrho(Q_i)$ or a class

of mathematical entities in $\mathbf{M}_0$; $\varrho(Q_i)$ is supposed to be associated with an algorithm determining the set $V(\varrho(Q_i))$ of all real numbers which correspond to all possible values that Q_i may assume as a result of a measurement.[4]

Clearly ϱ connects the *operational part* of $\mathbf{M}$ ($\langle S, Q_1, \ldots, Q_k \rangle$) with the mathematical part $\mathbf{M}_0$ of $\mathbf{M}$.

The formal first-order language $\mathscr{L}_{\mathbf{M}}$ associated with $\mathbf{M}$ includes the language $\mathscr{L}_{\mathbf{M}_0}$ of $\mathbf{M}_0$ (with names **a** for all elements a of $\mathbf{M}_0$); further it contains for any Q_i a countable set of special variables $\{q_{i_1}, q_{i_2}, \ldots \ldots, q_{i_n}, \ldots\}$ ranging over $V(\varrho(Q_i))$, and not admitting quantification. $\mathscr{L}_{\mathbf{M}}$ has no names for the elements of S. A sentence of $\mathscr{L}_{\mathbf{M}}$ is a well formed formula in which only special variables may occur free.

The definition of truth of a well formed formula α of $\mathscr{L}_{\mathbf{M}}$ in $\mathbf{M}(\underset{\mathbf{M}}{\models} \alpha)$ is as follows:

a) if α does not contain any variable q_i, then

$$\underset{\mathbf{M}}{\models} \alpha \quad \text{iff} \quad \underset{\mathbf{M}_0}{\models} \alpha$$

b) Suppose that α has the form

$$\beta(q_{i_1}, \ldots, q_{j_r}(q_{h_t}), \ldots, q_{l_n}, \ldots, q_{i_{tn}}, y_1, \ldots, y_m).$$

Let R_α be an n-ary relation representing the extension in $\mathbf{M}_0$ of the formula

$$\forall y_1, \ldots, y_m \beta(x_1, \ldots, x_n, y_1, \ldots, y_m).$$

Then $\underset{\mathbf{M}}{\models} \alpha$ iff for any physical situation s for which the quantities $Q_{i_1}, \ldots, Q_{j_r}, Q_{h_t} \ldots, Q_{l_n}$ are defined, there exists an n-ple of real numbers $r_1, \ldots, r_n$ such that

$$r_1 \in Q_{i_1}(s), \ldots, r_r \in [Q_{j_r}(Q_{h_t})](s), \ldots, r_n \in Q_{l_n}(s)$$

and $R_\alpha(r_1, \ldots, r_n)$ holds in $\mathbf{M}_0$.[5]

A physical theory $\mathscr{T}$ is now definable as a triple $\langle \mathbf{M}, A, R \rangle$ where

1) $\mathbf{M}$ is a physical structure.

2) A (the *axiom-system* of $\mathscr{T}$) is a *locally-consistent* set of sentences that are true in $\mathbf{M}$.

3) R is a system of rules of inference (for instance the rules of inference of classical *natural deduction*). By local consistency of a set of sentences we mean that for any α the set does not contain both α and $\neg\alpha$. We must explicitly require local consistency, for (as we shall see below) the set of all sentences true in a given physical structure will in general turn out to be inconsistent.

According to our definition, an axiom of a physical theory has always the form $\alpha(q_{i_1}, \ldots, q_{j_n})$ (and thus belongs to an $\mathscr{L}_{\mathbf{M}}$-language). One can find in the literature also a broader sense of the term 'axiom': in the sense of a metatheoretical condition on ϱ.[6]

In some cases ϱ may determine a mathematical realization not only for physical quantities but also for *physical states* which are members of the physical situations. If D is the set of all states of physical systems in S, $\varrho(D)$ will be a set of mathematical entities in $\mathbf{M}_0$ (for instance, $\varrho(D)$ may be, in classical mechanics the phase-space, and in quantum mechanics the set of all vectors in a Hilbert space).

II. THE TRIPLE CONNECTIVE-SYSTEM DEFINABLE IN A PHYSICAL STRUCTURE

The system of sentential connectives $\neg$ (not), $\wedge$ (and), $\vee$ (or), $\rightarrow$ (if...then) turns out to be *non-truth-functional* in $\mathbf{M}$ (whereas it is, obviously, truth-functional in $\mathbf{M}_0$). For, if α and β are sentences of $\mathscr{L}_{\mathbf{M}}$, there holds:

(1) $$\underset{\mathbf{M}}{\models} \alpha\wedge\beta \Rightarrow \underset{\mathbf{M}}{\models} \alpha \text{ and } \underset{\mathbf{M}}{\models} \beta$$

But the inverse relation does not generally hold:

$$\underset{\mathbf{M}}{\models} \alpha \text{ and } \underset{\mathbf{M}}{\models} \beta \not\Rightarrow \underset{\mathbf{M}}{\models} \alpha\wedge\beta$$

(2) $$\underset{\mathbf{M}}{\models} \alpha \text{ or } \underset{\mathbf{M}}{\models} \beta \Rightarrow \underset{\mathbf{M}}{\models} \alpha\vee\beta$$ But:

$$\underset{\mathbf{M}}{\models} \alpha\vee\beta \not\Rightarrow \underset{\mathbf{M}}{\models} \alpha \text{ or } \underset{\mathbf{M}}{\models} \beta$$

(3) $$\underset{\mathbf{M}}{\models} \neg\alpha \not\Rightarrow \text{not } \underset{\mathbf{M}}{\models} \alpha$$

$$\text{not } \underset{\mathbf{M}}{\models} \alpha \not\Rightarrow \underset{\mathbf{M}}{\models} \neg\alpha$$

(4) $$\models_{\mathbf{M}} \alpha \to \beta \not\Rightarrow (\models_{\mathbf{M}} \alpha \Rightarrow \models_{\mathbf{M}} \beta).$$
$$(\models_{\mathbf{M}} \alpha \Rightarrow \models_{\mathbf{M}} \beta) \not\Rightarrow \models_{\mathbf{M}} \alpha \to \beta.$$

Now in any logic where the connectives turn out to be non truth-functional (with respect to the bivalent system of truth-values) it is always possible to introduce trivially a new system of connectives $\langle \sqcap, \sqcup, \daleth, \rightharpoonup \rangle$ which is truth-functional. Let us set by means of a semantical definition:

(5) $$\models_{\mathbf{M}} \alpha \sqcap \beta \quad \textit{iff} \quad \models_{\mathbf{M}} \alpha \textit{ and } \models_{\mathbf{M}} \beta$$
$$\models_{\mathbf{M}} \alpha \sqcup \beta \quad \textit{iff} \quad \models_{\mathbf{M}} \alpha \textit{ or } \models_{\mathbf{M}} \beta$$
$$\models_{\mathbf{M}} \daleth\, \alpha \quad \textit{iff} \quad \textit{not} \models_{\mathbf{M}} \alpha.$$
$$\models_{\mathbf{M}} \alpha \rightharpoonup \beta \quad \textit{iff} \quad (\models_{\mathbf{M}} \alpha \Rightarrow \models_{\mathbf{M}} \beta).$$

There follows trivially (by (1)–(2)):

(6) $$\models_{\mathbf{M}} \alpha \wedge \beta \rightharpoonup \alpha \sqcap \beta. \qquad \textit{But not viceversa.}$$
$$\models_{\mathbf{M}} \alpha \sqcup \beta \rightharpoonup \alpha \vee \beta \qquad \textit{But not viceversa.}$$

At the same time $\rightharpoonup \alpha$ and $\daleth\, \alpha$, as well as $\alpha \to \beta$ and $\alpha \daleth \beta$ turn out to be incomparable with respect to . From an intuitive point of view, the two systems $\langle \wedge, \vee, \neg, \to \rangle$ and $\langle \sqcap, \sqcup, \daleth, \rightharpoonup \rangle$ distinguish, directly in the object-language, two different semantical ideas of logical operations (which in some cases may coincide): truth of the negation of a sentence and falsehood (i.e. non-truth) of the sentence; truth of the conjunction of two sentences and simultaneous truth of two sentences; similarly in the other cases.

If we compare this situation with the case of other logics that turn out to be non-extensional in a bivalent semantics (for instance, intuitionistic logic, probabilistic logic, quantum logic) we may observe that the only 'anomalies' arising in our empirical logic concern the violations of $\alpha \sqcap \beta \rightharpoonup \alpha \wedge \beta$ and $\neg \alpha \rightharpoonup \daleth\, \alpha$, which, on the contrary, hold in all the other cases.

The system $\langle \sqcap, \sqcup, \daleth, \rightharpoonup \rangle$ is, by definition, boolean. Also the system $\langle \wedge, \vee, \neg, \to \rangle$ is boolean; in the sense that one can prove that all classical laws (expressed in terms of $\wedge, \vee, \neg, \to$) are true in any $\mathbf{M}$.

Intuitively, it seems interesting to study also a third system of connectives $\langle \sqcap, \sqcup, \sim, \rightharpoonup \rangle$, where

$$\sim\alpha = \begin{cases} \neg\alpha & \textit{if } \alpha \textit{ does not contain any of the connectives } \sqcap, \\ & \sqcup, \urcorner, \rightharpoonup; \\ \urcorner\alpha & \textit{otherwise.} \end{cases}$$

Algebraically $\sqcap, \sqcup, \sim, \rightharpoonup$ determines only a distributive involutory lattice. Indeed $\langle \sqcap, \sqcup \rangle$ determines trivially a distributive lattice; on the other hand the negation $\sim$, while satisfying the involution law $(\models_{\mathbf{M}} \sim\sim\alpha \rightleftharpoons \alpha$, for any $\mathbf{M})$[7] fails to satisfy, for instance, the following boolean laws: *tertium non datur* $(\alpha \sqcup \sim \alpha)$; non-contradiction principle $(\sim(\alpha \sqcap \sim \alpha))$; De Morgan's laws; Philo's law $((\alpha \rightharpoonup \beta) \rightleftharpoons \sim \alpha \sqcup \beta)$; Chrisippus' law $((\alpha \rightharpoonup \beta) \rightleftharpoons \sim(\alpha \sqcap \sim \beta))$. Such a multiple logical situation (in which all the three systems $\langle \wedge, \vee, \neg, \rightarrow \rangle$, $\langle \sqcap, \sqcup, \urcorner, \rightharpoonup \rangle$, $\langle \sqcap, \sqcup, \sim, \rightharpoonup \rangle$ are considered) seems to represent a step towards an adequate formal description of concrete 'logical reasoning' that takes place in the case of empirical theories. For in this way one can meet at the same time the following requirements:

1) The mathematics is classical (obviously, the three connective-systems turn out to coincide in $\mathbf{M}_0$).
2) The semantics is bivalent. Thus, any physical law must be either true or false.
3) We have nevertheless at our disposal some 'more realistic' logical operations. For instance, we have a negation which admits of a certain kind of contradictions and violations of *tertium non datur* ($\alpha \sqcap \sim \alpha$ and $\sim(\alpha \sqcup \sim \alpha)$ may be true in $\mathbf{M}$).[8] Further, the intuitive meaning of our connective $\rightharpoonup$ seems to be very close to the informal concrete use of the notion of 'empirical implication'.

Let us state:

α *empirically implies* β iff for any physical structure $\mathbf{M}$ (belonging to a given class K), $\models_{\mathbf{M}} \alpha \rightharpoonup \beta$.

Most probably, this kind of empirical implication can avoid some classical difficulties which may arise (as is well known) when using material or strict implication in the case of empirical theories.[9]

III. A MODAL INTERPRETATION

Let us consider our language $\mathscr{L}_{\mathbf{M}}$ extended with the new connectives $\sqcap, \sqcup, ⅂, ⇀$. We shall indicate this extension of $\mathscr{L}_{\mathbf{M}}$ by $\mathscr{L}'_{\mathbf{M}}$. By definition of well formed formula of $\mathscr{L}'_{\mathbf{M}}$ the connectives $\wedge, \vee, \neg, \rightarrow$ can be applied only to *pure* formulas, i.e. to formulas that do not contain any of the connectives $\sqcap, \sqcup, ⅂, ⇀$. The intuitive meaning of such logical situation can be better clarified by means of a modal interpretation, stated in the style of the modal interpretations of intuitionistic logic and of quantum logic.[10]

Let $\mathscr{L}^*_{\mathbf{M}}$ be a modal extension of $\mathscr{L}_{\mathbf{M}}$ (with modal operators L (it is necessary that) and M (it is possible that)). In order to obtain a translation of $\mathscr{L}'_{\mathbf{M}}$ in $\mathscr{L}^*_{\mathbf{M}}$, we define, by simultaneous induction on the length of the formulas, two representations φ and ϱ of $\mathscr{L}'_{\mathbf{M}}$ in $\mathscr{L}^*_{\mathbf{M}}$:

$$(7) \qquad \varrho(\alpha) = \begin{cases} LM\varphi(\alpha) & \textit{if } \alpha \textit{ is pure;} \\ \varphi(\alpha) & \textit{otherwise} \end{cases}$$

$$\varphi(p^n_i t \dots t_n) = p^n_i t_1 \dots t_n$$

$$\begin{aligned}
\varphi(\neg\alpha) &= \neg\varphi(\alpha)\\
\varphi(\alpha\wedge\beta) &= \varphi(\alpha)\wedge\varphi(\beta)\\
\varphi(\alpha\vee\beta) &= \varphi(\alpha)\vee\varphi(\beta)\\
\varphi(\alpha\rightarrow\beta) &= \varphi(\alpha)\rightarrow\varphi(\beta)\\
\varphi(⅂\,\alpha) &= \neg\varrho(\alpha)\\
\varphi(\alpha\sqcap\beta) &= \varrho(\alpha)\wedge\varrho(\beta)\\
\varphi(\alpha\sqcup\beta) &= \varrho(\alpha)\vee\varrho(\beta)\\
\varphi(\alpha ⇀ \beta) &= \varrho(\alpha)\rightarrow\varrho(\beta).
\end{aligned}$$

One can prove that for any physical structure **M** there exists a Kripke S_4 model $\mathbf{M}^*$ such that:

$$(8) \qquad \underset{\mathbf{M}}{\models} \alpha \textit{ iff } \underset{\mathbf{M}^*}{\models} \varrho(\alpha).^{11}$$

This result represents a new example confirming the large-scale applicability of Kripke's Semantics (and modal conceptualization) to the logical analysis of empirical theories.[12]

A comparison between different modal interpretations of intuitionistic logic, quantum logic, and this kind of "multiple empirical logic" can clarify different intuitive ideas of logical operations. For instance, on this ground one can better understand the reason why negation, conceived as 'implication of an absurdity', appears to be meaningful in the case of an *epistemological* approach (intuitionistic and minimal logic); or in the case of a *descriptivistic-deterministic* approach (without approximation: classical logic).

However such a conception of negation may lose sense in the case of a *descriptivistic-indeterministic* approach (with approximation: quantum logic and empirical logic, where $\neg\alpha \rightarrow (\alpha \rightarrow \lambda)$; whereas obviously: $\neg\alpha \leftrightarrow (\alpha \rightarrow \lambda)$ and $\urcorner\alpha \rightleftharpoons (\alpha \rightharpoondown \lambda)$[13].

IV. QUANTUM LOGIC AND EMPIRICAL LOGIC

We want to study now some relations which hold between quantum logic and empirical logic. Apparently, most quantum logics studied in the literature, involve a very strong mathematical idealization and turn out to be totally determined by the mathematical part of our physical structures.

Let D be the set of all physical states of a given structure $\mathbf{M}$ and $\varrho(D)$ its mathematical realization. Any function $\tau: D \Rightarrow \varrho(D)$ represents an *ideal mathematical translation* of states. A single $\tau(D)$ determines univocally the values for all the Q_i's defined on $d \in D$.[14] Let us denote by $\tau^{Q_i}(d)$ the value for Q_i determined by $\tau(d)$.[15]

Without loss of generality we may suppose that a 'concrete' system of measurements carried out on a state d for all quantities Q_i, which are defined on d, is represented by a set $\{\tau_m\}$ of mathematical translations.

Let us now consider a special set of formulas of L_M, which we shall call *elementary*; they have the form

$$q_i = \mathbf{r} \ \mathit{or} \ q_i = x_j$$

Any elementary sentence $q_i = \mathbf{r}$ defines a mathematical property $P_r^{Q_i}$ of the elements of $\varrho(D)$:

$$(9) \qquad \models_{\mathbf{M}_0} P_r^{Q_i}\tau(d) \ \ \mathit{iff} \ \models_{\mathbf{M}_0} \tau^{Q_i}(d) = \mathbf{r}$$

By means of a trivial translation we may interpret the $\tau(d)$'s as 'possible worlds' and put:

$$(10) \qquad \models_{\mathbf{M}_0,\tau(d)} q_i = \mathbf{r} \ \ \mathit{iff} \ \models_{\mathbf{M}_0} P_r^{Q_i}\tau(d)$$

Let us now consider the complete boolean algebra that contains all subsets of $\varrho(D)$, and let E be the set of all (first-order) sentences of $\mathscr{L}_M$ which have as atomic subformulas only elementary formulas. We may define (in the standard way), by simultaneous induction, the concept of truth for any $\alpha \in E(\models_{\mathbf{M}_0,\psi} \alpha$, where $\psi \in \varrho(D))$ and the set X_α of all possible worlds in which α is true:

$$(11) \qquad X_\alpha = [\psi \in \varrho(D) / \models_{\mathbf{M}_0,\psi} \alpha]$$

$$\models_{\mathbf{M}_0,\psi} \neg\alpha \quad \textit{iff} \quad \psi \in \varrho(D) - X_\alpha$$

$$\models_{\mathbf{M}_0,\psi} \alpha \wedge \beta \quad \textit{iff} \quad \psi \in X_\alpha \cup X_\beta$$

$$\models_{\mathbf{M}_0,\psi} \alpha \vee \beta \quad \textit{iff} \quad \psi \in X_\alpha \cap X_\beta$$

$$\models_{\mathbf{M}_0,\psi\; x} \forall x \quad \textit{iff} \quad \psi \in \bigcap \{X_{\alpha(x/\mathbf{r})}\}_{r \in R}$$

$$\models_{\mathbf{M}_0,\psi\; x} \exists x \quad \textit{iff} \quad \psi \in \bigcup \{X_{\alpha(x/\mathbf{r})}\}_{r \in R}$$

Clearly the 'logic' arising in this way turns out to be classical logic. Suppose now that (as it is the case for quantum mechanics) a complete ortholattice is definable in $\varrho(D)$, and let $\underset{*}{-}, \underset{*}{\cap}, \underset{*}{\cup}, \underset{*}{\bigcap}, \underset{*}{\bigcup}$ be the ortholattice-operations on $\varrho(D)$. Exactly as in the previous case, we can define $\models_{\mathbf{M}_0,\psi} \alpha$ and X_α (let us indicate by $\underset{*}{\neg}, \underset{*}{\wedge}, \underset{*}{\vee}, \underset{*}{\forall}, \underset{*}{\exists}$ the *quantum connectives and quantifiers* corresponding to the ortho-operations). As a consequence of the mathematical formalism of quantum mechanics, the *truth* of an atomic sentence means:

(12) $\models_{\mathbf{M}_0,\psi} q_i = \mathbf{r}$ *iff there exists a* τ *such that:*

$\psi = \tau(d)$ *and* $\tau^{Q_i}(d) = r$, for a certain d.

iff ψ *is an eigenvector, with eigenvalue* r, *of the hermitian operator which represents the mathematical realization* $\varrho(Q_i)$ *of* Q_i.

As is well known, this gives rise to quantum logic. Thus apparently, quantum logic seems to be totally determined in the mathematical part of **M**. Let us try and extend this kind of logical construction also to the operational part of **M**. For any $\alpha \in E$, and $d \in S$, we can state:

$$\models_{\mathbf{M},d} \alpha \quad \textit{iff for at least a } \tau \in \{\tau_m\}, \qquad \models_{\mathbf{M}_0,\tau(d)} \alpha.$$

One easily verifies that we obtain in this way our empirical logic in the connectives $\wedge, \vee, \neg$ (the truth functional connectives, $\sqcap, \sqcup, \daleth$ being then definable in the obvious way).

An analogous definition of truth for α containing quantum connectives and quantifiers (instead of $\wedge, \vee, \neg, \forall, \exists$) gives rise to a "mixed quantum-empirical" logic. There holds then for any **M** and d:

$$\models_{\mathbf{M},d} \alpha \textit{ for any } \alpha \textit{ quantum valid.}$$

$$\models_{\mathbf{M},d} \alpha \underset{*}{\wedge} \beta \rightleftharpoons \alpha \wedge \beta \rightharpoonup \alpha \sqcap \beta. \textit{ But not viceversa.}$$

$$\models_{\mathbf{M},d} \alpha \sqcup \beta \rightharpoondown \alpha \vee \beta \rightharpoonup \alpha \underset{*}{\vee} \beta. \textit{ But not viceversa.}$$

$$\models_{\mathbf{M},d} \underset{*}{\neg} \alpha \rightharpoonup \neg \alpha.$$

$$\models_{\mathbf{M},d} \underset{*}{\neg} \alpha \not\rightharpoonup \daleth \alpha$$

$$\models_{\mathbf{M},d} \daleth \alpha \not\rightharpoonup \underset{*}{\neg} \alpha,$$

where α and β are supposed to contain, as logical constants, at most $\underset{*}{\wedge}, \underset{*}{\vee}, \underset{*}{\neg}, \underset{*}{\forall}, \underset{*}{\exists}$.

Therefore it may be possible that: $\models_{\mathbf{M},d} \alpha \sqcap \underset{*}{\neg} \alpha$.

V. A GENERALIZED EMPIRICAL STRUCTURE

The concept of physical structure $\mathbf{M} = \langle \mathbf{M}_0, S, Q_1, \ldots, Q_k, \varrho \rangle$ admits of an obvious generalization that may find applications also in other fields.

Let us suppose that:

a) $\mathbf{M}_0$ be an abstract structure.

b) S be a generic set of *empirical situations* (not necessarily belonging to physical science)

c) Any Q_i be a function which associates with each element s of S (for which it is defined) an element (or a system of elements) in $\mathbf{M}_0$. Probably, most theoretical fields which involve certain *abs-*

traction operations to be performed on a set of empirical data, can be formally described by means of general structures of this kind. Let us call such an **M** an *empirical structure.*

It may happen that S is not only a set, but a set with a structure:

$$\mathbf{S} = \langle S, R_1, \ldots, R_h \rangle$$

The R_i's may be conceived as intuitive relations expressed in the metatheory in which the system S is described. For instance, if **M** is a physical structure, R_i may represent an intuitive metatheoretical property which physicists use in their informal speech (such as 'atom', 'elementary particle' and so on).

In what sense could these relations R_i be formally defined in **M**? Let us suppose R_i to be a property. And let $\bar{R}_i$ be a name in L_S for R_i. Then an obvious answer is the following

R_i is *formally definable in* **M** iff there exists a formula $\alpha(q_{i_1}, \ldots, q_{i_n})$ of L_M such that: for any j $(1 \leqslant j \leqslant n)$ and any $s \in S$ (with name **s** in L_S) for which Q_{h_j} is defined, there exists a real number $r_j \in Q_{h_j}(s)$ and $\models_{\mathbf{S}} R_i s$ iff $\models_{\mathbf{M}_0} \alpha(r_1, \ldots, r_n)$

In such cases we can say that α *determines the physical meaning of* R_i.

VI. EXTENSIONAL OR INTENSIONAL SEMANTICS?

Our last example brings about some arguments for the discussion of a classical methodological problem: what theory of meaning represents an adequate semantical tool in the logical analysis of empirical theories?

The limits and inadequacy of extensional Semantics on this field have been pointed out on may occasions: how could we seriously claim that the physical meaning of a predicate such as 'elementary particle' is adequately represented by the set of *all* elementary particles in the Universe?

As is well known, some authors have, on this ground, reached the conclusion that the Semantics of some empirical theories is intrinsically *intensional*. However, the shortcoming of such a position is represented by the extremely vague character of the concept of 'intension'. And if we try and make it precise, on the lines of Kripke's Semantics,[16] we reach an 'extensional definition' of intension, which does not solve our previous difficulties. For, according to this approach, the intensional meaning of the predicate 'elementary particle' would become a function, which for any time-instant assumes as value the set of all elementary particles in the Universe at that time.

As we have seen, our semantical analysis, founded on the concept of 'physical structure', seems to lead, in a natural way, to the identification of the physical meaning of a predicate such as 'elementary particle' with its formal definition in terms of physical quantities. We are aware that such a proposal can meet, at least, the criticism of two kinds of objectors:

those who claim the impossibility of defining *all* physical predicates in terms of operationally defined quantities and on this ground reach the conclusion that operational definitions do not generally provide *meanings* but only *methods of use*.

2) those who refuse the identification of meanings with formulae (or sets of formulae): meanings cannot be dependent on languages; they must be extra-linguistic.

To the former objectors we observe that identifing meanings with formal definitions do not necessarily compel us to conclude that *all* predicates must be formally definable. As in the case of mathematical theories, some predicates may be *undefined* or even *undefinable* (with respect to a given structure **M**). In such cases we will say that they do not possess a definite meaning (in **M**).[17]

The second objection seems to be more critical. Indeed, a simple identification of meanings with formulae (i.e. with certain sets of signs) may generally lead to a number of counterintuitive consequence. Without attempting to solve this crucial question, we observe that most counterintuitive consequences seem to depend on the fact that one is searching for an absolute sense of 'meaning'. However, let us relativise meanings to languages, and let us limit ourselves to the question: what is 'determining a meaning' with respect to a given language"? In this case the proposal to characterize meanings as special kinds of translations seems to become much more intuitive and adequate.

After all, everybody accepts that the meaning of 'dog' in German is 'Hund', and everybody answers to a child's question 'what does this mean?' by a translation in a more familiar language.

BIBLIOGRAPHY

[1] Dalla Chiara Scabia, M. L. and Toraldo di Francia, G., 'A logical Analysis of Physical Theories', *Rivista del Nuovo Cimento, Serie* 2. *Vol.* **3** (1973), 1–20.
[2] Dalla Chiara, M. L., 'Some Logical Problems suggested by Empirical Theories', forthcoming in *Boston Studies in the Philosophy of Science.*
[3] Montague, R., 'Pragmatics', *Logic and Foundations of Mathematics*, Firenze, 1968.
[4] Van Fraassen, B. C., 'Semantic Analysis of Quantum Logic' in *Contemporary Research in the Foundations and Philosophy of Quantum Theory*, Dordrecht, 1973.
[5] Wójcicki, R., 'Set Theoretic Representations of Empirical Phenomena', in *Reports of the Seminar on Formal Methodology of Empirical Sciences*, Wrocław, November 1973.

NOTES

[1] As described in [1], the concept of 'operational definition of a physical quantity, admits of a rigorous characterization.
[2] Of course, instead of a real number interval, we may have in some cases the cartesian product of n real number intervals.
[3] For instance, Q_j may represent the quantity 'time' and Q_i may be any other quantity which is a function of time.
[4] As an example, in classical mechanics ϱ may associate with the quantity 'time' the set of real numbers, with the quantity 'length' a set of real-valued functions, and so on. In quantum mechanics ϱ associates with each quantity Q_i a hermitian operator in a Hilbert space. Respectively, in these cases $V(\varrho(Q_i))$ is represented by: the set of real numbers, the set of all possible values of all the real functions in $\varrho(Q_i)$, the set of all eigenvalues of $\varrho(Q_i)$. Note that not necessarily $\varrho(Q_i) \subset \mathbf{M}_0$. However, extending $\mathbf{M}_0$ to a higher-order structure, it is always possible to obtain trivially $\varrho(Q_i) \subset \mathbf{M}_0$.
[5] Our concept of physical structure is very close to the definition of *operational system* given by Wójcicki in [5]. Wójcicki's 'approximate truth' practically coincides with our 'truth in a physical structure'. Note that in Ref. [1] (p. 18, l. 12) due to an error of transcription, one should read '$s \in \bigcap_{j=1}^{m} S_{n_j}$ instead of 'm-ple $s_{h_1}, \ldots, s_{h_m}$ $(s_{h_j} \in S_{h_j})$'.
[6] As an example we may refer to the 'axiom of quantum mechanics' according to which any physical observable quantity is associated with a hermitian operator in a Hilbert space.

[7] $\rightleftharpoons$ is defined, in a obvious way, by means of $\sqcap$ and $\rightharpoonup$.
[8] In a certain sense, we could say that we are assuming a 'dialectical principle'.
[9] For example, when one maintains that '*if* there are certain initial and boundary conditions, *then* there are certain results'.
[10] A modal interpretation of quantum logic is described in [2].
[11] Using Wójcicki's terminology, we could say that **M*** consists of all idealizations of **M** (for any physical situation).
[12] As observed by Van Fraassen in [4], the notion of 'possible world', in spite of its metaphysical appearance, allows a number of 'empirical applications'. For instance it may be interpreted on some occasions as 'physical state' or as 'physical situation' etc. etc.
[13] λ represents an absurdity.
[14] These values are in classical mechanics real numbers (or *n*-ples of real numbers); in quantum mechanics probability distributions of real numbers.
[15] If $\tau^{Q_i}(d)$ consists of a single value r with probability 1, we shall simply write $\tau^{Q_i}(d) = r$
[16] See for instance [3].
[17] It is worthwhile to note that some stronger difficulties arise only if one intends 'operational definition' in a very narrow sense (as the set of 'concrete material operations', without theoretical import). As described in [2], the characterization of 'operational definition' involved in our physical structure is not so narrow.

JAAKKO HINTIKKA AND ILKKA NIINILUOTO

AN AXIOMATIC FOUNDATION FOR THE LOGIC OF INDUCTIVE GENERALIZATION

I. THE REPRESENTATIVE FUNCTION OF AN INDUCTIVE METHOD

One of the most interesting viewpoints from which inductive logic can be looked at is to ask what the different factors are that must be taken into account in singular inductive inference, i.e., in the usual technical jargon, what the arguments of the representative functions of one's system of inductive methods are. It is well known that Carnap's λ-continuum of inductive methods can be derived from essentially one single assumption concerning these arguments of the representative function.[1] It is shown in this paper that a logic of inductive generalization is obtained if this assumption is weakened in a natural way.

Representative function is a numerical function which 'represents' an inductive probability measure defined for the statements of the underlying language. Suppose that the individuals of a universe U are classified in K exclusive classes into one of which each individual of U must fall, i.e., that universe U is partitioned into K 'cells'. Given a sample of n individuals of which n_i belong to the i:th cell ($i = 1, \ldots K$), the representative function f specifies the probability that an unknown further individual ('next individual to be observed') falls into a given cell. (We may think of this probability as a betting ratio for the bet that the next individual belongs to the i:th cell.) The *symmetry assumption*, de Finetti's 'exchangeability'[2], says that f depends on the sample only through the numbers $n_1, n_2, \ldots, n_K$, not on the order in which the several individuals in the sample were observed, and that the probability of the transition from any sequence of n_i's to another one is independent of the path. (It also implies that the probability does not depend on the new 'unknown' individual.) Thus it is already one of the kinds of assumptions that limits the choice of the arguments of the representative function.

II. CARNAP'S λ-CONTINUUM

The assumption that Carnap employs for the derivation of his λ-continuum goes a long way further than the symmetry assumption and says that f depends on the sample only through its size n (i.e., the sum $n_1 + + \ldots + n_K$) and the number n_i of sample individuals in the cell in question. Thus the probability that the next individual belongs to a cell with n' individuals so far observed in it will be of the form $f(n', n)$. (Function f depends also on the classification system through the fixed number K of cells.) It is course required also that this representative function must give rise to a symmetric probability distribution, that is, to a probability measure satisfying probability calculus. From this assumption and the assumption concerning the arguments of f it follows that f must be of the form

$$(1) \qquad f(n', n) = \frac{n' + \dfrac{\lambda}{K}}{n + \lambda}$$

where λ is a positive, real-valued constant. The parameter λ is determined by the equation

$$f(0, 1) = \frac{\dfrac{\lambda}{K}}{1 + \lambda},$$

which yields as a solution

$$(2) \qquad \lambda = \frac{K \cdot f(0, 1)}{1 - K \cdot f(0, 1)}$$

Strictly speaking, (1) follows only if $K > 2$. In the case $K = 2$ it follows only in conjunction with the additional assumption that f be a linear function of n'.

The parameter λ can be interpreted as an index of caution with respect to singular inductive inference, or alternatively as a measure of the disorder (randomness) in the universe U.[3] It can also be interpreted as the size of an imaginary homogeneous sample of λ individuals which is combined with the real sample of n individuals in estimating the frequency of different kinds of individuals in the whole universe U.

III. THE PROBLEM OF INDUCTIVE GENERALIZATION

The main disadvantage with Carnap's λ-continuum is that it assigns zero prior probabilities to all nontrivial generalizations in an infinite universe, and hence likewise assigns zero probabilities to them a posteriori on any finite evidence. Moreover, in a finite universe appreciable posterior probabilities are associated with nontrivial generalizations only when our sample is of the same order of magnitude as the whole universe. This makes Carnap's λ-continuum useless as a logic of inductive generalization.

Inductive logics which are free of this defect have been proposed in the literature. A case in point is Hintikka's *α-λ-system.*[4] However, these systems have so far not been motivated on the basis of simple assumptions concerning the representative function.

IV. A NEW ARGUMENT FOR THE REPRESENTATIVE FUNCTION

In this paper, a new type of system of inductive logic is examined. It is based on an assumption which is the most natural way of loosening the axioms of the λ-system. Instead of assuming that the representative function depends only on n' and n, we assume that it may also depend on the number c of cells instantiated in the sample (i.e., on the number of i's such that $n_i > 0$). The importance of this parameter for inductive generalization is shown already by the fact that it determines the number of nonequivalent generalizations compatible with the sample. Thus our working hypothesis in this paper is that the representative function is of the form $f(n', n, c)$ and that it gives rise to a symmetric probability distribution. These two assumptions are the 'axioms' of the new system studied in this paper.

Certain members of the α-λ-continuum fall within the scope of the new assumption, but others do not. An instance of the former is the generalized 'combined system' of Hintikka.[5] Its representative function is, for $n' > 0$,

$$(3) \qquad (n'+1)\frac{\displaystyle\sum_{i=0}^{K-c}\binom{K-c}{i}\frac{(\alpha+c+i-1)!}{(n+c+i)!}}{\displaystyle\sum_{i=0}^{K-c}\binom{K-c}{i}\frac{(\alpha+c+i-1)!}{(n+c+i-1)!}}$$

which is seen to depend only on n', n, c, and K. Conversely, the new assumption allows for inductive strategies not covered by the α-λ-continuum.

The consequences of the new assumption will not receive an exhaustive analysis in this paper. What is attempted here is an exploration of some of the most important qualitative (asymptotic) consequences of this assumption, which will in the sequel be referred to as our *Basic Assumption*. In particular we shall show that it behaves *vis-à-vis* inductive generalization in the way we are entitled to expect, i.e., that it assigns nonzero prior probabilities to generalization even in an infinite universe and that the posterior probabilities of generalization (including constituents, that is, statements which specify the empty and the nonempty cells of the universe) in the right way in most cases. This means that the simplest constituent compatible with evidence will be highly confirmed in the long run when the evidence grows without limit. (The evidence is assumed to consist of a number, say n, individuals located in specified cells of the universe).

We shall begin by establishing a number of separate results on the representative function $f(n', n, c)$. Later, some of the more general consequences of our observations, and their relations to Carnap's λ-continuum, will be discussed.

V. BASIC THEOREMS

It is useful to introduce a few defined symbols.

Definition 1. $h(n, c) = (K-c)f(0, n, c)$.

Definition 2. $g(n, c) = 1-h(n, c)$.

The intuitive meaning of the functions h and g is clear: while g gives the probability that an unknown new individual belongs to a cell already instantiated in the sample (i.e., is of the same kind as the individuals already observed), h gives the probability that it does not (i.e., that it belongs to a new cell.)

Theorem 1. If $n_1', n_2', \ldots, n_c'$ are positive integers $\geqslant 1$ whose sum is n, and if $0 \leqslant c \leqslant K$, then

$$\sum_{i=1}^{c} f(n_i', n, c) + (K-c)f(0, n, c) = 1.$$

Proof. Follows immediately from the assumption that f defines a probability distribution.

Corollary 1.1. $f(0, 0, 0) = \frac{1}{K}$.

Corollary 1.2. $f(1, K, K) = \frac{1}{K}$.

Corollary 1.3. $f(n, n, 1) + (K-1)f(0, n, 1) = 1$

Corollary 1.4. $f(n', n, 2) + f(n-n', n, 2) + (K-2)f(0, n, 2) = 1$, provided that $n > n' > 0$.

Corollary 1.5. $f(n', cn', c) = \frac{1}{c} g(cn', c)$, provided that $c > 0$.

Corollary 1.6. $\sum_{i=1}^{c} f(n_i', n, c) = g(n, c)$, on the assumptions of Theorem 1.

The next theorem is based on the assumption that f defines a symmetric probability distribution.

Theorem 2. $f(n', n, c) \cdot f(n'', n+1, c) = f(n'', n, c) \cdot f(n', n+1, c)$, provided that $n - n' - n'' \geqslant c-2$, $n' \geqslant 1$, $n'' \geqslant 1$, $K \geqslant c \geqslant 2$, and that $n = n' + n''$ if $c = 2$.

Proof. This theorem is proved by considering the two orders in which two new individuals can be found, one falling into a cell with n' individuals ($n' > 0$) and the other into a cell with n'' individuals ($n'' > 0$), and by identifying the two transition probabilities.

Theorem 2 says essentially that $f(n', n, c)$ is a linear function of $n' > 0$. Later, a more explicit form of this fact will be found.

Corollary 2.1. $f(n', n+1, c) = f(n'', n+1, c) \cdot \frac{f(n', n, c)}{f(n'', n, c)}$ on the assumptions of Theorem 2.

Corollary 2.2. $f(n', n, 2) = f(n-n'-1, n, 2) \cdot \frac{f(n', n-1, 2)}{f(n-n'-1, n-1, 2)}$ provided that $n-1 > n' > 1$.

Because of our Basic Assumption, Theorem 2 will generally be false in the case $n'' = 0$ (and $c < K$). Instead of Theorem 2, the following result is obtained for this case.

Theorem 3. $f(n', n, c) \cdot f(0, n+1, c) = f(0, n, c) \cdot f(n', n+1, c+1)$, provided that $n-n' \geqslant c-1, n' \geqslant 1, K > c \geqslant 1$, and that $n' = n$ if $c = 1$.

Proof. This theorem may be proved by considering the two orders in which two new individuals can make their appearance, one belonging to an already instantiated cell with n' individuals and the other to a given empty cell, and by identifying the two transition probabilities.

Corollary 3.1. $f(n', n, c) = \dfrac{f(0, n, c-1)}{f(0, n-1, c-1)} f(n', n-1, c-1)$, provided that $n-n' \geqslant c-1, n' \geqslant 1, K \geqslant c \geqslant 2$, and that $n-1 = n'$ if $c = 2$.

Corollary 3.2.
$$f(n', n, c) = \frac{f(0, n, c-1)}{f(0, n-1, c-1)} \cdot \frac{f(0, n-1, c-2)}{f(0, n-2, c-2)} \cdot \ldots$$
$$\ldots \frac{f(0, n-c+3, 2)}{f(0, n-c+2, 2)} f(n', n-c+2, 2),$$
with the assumptions of Corollary 3.1.

Corollary 3.3. $f(n', n, c) = \dfrac{f(0, n, c)}{f(0, n+1, c)} f(n', n+1, c+1)$ provided that $n-n' \geqslant c-1, n' \geqslant 1, K > c \geqslant 1$, and that $n = n'$ if $c = 1$.

Corollary 3.4.
$$f(n', n, c) = \frac{f(0, n, c)}{f(0, n+1, c)} \cdot \frac{f(0, n+1, c+1)}{f(0, n+2, c+1)} \cdot \ldots$$
$$\ldots \frac{f(0, n+K-c-1, K-1)}{f(0, n+K-c, K-1)} f(n', n+K-c, K),$$
provided that $n-n' \geqslant c-1, n' \geqslant 1$, and that $n' = n$ if $c = 1$.

Corollary 3.4 shows that $f(n', n, c)$ can be computed by multiplying $f(n', n+K-c, K)$ with a factor which does not depend on n'. On the other hand, when each of the K cells has been exemplified, the representative function behaves essentially in the same way as in Carnap's λ-continuum.

Theorem 4. $$f(n', n, K) = \frac{n' + \dfrac{\lambda_K}{K}}{n + \lambda_K},$$
where λ_K is a constant, $\lambda_K > -K$,

$$\lambda_K = \frac{Kf(1, K+1, K)}{1 - Kf(1, K+1, K)} - K,$$

provided that $n - n' \geqslant K-1$, and $n' \geqslant 1$.

Proof. This theorem can be proved essentially in the same way as the main result for Carnap's λ-continuum. The argument is given in Appendix 1.

If $K = 2$, Theorem 4 depends on the auxiliary assumption as Carnap's λ-continuum in the analogous case.

By combining Corollary 3.4 and Theorem 4, the following theorem is immediately obtained.

Theorem 5. $f(n', n, c)$ is of the form

$$\mu(n, c) \frac{n' + \dfrac{\lambda_K}{K}}{n + K - c + \lambda_K},$$

provided that $n - n' \geqslant c-1$, and $n' \geqslant 1$.

Corollary 5.1. $f(n', n, c)$ is a linear function of $n' > 0$, provided that $n - n' \geqslant c-1$,

If we introduce the notation

$$\alpha(n, c) = \frac{\mu(n, c)}{n + K - c + \lambda_K},$$

then the following result is obtained by Theorem 5, Corollary 1.6, and Definitions 1 and 2.

Theorem 6. (a) $$f(n', n, c) = \alpha(n, c)\left(n' + \frac{\lambda_K}{K}\right),$$

provided that $n - n' \geqslant c-1$, and $n' \geqslant 1$,

(b) $$g(n, c) = \alpha(n, c)\left(n + \frac{c\lambda_K}{K}\right), \text{ if } n \geqslant c,$$

$$\text{(c)}\quad f(0, n, c) = \frac{1-\alpha(n, c)\left(n+\dfrac{c\lambda_K}{K}\right)}{K-c}, \text{ if } n \geqslant c.$$

In this theorem, we have separated the role of n' in f from those of n and c.

In the sequel, we shall write simply λ for the parameter λ_K given in Theorem 4.

Theorem 7. When $n \geqslant c$ and $K > c \geqslant 1$.

$$\alpha(n+1, c+1)\left[n+\frac{c\lambda}{K}-\frac{1}{\alpha(n, c)}\right]-$$

$$-\alpha(n+1, c)\left[n+\frac{c\lambda}{K}+1\right]+1 = 0.$$

Proof. Substitute values given by Theorem 6 (a) and (c) into Theorem 3.

Corollary 7.1. When $n \geqslant c$ and $K > c \geqslant 1$.

$$\alpha(n+1, c+1)\left[n+\frac{c\lambda}{K}-\frac{1}{\alpha(n, c)}\right] =$$

$$= \alpha(n+1, c)\left[n+1+\frac{c\lambda}{K}-\frac{1}{\alpha(n+1, c)}\right].$$

Corollary 7.2. When $n \geqslant K-1$,

$$\alpha(n+1, K-1) = \frac{2(n+\lambda)+1-\dfrac{\lambda}{K}-\dfrac{1}{\alpha(n, K-1)}}{(n+\lambda+1)\left(n+\lambda+1-\dfrac{\lambda}{K}\right)}.$$

Proof: Substitute $K-1$ for c in Theorem 7, and recall that

$$\alpha(n, K) = \frac{1}{n+\lambda}.$$

Corollary 7.2 expresses a nonlinear difference equation for $\alpha(n, K-1)$. Its solution, which can be written as a continued fraction, gives the value of $\alpha(n, K-1)$ as a function of n, λ, K, and $\alpha(K-1, K-1)$. When the values $\alpha(n, K-1), n \geqslant K-1$, are known, a difference equation for $\alpha(n, K-2)$ is obtained from Theorem 7, and similarly for $\alpha(n, c)$ for all $c = 1, \ldots, K-1$.

VI. THE PARAMETERS OF THE NEW SYSTEM

Theorem 8. $f(0, c, c), c = 1, 2, \ldots, K-1$, and $f(1, K+1, K)$ can be chosen as the independent free parameters of the new system.

Proof. Note first that we can determine $f(n', n, c)$ and $g(n, c)$, by Theorems 4 and 6, if we know $\alpha(n, c)$ and $f(1, K+1, K)$. What we are claiming is that $f(n', n, c)$ is determined for arbitrary values of n', n, c $(K \geqslant c \geqslant 1, n-n' \geqslant c-1)$ if $f(0, c, c)$ $(c = 1, 2, \ldots, K-1)$ and $f(1, K+1, K)$ are fixed.

This claim holds for $c = K$ by Theorem 4. Assume now that it holds when $c > c_0$ and $n < n_0$. Then Theorem 7 shows that it holds when $c = c_0$ and $n = n_0$. This amounts to an inductive proof of Theorem 8, for the smallest value that can be used in Theorem 7 is just $n = c$, which corresponds to $\alpha(c, c)$ or $f(0, c, c)$.

Corollary 8.1. There are K free parameters in the new system.

The values of the parameters $f(0, c, c)$ $(c = 1, 2, \ldots, K-1)$ are not determined by our Basic Assumption, but we shall see later that there is a constraint imposed upon them in that they cannot be chosen so that they yield more 'pessimistic' probabilities than the corresponding Carnapian ones. The values of $f(0, c, c)$ thus belong to certain subintervals of $[0, 1]$ with 0 as the left end point. The point 0 is excluded if f is to define a regular probability measure. The same holds for the parameter $f(1, K+1, K)$. If a principle of 'instantial relevance' to the effect that

$$f(2, K+1, K) > f(1, K, K)$$

is assumed, then it follows that

$$f(1, K+1, K) < \frac{1}{K}.$$

Note also that if $f(1, K+1, K)$ is determined in Carnap's λ-continuum with *his* $\lambda = \lambda_0$, i.e., if

$$f(1, K+1, K) = \frac{1 + \dfrac{\lambda_0}{K}}{K+1+\lambda_0},$$

then by Theorem 4

$$\lambda_K = \frac{K+\lambda_0}{K+1+\lambda_0}(K+1+\lambda_0)-K = \lambda_0 .$$

In other words, if parameter $f(1, K+1, K)$ is given its Carnapian value, then *our* λ (i.e., λ_K) equals the λ of Carnap's λ-continuum.

As soon as the value of $f(1, K+1, K)$ is fixed, $f(0, c, c)$, $\alpha(c, c)$, and $f(1, c, c)$ determine each other through the equations

$$f(0, c, c) = \frac{1-c\left(1+\frac{\lambda}{K}\right)\alpha(c, c)}{K-c}$$

and

$$f(1, c, c) = \left(1+\frac{\lambda}{K}\right)\alpha(c, c).$$

Thus, instead of $f(0, c, c)(c = 1, 2, ..., K-1)$ and $f(1, K+1, K)$ we could choose as the free parameters of the new system $f(1, c, c)(c = = 1, 2, ..., K-1)$ and $f(1, K+1, K)$.

For example, in the case $K = 2$, the free parameters can be taken to be $f(1, 3, 2)$ and either $f(0, 1, 1)$ or $f(1, 1, 1)$.

The intuitive interpretation of the parameters of the new system will be discussed later when we have seen what their role in inductive generalization is.

VII. POSTERIOR PROBABILITIES FOR CONSTITUENTS: CASE $c = K-1$

After these preliminary results, we shall address ourselves to the most central task of this paper. One of the most important things we can hope to find out about the new system of inductive logic is how the posterior probabilities of generalizations behave when evidence for them accumulates. In certain cases, the calculation of these probabilities can be carried out explicitly.

Let us consider the case in which the evidence exemplifies $K-1$ cells, leaving just one of the cells empty. Let us assume that the evidence e consists of a complete classification of n individuals. Then the posterior probability of the constituent $C^{(K-1)}$ which says that all the $K-1$ cells exemplified in the evidence are instantiated, but leaves the one cell empty that was not examplified in the sample, either, is

$$\text{(4)} \qquad P(C^{(K-1)}/e) = \prod_{i=0}^{\infty} g(n+i,\, K-1).$$

This formula holds whenever e is a sample from an infinite (denumerable or nondenumerable) universe. Assuming e, constituent $C^{(K-1)}$ is equivalent to a universal statement claiming that all individuals of the universe belong to the $K-1$ cells already exemplified in e. According to the results of Gaifman, Scott and Krauss[6], the probability $P(C^{(K-1)}/e)$ equals the infimum of the probabilities of this universal statement on finite subsets of the universe (containing the sample e). This is precisely what we get from formula (4).

The value of (4) can be calculated by considering the transition given in Figure 1 as being accomplished in two different ways and and equating the resulting probabilities.

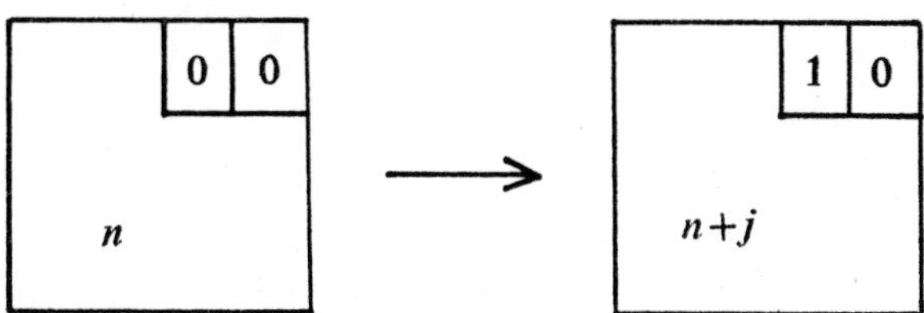

Fig. 1.

This yields

$$\text{(5)} \qquad \begin{aligned} &g(n,K-1)g(n+1,K-1)\ldots g(n+j-1,K-1)(1-\\ &-g(n+j,K-1)) = (1-g(n,K-1))(1-f(1,n+1,K))\ldots\\ &+(1-f(1,n+j,K)). \end{aligned}$$

Here

$$\text{(6)} \qquad 1-f(1,n+i,K) = 1-\frac{1+\dfrac{\lambda}{K}}{n+i+\lambda} = \frac{n+i-1+\dfrac{K-1}{K}\lambda}{n+i+\lambda}.$$

From (5) and (6) we get

$$(7)\qquad \prod_{i=0}^{j-1} g(n+i, K-1) - \prod_{i=0}^{j} g(n+i, K-1) = $$

$$= (1-g(n, K-1)) \prod_{i=1}^{j} \frac{n+i-1+\frac{K-1}{K}\lambda}{n+i+\lambda}$$

Taking the sum over values $j = 1, 2, \ldots, m$ on both sides of (7) we obtain the partial product

$$(8)\qquad \prod_{i=0}^{m} g(n+i, K-1) = g(n, K-1) - (1-g(n, K-1)) \times$$

$$\times \sum_{j=1}^{m} \prod_{i=1}^{j} \frac{n+i-1+\frac{K-1}{K}\lambda}{n+i+\lambda}$$

Hence, the desired probability equals

$$(9)\qquad g(n, K-1) - (1-g(n, K-1)) \frac{\Gamma(n+1+\lambda)}{\Gamma\left(n+\frac{K-1}{K}\lambda\right)} \times$$

$$\times \sum_{j=1}^{\infty} \frac{\Gamma\left(n+j+\frac{K-1}{K}\lambda\right)}{\Gamma(n+j+1+\lambda)}.$$

Here Γ is the Gamma function. The infinite sum in (9) can be evaluated when λ is a multiple of K. The calculation is given in Appendix 2. It yields the following result.

Theorem 9. If λ is a multiple of K, then the posterior probability (4) of constituent $C^{(K-1)}$ is

$$(10)\qquad \prod_{i=0}^{\infty} g(n+i, K-1) = 1 - \frac{n+\lambda}{\frac{\lambda}{K}} f(0, n, K-1).$$

The posterior probability (10) is greater than zero if and only if

$$f(0, n, K-1) < \frac{\frac{\lambda}{K}}{n+\lambda},$$

i.e., if and only if the inductive method determined by f is more optimistic than the methods in Carnap's λ-continuum. When $f(0, n, K-1)$ equals the Carnapian value, the conditional (a posteriori) probability of $C^{(K-1)}$ equals zero. In other words, in the special case at hand, Carnap's λ-continuum represents the most pessimistic possible strategy of inductive generalization.

Notice, incidentally, that for λ which is a multiple of K we obtain as the probability of $C^{(K-1)}$ on evidence which consists of $K-1$ individuals in $K-1$ cells

$$1-\frac{(K-1+\lambda)K}{\lambda}f(0, K-1, K-1)$$

which is greater than zero if and only if $f(0, K-1, K-1)$ is chosen more optimistically than by Carnap. It equals zero for the Carnapian value of $f(0, K-1, K-1)$, and is smaller than zero for a more pessimistic value of $f(0, K-1, K-1)$. This shows that the parameter $f(0, K-1, K-1)$ cannot be chosen more pessimistically than by Carnap.

It remains to estimate the asymptotic behavior of $f(0, n, K-1)$ when n grows without limit. For the purpose, note first that if we are more optimistic than Carnap with respect to the probability $f(1, K-1, K-1)$, we must ever since be more optimistic than Carnap:

Theorem 10. $f(n', n, K-1) > \dfrac{n'+\dfrac{\lambda}{K}}{n+\lambda}$,

if

$$f(1, K-1, K-1) > \frac{1+\dfrac{\lambda}{K}}{K-1+\lambda}.$$

Proof: By induction on n by means of Corollary 7.2.

The condition for $f(1, K-1, K-1)$ in Theorem 10 is equivalent to

$$f(0, K-1, K-1) < \frac{\dfrac{\lambda}{K}}{K-1+\lambda}. \tag{11}$$

N te further that

$$f(0, n+1, K-1) = f(0, n, K-1)\frac{f(n', n+1, K)}{f(n', n, K-1)}$$

by Theorem 3, where

$$f(n', n+1, K) = \frac{n' + \dfrac{\lambda}{K}}{n+\lambda}$$

by Theorem 4. A simple inductive argument shows then that

$$f(0, n+1, K-1) < \frac{\dfrac{\lambda}{K}}{n+1+\lambda}$$

for a suitable value of $f(0, K-1, K-1)$, that is, for a value satisfying (11). Hence the posterior probability (10) will be greater than zero for a suitable initial value of $f(0, K-1, K-1)$, viz. for those values of this parameter which are more optimistic than Carnap's value. A closer analysis shows that when $n \to \infty$, in this case, the posterior probability (10) will approach one as the limit.

A sketch of such an argument can be given as follows. Suppose that $f(0, K-1, K-1)$ satisfies the condition (11). By Theorem 10, for any $m \geqslant K-1$ there is a $\mu > 1$ such that $\alpha(m, K-1)$ is μ times better than Carnap's value, i.e.,

$$\alpha(m, K-1) = \frac{\mu}{m+\lambda}.$$

Then by Corollary 7.2

$$\text{(12)} \qquad \alpha(m+1, K-1)\,(m+\lambda+1) = 1 + \frac{m+\lambda - \dfrac{m+\lambda}{\mu}}{m+\lambda+1-\dfrac{\lambda}{K}} =$$

$$= \frac{\mu-1}{\mu} \cdot \frac{m+\lambda}{m+\lambda+1-\dfrac{\lambda}{K}} + 1.$$

By Theorems 3, 4 and 6, we obtain

$$(m+\lambda)f(0, m, K-1) = (m+\lambda)f(0, m+1, K-1) \times$$

$$\times \frac{f(n', m, K-1)}{f(n', m+1, K)} = (m+\lambda)f(0, m+1, K-1)\alpha(m, K-1) \times$$

$$\times (m+1+\lambda) = (m+j+\lambda) f(0, m+j, K-1)$$

$$\prod_{i=0}^{j-1} \alpha(m+i, K-1)\,(m+i+\lambda)$$

It thus follows that

$$\text{(13)} \qquad \lim_{j\to\infty} \frac{f(0, m+j, K-1)}{\dfrac{\dfrac{\lambda}{K}}{m+j+\lambda}} = \frac{\dfrac{K}{\lambda}(m+\lambda) f(0, m, K-1)}{\lim\limits_{j\to\infty} \prod\limits_{i=0}^{j-1} \alpha(m+i, K-1)\,(m+i+\lambda)}$$

For sufficiently large values of m, the infinite product in the denominator of (13) is, by formula (12), approximately

$$\mu\left(1+\frac{\mu-1}{\mu}\right)\left(1+\frac{1+\dfrac{\mu-1}{\mu}-1}{1+\dfrac{\mu-1}{\mu}}\right)\cdots$$

$$= \mu\left(1+\frac{\mu-1}{\mu}\right)\left(1+\frac{\mu-1}{2\mu-1}\right)\cdots\left(1+\frac{\mu-1}{i\mu-i+1}\right)\cdots$$

The general term here is

$$1+\frac{\mu-1}{i\mu-i+1} = \frac{i(\mu-1)+\mu}{i(\mu-1)+1}\,.$$

Hence the product up to $j-1$ grows without limit when $j \to \infty$. It follows that formula (13) receives the value zero.

This argument shows that the posterior probability (10) approaches one when $n \to \infty$ and condition (11) holds. We have thus established the following theorem.

Theorem 11. Whenever the parameter $f(0, K-1, K-1)$ is more optimistic than Carnap's value (i.e., satisfies condition (11)), the posterior probability (10) of constituent $C^{(K-1)}$ receives asymptotically the value one.

VIII. POSTERIOR PROBABILITIES FOR CONSTITUENTS: GENERAL CASE

Although the results of Section 7 concern directly only the special case in which $c = K-1$ and λ is a multiple of K, they can be generalized.

It can be seen that the product

$$(14) \qquad \prod_{i=0}^{j} g(n+i, K-1)$$

is a monotonic function of λ, when $\lambda \geqslant K$. Hence we can see qualitatively the behavior of the product (14) at large ($\lambda \geqslant K$) by interpolating from our special result (Theorem 11). For all values of $\lambda \geqslant K$ the product (14), and hence the posterior probability of constituent $C^{(K-1)}$, approaches one when $j \to \infty$, provided that $f(0, K-1, K-1)$ is smaller (i.e., more optimistic) than its Carnapian value.

This result can be extended to other values of c, $c < K-1$, by considering the transition given in Figure 2 and by identifying the probabilities of two special ways in which it can be accomplished.

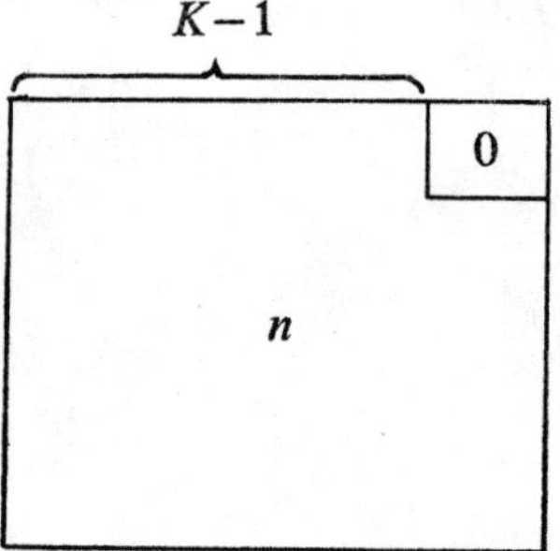

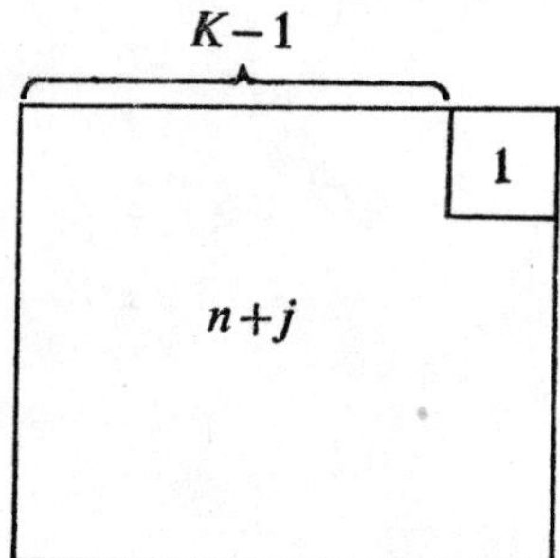

Fig. 2.

This yields

$$f(0, n, K-2) \prod_{i=0}^{j-1} [g(n+1+i, K-1) - f(1, n+1+i, K-1)] =$$
$$= \prod_{i=0}^{j-1} g(n+i, K-2) f(0, n+j, K-2).$$

Since $f(0, n, K-2) > f(0, n+j, K-2)$ for sufficiently large values of j, we have, by Theorem 6 (a) and (b), for non-Carnapian values

$$\prod_{i=1}^{j-1} g(n+i, K-2) = \prod_{i=0}^{j-1} \left(1 - \frac{1 + \dfrac{\lambda}{K}}{n+1+i+\dfrac{(K-1)\lambda}{K}} \right) \times$$
$$\times g(+1+i, K-1).$$

Since the right-hand expression in (15) approaches one when $j \to \infty$ (for non-Carnapian parameter values), the left-hand expression must do likewise.

By putting these results together, the following theorem is obtained.

Theorem 12. Whenever $\lambda \geqslant K$, $1 \leqslant c < K$, and the parameters $f(0, c, c)$ are chosen more optimistically than their Carnapian values, the posterior probability on increasing evidence is asymptotically one for the simplest constituent $C^{(c)}$ compatible with evidence and asymptotically zero for other constituents compatible with evidence.

This theorem shows that we in fact obtain, from our Basic Assumption, a qualitatively (asymptotically) satisfactory treatment of inductive generalization.

IX. INTERRELATIONS OF THE NEW SYSTEM AND CARNAP'S λ-CONTINUUM

In earlier sections, we have established a number of results which illustrate the relation of our Basic Assumption to Carnap's λ-continuum. Theorem 10 which concerns the case $c = K-1$ can be generalized other values of c by a simple inductive argument.

Theorem 13. Assuming

$$f(0, c+i, c+i) < \frac{\frac{\lambda}{K}}{c+i+\lambda}$$

for $i = 0, 1, \ldots, K-c-1$, we have

$$f(n', n, c) > \frac{n' + \frac{\lambda}{K}}{n+\lambda}.$$

Proof. Assume that the claim of this theorem holds when $c > c_0$ and when $n < n_0$. Then from Theorem 7 it follows that it holds for c_0 and n_0, i.e., that it holds for c_0 if it holds for $c > c_0$. The rest follows now from Theorem 10.

We have also seen that the only exception to Theorem 12 concerning inductive generalization is the case when the free parameters of our system have their Carnapian values. In brief, it can be said that Carnap's λ-continuum represents the worst possible case in our wider continuum.

The parameters $f(0, c, c)$ $(c = 1, 2, \ldots, K-1)$ of the new system can be chosen, in several ways, more optimistically than their corresponding Carnapian values. One such possibility which seems interesting is to choose them in Carnapian way, but with a value λ_0 of λ which yields more optimistic values than the λ corresponding to the case $c = K$, i.e.,

$$(16) \qquad f(0, c, c) = \frac{\frac{\lambda_0}{K}}{c+\lambda_0} < \frac{\frac{\lambda}{K}}{c+\lambda}.$$

Here $0 < \lambda_0 < \lambda$. The difference $\lambda - \lambda_0$ in a sense measures the 'distance' of the choice (16) of the parameters from Carnap's λ-continuum, and λ_0 serves as a kind of an index of caution with respect to inductive generalization.

X. THE INTERPRETATION OF THE PARAMETERS

A feeling for the intuitive interpretation of the parameters of our system is obtained from the simplest case $K = 2$. In this case, there are two free parameters for which we can choose $f(0, 1, 1)$ and $f(1, 3, 2)$. They behave somewhat like parameters α and λ, respectively, within Hintikka α-λ-continuum. In fact, by Theorem 4

$$f(n', n, 2) = \frac{n' + \frac{\lambda_2}{2}}{n+\lambda_2},$$

where λ_2 is uniquely determined by $f(1, 3, 2)$. When λ_2 is an even positive integer, Theorem 9 yields as the a priori probability of constituent $C^{(1)}$

$$(17) \qquad \frac{1}{2}\left(1 - \frac{2(\lambda_2+1)}{\lambda_2} f(0, 1, 1)\right) = \frac{1}{2} - \frac{\lambda_2+1}{\lambda_2} f(0, 1, 1).$$

Since $f(1, 1, 1) = 1 - f(0, 1, 1)$, in this case, (17) is equal to

$$(18) \qquad \frac{1}{2}\left[f(1, 1, 1) - (1 - f(1, 1, 1))\frac{\lambda_2+2}{\lambda_2}\right].$$

Here $C^{(1)}$ is a constituent which claims that precisely one particular cell of the partition is instantiated. The a priori probability of the constituent $C^{(2)}$ which claims that both cells are instantiated can be evaluated as follows:

$$(19) \qquad 2g(0,0)\,(1-g(1,1)) + \sum_{m=1}^{\infty} 2g(0,0)g(1,1) \dots g(m,1) \times$$

$$\times (1-g(m+1,1) = 1-g(1,1)+g(1,1) - \prod_{i=1}^{\infty} g(i,1) =$$

$$= 1 - \prod_{i=1}^{\infty} g(i,1) = \frac{2(\lambda+1)}{\lambda} f(0,1,1).$$

It follows that the a priori probability of the constituent $C^{(0)}$ which claims that both cells are empty is zero. (This fact is already implicit in Corollary 1.1.)

The a priori probability of $C^{(2)}$ is thus proportional to $f(0,1,1)$, while the probability of $C^{(1)}$ is inversely related to $f(0,1,1)$. When $f(0,1,1)$ grows towards the Carnapian value

$$\frac{\frac{\lambda_2}{2}}{1+\lambda_2},$$

the probability of $C^{(1)}$ approaches zero and the probability of $C^{(2)}$ approaches one. This corresponds to the case $\alpha \to \infty$ within Hintikka's α-λ-continuum. When $f(0,1,1) = 0$, the probabilities (17) and (19) reduce to 1/2 and 0, respectively. This case falls outside the scope of Hintikka's α-λ-continuum: here the a priori probability is one for the hypothesis that there exists individuals in one and only one cell of the universe, so that already one observation is enough to raise the posterior probability of the corresponding constituent $C^{(1)}$ to one.

When $\lambda_2 \to \infty$, the probabilities (17) and (19) approach the values $1/2 - f(0,1,1)$ and $2f(0,1,1)$, respectively. These probabilities are equal, in this case, if $f(0,1,1) = 1/6$.

In the general case ($K \geqslant 2$), parameter $f(1, K+1, K)$ helps to govern the behavior of our system with respect to singular inductive inference, while parameters $f(0, c, c)$, $c = 1, 2, \dots, K-1$, help to govern its behavior both with respect to singular inductive inference and with respect to inductive generalization. The two kinds of inference are thus less sharply distinguished from one another than in Hintikka's α-λ-system, and their intuitive content accordingly is less clear.

The interpretation of our parameters along these lines can be further illustrated by pointing out that when $\lambda = K$ and when the parameters $f(0, c, c)$, $c = 1, 2, \ldots, K-1$, are chosen in a suitable way we obtain Hintikka's generalized combined system, which likewise results from setting the λ of Hintikka's α-λ-continuum equal to K. This possibility of deriving the generalized combined system in two essentially different ways seems to be a strong argument for its naturalness.

The representative function of Hintikka's generalized combined system was given above in formula (3). It follows from (3) that

$$f(n', n, K) = \frac{n'+1}{n+K},$$

which corresponds to Theorem 4 with $\lambda_K = K$. By (3) we obtain further

$$f(n', n, K-1) = \frac{(n'+1)(n+2K+\alpha-1)}{(n+K)(n+2K+\alpha-2)}.$$

Hence

$$f(0, n, K-1) = \frac{K+\alpha-1}{(n+K)(n+2K+\alpha-2)}.$$

and, in particular,

$$f(0, K-1, K-1) = \frac{1}{(2K-1)\left(1+\dfrac{2K-2}{K+\alpha-1}\right)} \tag{20}$$

From formula (20) and Theorem 9 we see that the posterior probability of constituent $C^{(K-1)}$ on evidence exemplifying $K-1$ cells is

$$1-\frac{K+\alpha-1}{n+2K+\alpha-2}.$$

Formula (20) defines a one-to-one correspondence between certain values of the parameter $f(0, K-1, K-1)$ of the new system and the parameter α of Hintikka's generalized combined system. When α grows from 0 to ∞, $f(0, K-1, K-1)$ grows from $\dfrac{1}{3(2K-1)}$ to $\dfrac{1}{2K-1}$. The latter equals the corresponding Carnapian value, in this case. On the other hand, the values

$$f(0, K-1, K-1) < \frac{1}{3(2K-1)}$$

fall outside the scope of the combined system, i.e., they are more optimistic than the inductive strategies within Hintikka's α-λ-continuum with $\lambda(w) = w$.

XI. CONCLUSIONS

What are the philosophical implications of our results? The following are among the general points that can be made on the basis of our observations.

(1) It is possible to build on a simple axiomatic basis a logic of inductive generalization which is at least qualitatively adequate and which allows for a considerable flexibility in the choice of its characteristic parameters. This can be done in such a way that we obtain Carnap's λ-continuum as a limiting case and Hintikka's generalized combined system as another special case.

Thus our results serve to emphasize the naturalness of a logic of inductive *generalization*. It is only when the representative function is allowed to disregard completely the generalization aspect (i.e., disregard c) that one does not assign nonzero probabilities to generalizations also in an infinite universe.

(2) On earlier occasions, Hintikka has emphasized that if the symmetry (exchangeability) assumption is adhered to, to 'bet' on (associate behavioristically detectable probabilities with) *strict generalizations* (in an infinite universe) amounts to betting in a certain manner on singular events (on the basis of finite evidence).[7] How one is willing to bet on generalizations nevertheless is usually shown by no particular finite bets one makes but rather one's whole strategy of betting on singular events (on finite evidence).

However, if the Basic Assumption is made, we can go further. Given the Basic Assumption, how I am going to bet on generalizations is shown by the finite set of values I give to $f(1, c, c)$ $(c = 1, 2, \ldots, K-1)$

and $f(1, K+1, K)$, i.e., is shown by a finite number of odds for betting on singular event on finite (and in fact quite small) evidence. This demonstrates strikingly the possibility of connecting closely bets on generalizations and bets on finite events.

APPENDIX 1

A proof for Theorem 4 is given in this appendix. The following lemmas are immediate corollaries of Theorem 1.

Lemma 1. $f(n-K+1, n, K) = 1-(K-1)f(1, n, K)$, *if* $n \geqslant K$.

Lemma 2. $f(n-K+1, n+1, K)+f(2, n+1, K)+(K-2)f(1, n+1, K) = 1$, *if* $n \geqslant K$.

As a special case of Corollary 2.1 we obtain

Lemma 3. $f(n', n+1, K) = f(1, n+1, K)\dfrac{f(n', n, K)}{f(1, n, k)}$, *provided that* $0 < n' \leqslant n-K+1$.

By combining Lemma 2 and Lemma 3, we obtain

Lemma 4. $$f(1, n+1, K)\left[\frac{f(n-K+1, n, K)}{f(1, n, K)} + \frac{f(2, n, K)}{f(1, n, K)} + (K-2)\right] = 1.$$

Lemma 1 and Lemma 4 together yield

Lemma 5. $$\frac{f(1, n+1, K)}{f(1, n, K)} = \frac{1}{1-f(1, n, K)+f(2, n, K)}.$$

We are now ready to prove the following result.

Claim. $$f(n', n, K) = \frac{(n'-1)-(K(n'-1)-1)\beta}{(n-K)-(n-K-1)K\beta}$$

where $\beta = f(1, K+1, K)$, $0 \leqslant \beta \leqslant 1$, $n' > 0$, *and* $n \geqslant K > 2$.

Proof. By induction on $n \geqslant K$. Note first that when $n = K$, then $n' = 1$ and $f(1, K, K) = \dfrac{1}{K}$. When $n = K+1$ and $n' = 1$,

$$f(1, K+1, K) = \frac{0-(K\cdot 0-1)\beta}{1-0\cdot K\beta} = \beta.$$

Inductive step:

$$f(n', n+1, K) = f(n', n, K)\frac{f(1, n+1, K)}{f(1, n, K)} \qquad \text{(by Lemma 3)}$$

$$= \frac{f(n', n, K)}{1-f(1, n, K)+f(2, n, K)} \qquad \text{(by Lemma 5)}$$

$$= \frac{[(n'-1)-(K(n'-1)-1)\beta]\times [(n-K)-(n-K-1)K\beta]}{[(n-K)-(n-K-1)K\beta]\,[(n-K)-(n-K-1)K\beta-\beta+1-(K-1)\beta]} \qquad \text{(by inductive hypothesis)}$$

$$= \frac{(n'-1)-(K(n'-1)-1)\beta}{(n-K+1)-(n-K)K\beta}.$$

Theorem 4 follows from the above Claim by putting

$$\lambda_K = \frac{K\beta}{1-K\beta} - K.$$

APPENDIX 2

A proof of Theorem 9 is given in this appendix. We shall first prove the following result.

Lemma 6. If a and b are positive integers, then

$$\sum_{j=1}^{\infty} \frac{\Gamma(j+a)}{\Gamma(j+a+b+1)} = \frac{\Gamma(a+1)}{b\Gamma(a+b+1)}.$$

Proof. Let

$$r(j, b) = \frac{\Gamma(j+a)}{\Gamma(j+a+b+1)} = \frac{1}{(j+a)\,(j+a+1)\ldots(j+a+b)}.$$

Then $r(j, b)$ can be expressed as a sum

$$r(j, b) = \sum_{k=0}^{b} \frac{c_k}{j+a+k}$$

where the coefficients c_k, $k = 0, 1, \ldots, b$, are determined by

$$c_k = \lim_{j \to -(a+k)} \frac{1}{(j+a) \ldots (j+a+k-1)(j+a+k+1) \ldots (j+a+b)}$$

$$= \frac{(-1)^k}{k!(b-k)!} = \frac{(-1)^k \binom{b}{k}}{b!}$$

A simple inductive argument shows that

$$d_k = \sum_{i=0}^{k} c_i = \frac{(-1)^k}{k!(b-k-1)!\,b} = \frac{(-1)^k \binom{b-1}{k}}{b!}$$

for $k = 0, 1, \ldots, b-1$, and

$$d_b = \sum_{i=0}^{b} c_i = 0.$$

Hence the infinite sum of $r(j, b)$, $j = 1, 2, \ldots$, reduces to a finite sum, and can be evaluated as follows:

$$\sum_{j=1}^{\infty} r(j, b) = \sum_{k=0}^{b-1} \frac{d_k}{a+k+1} = \sum_{k=0}^{b-1} \frac{(-1)^k \binom{b-1}{k}}{b!(1+a+k)} =$$

$$= \frac{(b-1)!}{b!} r(1, b-1) = \frac{r(1, b-1)}{b} = \frac{\Gamma(a+1)}{b\Gamma(a+b+1)}$$

This completes the proof of Lemma 6.

Suppose now that $\lambda(= \lambda_K)$ is a multiple of K, i.e., $\dfrac{\lambda}{K} = b$ is a positive integer. Let $a = n + (K-1)b$. Then, by Lemma 6,

$$\frac{\Gamma(n+1+\lambda)}{\Gamma\left(n + \dfrac{K+1}{K}\lambda\right)} \sum_{j=1}^{\infty} \frac{\Gamma\left(n+j+\dfrac{K-1}{K}\lambda\right)}{\Gamma(n+j+1+\lambda)} =$$

$$= \frac{\Gamma(a+b+1)}{\Gamma(a)} \sum_{j=1}^{\infty} \frac{\Gamma(j+a)}{\Gamma(j+a+b+1)} =$$

$$= \frac{\Gamma(a+b+1)\Gamma(a+1)}{b\Gamma(a)\Gamma(a+b+1)} = \frac{a}{b} = \frac{K}{\lambda}\left(n + \frac{K-1}{K}\lambda\right).$$

By formula (9), we obtain as the a posteriori probability of constituent $C^{(K-1)}$

$$g(n, K-1) - (1 - g(n, K-1))\frac{K}{\lambda}\left(n + \frac{K-1}{K}\lambda\right)$$

which equals

$$1 - f(0, n, K-1)\frac{K}{\lambda}(n+\lambda),$$

since $g(n, K-1) = 1 - f(0, n, K-1)$ by Theorem 6.

REFERENCES

[1] This assumption is Carnap's 'Axiom of Predictive Irrelevance'. For the derivation of Carnap's λ-continuum, see J. G. Kemeny, 'Carnap's theory of probability and induction', in P. A. Schilpp (ed.), *The Philosophy of Rudolf Carnap*, Open Court, LaSalle, Ill., 1963, pp. 711–738, R. Carnap and W. Stegmüller, *Induktive Logik und Wahrscheinlichkeit*, Springer-Verlag, Wien, 1958, pp. 242–249; and R. Carnap, 'Notes on probability and induction', *Synthese* **25** (1973), pp. 286–292. This assumption is termed 'W. E. Johnson's sufficiency postulate' in I. J. Good, *The Estimation of Probabilities*, Research Monograph No. 30, The MIT Press, Cambridge, Mass., 1965, p. 17.

[2] See B. de Finetti, 'Foresight: Its logical laws, its subjective sources', in H. E. Kyburg and H. E. Smokler (eds.), *Studies in Subjective Probability*, John Wiley, New York, 1964, pp. 93–158.

[3] For a brief discussion on the development of Carnap's own views about the intuitive interpretation of λ, see R. Hilpinen, 'Carnap's new system of inductive logic', *Synthese* **25** (1973), pp. 311–313.

[4] See J. Hintikka, 'A two-dimensional continuum of inductive methods', in J. Hintikka and P. Suppes (eds.), *Aspects of Inductive Logic*, North-Holland, Amsterdam 1966, pp. 113–132.

[5] The 'combined system' was originally introduced in J. Hintikka, 'On a combined system of inductive logic', in *Studia Logico-Mathematica et Philosophica in Honorem Rolf Nevanlinna*, *Acta Philosophica Fennica* **18** (1965), pp. 21–30. For its generalization (to positive values of α), see the work referred to in Note 4.

[6] See H. Gaifman, 'Concerning measures on first-order calculi', *Israel Journal of Mathematics* **2** (1964), 1–18; and D. Scott and P. Krauss, 'Assigning probabilities to logical formulas', in J. Hintikka and P. Suppes (eds.), *Aspects of Inductive Logic*, North-Holland, Amsterdam, 1966, p. 224.

[7] Cf. J. Hintikka, 'Unknown probabilities, Bayesianism, and de Finetti's Representation Theorem', in R. C. Buck and R. S. Cohen (eds.), *Boston Studies in the Philosophy of Science VIII*, D. Reidel, Dordrecht, 1971, pp. 325–341.

THEO A. F. KUIPERS

A TWO-DIMENSIONAL CONTINUUM OF A PRIORI PROBABILITY DISTRIBUTIONS ON CONSTITUENTS

I

Hintikka has defined a *one-dimensional* continuum of a priori probability distributions on constituents and has built on it a two-dimensional continuum of inductive methods with the aid of Carnap's λ-continuum ([1]), which plays also a fundamental role in his continuum of a priori distributions, and the formula of Bayes ([2]). Here a *two-dimensional* continuum of a priori probability distributions will be introduced. On its base a three-dimensional continuum of inductive methods can be constructed in the same way as Hintikka has done. The importance of the new continuum of a priori distributions, which is also based on Carnap's λ-continuum but in a completely different way, is that it leaves room for almost all kinds of a priori considerations, whereas Hintikka's continuum admits only considerations that lead to increasing probability for the constituents by increasing size.

In this article we will show furthermore (in section 10) the mathematical equivalence between a *generalized carnapian continuum* and the corresponding (*generalized*) so-called *Polya-distribution.*

II

There is given a set of K ($\geqslant 2$) so-called *Q-predicates*. They are supposed to be mutually exclusive and together exhaustive with respect to the universe considered. The *constituent-structure* S_w tells us that exactly w Q-predicates are exemplified in our universe (this concept was introduced by Carnap in [3]). A *constituent of size* w, abbreviated by C_w, tells us not only that there are w Q-predicates exemplified (i.e. that S_w holds) but it specifies also which ones. In the context of P (probability)--values for constituents we do not make any difference between constituents of the same size. Therefore C_w refers to an arbitrary constituent of size w and we have therefore: $P(S_w) = \binom{K}{w}\ P(C_w)$.

III

Hintikka defines $P(C_w)$ proportional to the probability, according to Carnap's λ-continuum that α individuals turn out to be compatible with C_w, hence $P(C_w)$ is equal to

$$\pi(\alpha, w \cdot \lambda/K) \sum_{v=0}^{K} \binom{K}{v} \pi(\alpha, v \cdot \lambda/K), \; \alpha \geqslant 0, \; \lambda > 0,$$

where $\pi(n, x) =_{\mathrm{df}} x \cdot (x+1) \ldots (x+n-1)$ if $n = 1, 2, 3, \ldots$ and $\pi(0, x) =_{\mathrm{df}} 1$.
It is easily seen that this distribution is directed to disorder, i.e. it satisfies the inequality $P(C_w) < P(C_v)$ if $0 \leqslant w < v \leqslant K$, as soon as $\alpha > 0$. If $\alpha = 0$ all constituents get the same probability, 2^{-K}.

IV

According to Hintikka a priori considerations of symmetry suggest that P has to satisfy the inequality just mentioned ([2], p. 117). It is not clear whether Hintikka would come to the same conclusion if he did not have assumed that the Q-predicates are defined in terms of a number of primitive, logically independent predicates. How this may be, we only assume that the Q-predicates are mutually exclusive and together exhaustive (on logical, linguistic or empirical grounds) with respect to our universe and we do not even assume that the universe is non--empty. To our opinion we have therefore to leave room for as many as possible different conclusions based on different a priori logical, statistical or metaphysical considerations.

V

In order to obtain such a rich continuum of distributions we start to look more closely to the relation between the universe of individuals, U, and our set of Q-predicates. The latter is in fact a second order universe, U_Q, with K elements and we have that for all $u \in U$ there is exactly one $Q \in U_Q$ such that Qu holds. Now we define two *metapredicates* E and N as follows: if $Q \in U_Q$ then

$$E(Q) \quad \textit{if and only if} \quad \exists_{u \in U} Qu$$

$$N(Q) \quad \textit{if and only if} \quad \sim \exists_{u \in U} Qu .$$

It is easily seen that these two predicates are mutually exclusive and together exhaustive with respect to U_Q, given U.

Suppose now that we were going to investigate the elements of U_Q, i.e. observe for each Q in U_Q, whether $E(Q)$ or $N(Q)$, then we would certainly use a generalized carnapian continuum for a family with two predicates in order to calculate a priori and a posteriori probabilities. By a *generalized carnapian continuum* we mean a λ-continuum with not necessarily equal logical weights for the members of the family of predicates and with room for 'negative induction', i.e. negative values for λ are permitted as far as they do not lead to negative 'probabilities'.

The foregoing includes, among others, an assignment of a priori probabilities to constituents and therefore to constituent-structures. However, in fact we are not going to investigate U_Q, at least not in a direct way, but U. But we need a priori probabilities for constituents and our proposal is to take the same values as those which we would take if we were going to investigate U_Q. Because there is no logical difference before we start our investigations this proposal seems to be reasonable.

VI

The assignment proposed in section V is the following. Assign weights p and q to E ('to be exemplified') and N ('not to be exemplified') resp. such that $p = 1-q$ and $0 < p < 1$, and therefore $0 < q < 1$. If exactly m_E of the Q-predicates $Q_1, Q_2, \ldots, Q_m$ $(m < K)$ turn out to be exemplified in U then the probability that Q_{m+1} will also be exemplified is, according to Carnap's λ-continuum (with $\lambda > 0$), equal to $(m_E+p\lambda)/(m+ +\lambda)$ and therefore we get, by applying the product rule, as the a priori probability of a constituent C_w:

$$\frac{p\lambda\cdot(p\lambda+1)\ldots(p\lambda+w-1)\cdot q\lambda(q\lambda+1)\ldots(q\lambda+K-w-1)}{\lambda(\lambda+1)(\lambda+2)\ldots(\lambda+K-1)}$$

which is equal to:

(1) $\qquad \pi(w, p\lambda)\cdot\pi(K-w, q\lambda)/\pi(K, \lambda), \ \lambda > 0.$

We have assumed for a moment that λ is positive, but (1) is defined for all real values of λ except for $\lambda = 0, -1, -2, \ldots, -K+1$. However, not all these permitted negative values for λ lead to non-negative values for (1) and this is necessary if we want to have a genuine probability distribution. Now if λ is negative, say $-\lambda = \gamma > 0$, we get positive values by (1) as soon as $p\gamma > K-1$ and $q\lambda > K-1$ (and therefore $\gamma > 2K-2$), which is easily seen by rewriting (1) in terms of γ,

$$\frac{p\gamma(p\gamma-1) \ldots (p\gamma-w+1)\cdot q\gamma(q\gamma-1) \ldots (q\gamma-K+w+1)}{\gamma(\gamma-1) \ldots (\gamma-K+1)},$$

or in terms of $\varrho(n, x) =_{\mathrm{df}} x\cdot(x-1) \ldots (x-n+1)$, if $n = 1, 2, 3, \ldots$ and $\varrho(0, x) =_{\mathrm{df}} 1$.

(2) $\qquad \varrho(w, p\gamma)\cdot\varrho(K-w, q\gamma)/\varrho(K, \gamma), \; -\lambda = \gamma > 0.$

As we will see at the end of section 8 there are included in (2) non-negative assignments for which $p\gamma$ and/or $q\gamma$ do not exceed $K-1$.

VII

It is possible to reformulate (1) and (2) in terms of the *gamma-function* and in terms of the *beta-function*. If $\lambda > 0$ then (1) is equivalent to

(1*) $$\frac{\Gamma(p\lambda+w)\cdot\Gamma(q\lambda+K-w)}{\Gamma(\lambda+K)} \bigg/ \frac{\Gamma(p\lambda)\cdot\Gamma(q\lambda)}{\Gamma(\lambda)}$$

which is also equivalent to

(1**) $\qquad B(p\lambda+w, q\lambda+K-w)/B(p\lambda, q\lambda).$

If $0 > \lambda = -\gamma$ and $p\gamma > K-1$ and $q\gamma > K-1$ then (2) is equivalent to

(2*) $$\frac{(\gamma+1)\Gamma(p\gamma+1)\cdot\Gamma(q\gamma+1)}{\Gamma(\gamma+2)} \bigg/ \frac{(\lambda-K+1)\Gamma(p\gamma-w+1)\cdot\Gamma(q\gamma-K+w+1)}{\Gamma(\gamma-K+2)}$$

which is also equivalent to

(2**) $\qquad (\gamma+1)\cdot B(p\gamma+1, q\gamma+1)/(\gamma-K+1)B(p\gamma-w+1, q\gamma-$
$-K+w+1).$

In order to obtain these alternative formulations one only needs the following theorem

$$\Gamma(x+1) = x\cdot\Gamma(x), x > 0$$

and the definition of the beta-function:

$$B(x,y) =_{\mathrm{df}} \Gamma(x)\Gamma(y)/\Gamma(x+y),\ x > 0,\ y > 0.$$

VIII

Now we will consider some particular symmetric distributions ($p = q$) that result if λ assumes a particular value. Note first that $p = q$ implies $P(C_w) = P(C_{K-w})$ ($w = 0, 1, 2, \ldots, K$).

$\lambda \to \pm\infty$ $\quad P(C_w) = 2^{-K},\ w = 0, 1, 2, \ldots, K.$
here all constituents get the same probability and therefore we have that the constituent-structure of size w gets a probability exactly proportional to the number of constituents of size w; this assignment corresponds to *Hintikka's assignment* (see section 3) with $\alpha = 0$.

$\lambda = 2$: $\quad P(S_w) = (K+1)^{-1},\ w = 0, 1, 2, \ldots, K.$
here all constituent-structures get the same probability; this assignment has been suggested by Carnap in [3].

$-\lambda = \gamma = 2K$: $\quad P(C_w) = \binom{K}{w} \Big/ \binom{2K}{K},\ w = 0, 1, \ldots, K.$
here the probability of a constituent of size w is exactly proportional to the number of constituents with size w.

The last assignment ($\gamma = 2K$) is an extreme case of the following two generalizations for which a satisfactory interpretation has still to be found:

$-\lambda = \gamma = 2rK$:
$r = 1, 2, 3, \ldots$ $\quad P(S_w) = \binom{rK}{w}\binom{rK}{K-w} \Big/ \binom{2rK}{K},\ w = 0, 1, 2, \ldots, K.$

$-\lambda = \gamma = 2K - 2\dot{Z}$: $\quad P(C_{\dot{Z}-i}) = 0 = P(C_{K-(\dot{Z}-i)}),\ i = 1, 2, \ldots, \dot{Z}$

$K = 2L$ *or* $2L+1$, *and*

$$\dot{Z} = 0, 1, 2, \ldots, L. \qquad P(C_{\dot{Z}+i}) = P(C_{K-(\dot{Z}+i)}) = = \binom{K-2\dot{Z}}{i} \Big/ \binom{2K-2\dot{Z}}{K-\dot{Z}}, \quad i = 0, 1, 2, \ldots, L-\dot{Z}.$$

IX

In the previous section we have seen that the proposed continuum of distributions includes a great variety of symmetric distributions. As to asymmetric distributions we will restrict our attention to the case $\lambda > 0$. Our aim is to indicate that under certain restrictions on the possible values of the parameters p and λ the a priori distribution satisfies certain relations of monotony which may be seen as reflecting specific a priori considerations.

We will only consider $p > \frac{1}{2}$ because all relations to be given change in the opposite one if $p < \frac{1}{2}$ and if p and q are interchanged in the conditions.

I *for all* $w < K-w$: $P(C_w) < P(C_{K-w})$ *and therefore* $P(S_w) < < P(S_{K-w})$.

II *if* $\lambda(p-q) > K-1$ *then for all* $w < v$: $P(C_w) < P(C_v)$ (as we have seen in section 3 this relation holds generally in Hintikka's distribution as soon as $\alpha > 0$)

III *if* $q\lambda - 1 < 0 < p\lambda - 1$ *or* $0 \leqslant q\lambda - 1 < (p\lambda - 1)/K$ *or* $0 \leqslant 1 - p\lambda < (1 - q\lambda)/K$ *then for all* $w < v$: $P(S_w) < P(S_v)$

IV *if* $qK < \lambda q(p-q) + q < p$ *or* $K - p\lambda \leqslant Kq\lambda - p\lambda < K-1 < < \lambda(p-q)$ *then for all* $w < v$: $P(C_w) < P(C_v)$ *and* $P(S_w) < < P(S_v)$ (the first disjunctive condition combines the (sole) condition of II with the first of III, the second combines that of II with the second of III; the condition of II is incompatible with the third of III)

The proofs for all these theorems are elementary: the simplest way to start is to write $P(C_w) < P(C_{w+1})$ or $P(S_w) < P(S_{w+1})$ in terms of the gamma-function (see section 7).

From these theorems we may conclude that the parameters p (if $\neq q$) and λ determine in a complex way the character of an asymmetric assignment. In section 11 we will see that $p \cdot K$ is the a priori expectation of the size of the true constituent whereas λ determines the a priori correlation coefficient between the possibilities of being exemplified of two different Q-predicates.

X

In section V we have defined a generalized carnapian continuum as a λ-continuum with suitable real values for λ and a logical weight for each Q-predicate. In order to give some a priori and limit properties of our particular continuum of distributions we prove in this section the mathematical equivalence between such a generalized carnapian continuum and the (generalized) so-called *Polya-distribution* ([4], [5], [6]).

Suppose we have an urn containing N balls such that there are N_i with property (colour) $Q_i (i = 1, 2, \ldots, K^*)$, $\sum_{i=1}^{K^*} N_i = N$. Let p_i be equal to N_i/N, $\sum_{i=1}^{K^*} p_i = 1$.

Our experiments consist out of succesive random selections of one ball with replacement of $\Delta+1$ balls with the same colour. Here Δ is some integer; if Δ is negative then there are some obvious restrictions to the total number of experiments. Note that if $\Delta = 0$ we have drawings with replacement and that if $\Delta = -1$ we have drawings without replacement. Let N/Δ be equal to λ.

The result of the first m trials is indicated by e_m. Then e_m specifies for each i how many trials have resulted in Q_i, say m_i. Let h_i indicate the hypothesis that the next (the $m+1-$th) trial will result in Q_i. Now it is easy to prove that the objective conditional probability of h_i, given e_m, is equal to

$$\frac{m_i \cdot \Delta + N_i}{m \cdot \Delta + N} = \frac{m_i + N_i/\Delta}{m + N/\Delta} = \frac{m_i + p_i \lambda}{m + \lambda},$$

which is equivalent to the corresponding 'special value' in a generalized carnapian continuum.

By applying the product rule we obtain as objective unconditional probability of e_m: $\prod_{i=1}^{K^*} \pi(m_i, p_i \lambda)/\pi(m, \lambda)$
and this is equivalent to the a priori probability of the state-description corresponding to e_m in a generalized carnapian continuum.

XI

Friedman has given in [6], a number of a priori properties of the (actual) Polya distribution in which $K^* = 2$ and this corresponds therefore to our continuum defined by (1). Let x_m be a random variable that equals 1 if Q_m tourns out to be exemplified and 0 if not. Let the random variable X be defined as $x_1 + x_2 + \dots + x_K$. In the following E indicates the (mathematical) expectation operator.

According to [6] we have the following a priori properties of our continuum of distributions defined by (1):

$$\mathrm{E}(x_m) = p$$

i.e. *the a priori expectation of* x_m, *or the a priori probability that* Q_m *appears to be exemplified*; note the independence (at least explicitely) *of* K, λ and m.

$$\mathrm{E}(X) = Kp = \sum_{w=0}^{K} wP(S_w)$$

i.e. *the a priori expectation of* X, *or the a priori expectation of the size of the actual constituent*: E (*size*); note the independence of λ.

$$\mathrm{E}(x_m - \mathrm{E}(x_m))^2 = p \cdot q$$

i.e. *the a priori variance of* x_m;
note the independence of K, λ and m.

$$\mathrm{E}\{X - \mathrm{E}(X)\}^2 = Kpq(\lambda + K)/(\lambda + 1) =$$

$$= \sum_{w=0}^{K} P(S_w) \left\{ w - \sum_{w=0}^{K} wP(S_w) \right\}^2$$

i.e. *the a priori variance of* X: VAR (*size*);
note the independence on K, λ *and* p.

$$\mathrm{E}\{(x_m - \mathrm{E}(x_m))\,(x_n - \mathrm{E}(x_n))\} = p \cdot q/(\lambda + 1)$$

i.e. *the a priori covariance of* x_m *and* x_n $(m \neq n)$.

$$\frac{\mathrm{E}\{(x_m - \mathrm{E}(x_m))\,(x_n - \mathrm{E}(x_n))\}}{\sqrt{\mathrm{E}(x_m - \mathrm{E}(x_m))^2 \cdot \mathrm{E}(x_n - \mathrm{E}(x_n))^2}} = 1/(\lambda+1) =_{\mathrm{df}} c.c.$$

i.e. *the a priori correlation coefficient between* x_m *and* x_n $(m \neq n)$.

$$\mathrm{E}(X - \mathrm{E}(X))^2 \Big/ \sum_{m=1}^{K} \mathrm{E}(x_m - \mathrm{E}(x_m))^2 = (\lambda+K)/(\lambda+1) =_{\mathrm{df}} Q^2$$

i.e. *the a priori dispersion coefficient of X in relation to the* x_m*s*.

XII

In [5] Polya has presented the limit-distributions of the Polyadistribution. Here we are only interested in discrete limit-distributions.

K *finite*, $\lambda \to \infty, 0 < p < 1$

$$P(S_w) = \binom{K}{w} p^w q^{K-w} \qquad \textit{(binominal distribution)}$$

$\mathrm{E}(\textit{size}) = Kp$; VAR $(\textit{size}) = Kpq$; $Q^2 = 1$; $c.c. = 0$

$K \to \infty, \lambda \to \infty, p \to 0$, such that $Kp \to h > 0$ and $K/\lambda \to 0$

$$P(S_w) = e^{-h} h^{w}/w! \qquad \textit{(Poisson distribution)}$$

E $(\textit{size}) = h$; VAR $(\textit{size}) = h$; $Q^2 = 1$; $c.c. = 0$

$K \to \infty, \lambda \to \infty, p \to 0$, such that $Kp \to h > 0$ and $K/\lambda \to$ $\to d > 0$

$$P(S_w) = \binom{h/d + w - 1}{w} \left(\frac{1}{1+d}\right)^{h/d} \left(\frac{d}{1+d}\right)^{w}$$

(*a special case of the negative binominal distribution*)

E $(\textit{size}) = h$; VAR $(\textit{size}) = h(1+d)$; $Q^2 = 1+d$; $c.c. = 0$

From the last two results we may conclude that our proposal leaves room for a denumerably infinite number of Q-predicates. In this case we get zero probability for all constituents, but positive probabilities for constituent-structures.

XIII

In this section we will present the a posteriori distribution on constituent--structures that results if we substitute our a priori continuum of distributions in *Hintikka's frame-work* ([2]) based on Bayes' formula. Suppose we are investigating the individuals of U one by one, with or without replacement. Let e_n be a state-description of n investigations, according to which there are n_i exemplifications of Q_i and c is the number of Q_i' s such that $n_i > 0$. In the following we assume that e_n is compatible with C_w in the expression $P(e_n/C_w)$.

According to Bayes' formula the a posteriori probability of S_w ($w \geqslant c$), given e_n:

$$P(S_w/e_n) = \binom{K-c}{w-c} P(C_w)P(e_n/C_w) \Big/ \sum_{v=c}^{K} \binom{K-c}{v-c} P(C_v)P(e_n/C_v).$$

In the same way as Hintikka has done we take $P(e_n/C_w)$ equal to the probability of e_n in a carnapian continuum with w Q-predicates. Because the parameter λ occurs already in our a priori distribution we replace here however $\lambda(\lambda(w))$ by $\eta(\eta(w))$. Then we get

$$P(e_n/C_w) = \prod_{i-1}^{K} \pi(n_i, \eta(w)/w)/\pi(n, \eta(w))$$

For convenience we strict our attention to the case that $\eta(w) = w$. Then $P(e_n/C_w)$ is equal to

$$\prod_{i=1}^{K} n_i!/\pi(n, w) \tag{4}$$

Replacing $P(C_w)$ by (1) (section 6) and $P(e_n/C_w)$ by (4) in (3) leads to:

$$P(S_w/e_n) = \frac{\binom{K-c}{w-c} \pi(w, p\lambda)\cdot\pi(K-w, q\lambda)/\pi(n, w)}{\sum_{v=c}^{K} \binom{K-c}{v-c} \pi(v, p\lambda)\cdot\pi(K-v, q\lambda)/\pi(n, v)}$$

In case of symmetric a priori distributions $\left(p = q = \frac{1}{2}\right)$ we get e.g.

$$\textit{if } \lambda \to \pm\infty: \quad P(S_w/e_n) = \binom{K}{w}\binom{w}{c}\Big/\pi(n, w)\Big/ \left\{\sum_{v=c}^{K}\binom{K}{v}\binom{v}{c}/\pi(n, v)\right\}$$

$$\textit{if } \lambda = 2: \quad P(S_w/e_n) = \binom{w}{c}\Big/\pi(n, w)\Big/\left\{\sum_{v=c}^{K}\binom{v}{c}\Big/\pi(n, v)\right\},$$

note that $P(S_w/e_1) = 1/K(w = 1, 2, ..., K)$

$$\textit{if } \lambda = -2K: \quad P(S_w/e_n) = \binom{K-c}{w-c}\binom{K}{w}\Big/\pi(n, w)\Big/ \left\{\sum_{w=c}^{K}\binom{K-c}{v-c}\binom{K}{v}\Big/\pi(n, v)\right\}$$

XIV

In summary we hope to have shown that the proposed two-dimensional continuum of a priori distributions leaves indeed room for many kinds of considerations that we may have in a particular inductive situation. However, the problem of finding an acceptable a priori distribution is not a simple question of deductive nature, for any distribution includes e.g. an a priori expectation of the size of the true constituent and it is clear that such an expectation cannot be deductively derived from what we actually know about the universe of individuals and the set of Q-predicates if we are confronted with an inductive situation.

BIBLIOGRAPHY

[1] Carnap, R., *The Continuum of Inductive Methods*, The University of Chicago Press, Chicago 1952.

[2] Hintikka, J., 'A two-dimensional continuum of inductive methods', in *Aspects of Inductive Logic* (ed. by J. Hintikka and P. Suppes), North-Holland Publ. Comp. Amsterdam 1966, pp. 113–132.

[3] Carnap, R., 'The concept of constituent-structure', in *The Problem of Inductive Logic* (ed. by I. Lakatos), North-Holland Publ. Comp., Amsterdam 1968, pp. 218–220.

[4] Eggenberger, F. and Polya, G., 'Ueber die Statistik verketterter Vorgänge'; *Zeitschrift für Angewandte Mathematik und Mechanik*, vol. **3**, pp. 279–289 (1923).

[5] Polya, G., 'Sur quelques points de la théorie des probabilités' *Annales de l'Institut Henri Poincaré*, vol. **1**, pp. 117–161 (1931).

[6] Friedman, B., 'A simple urn model'; *Communications on Pure and Applied Mathematics*, vol. **2** pp. 59–70 (1949).

ILKKA NIINILUOTO

INDUCTIVE LOGIC AND THEORETICAL CONCEPTS

I. INDUCTIVE LOGIC: AIMS AND DEVELOPMENTS

The first serious attempts to base the theory of non-demonstrative scientific inference upon the notion of probability were presented by philosophers of science in the 1870's. In his *Principles of Science* (1874), Stanley Jevons argued that the Laplacean doctrine of inverse probability could be applied to determine the probability of a scientific hypothesis relative to the data or evidence supporting it.[1] In 'The Doctrine of Chance' (1878), Charles S. Peirce defined—referring to Locke the probability associated with a mode of argument, as the proportion of cases in which it carries truth with it. While Jevons attributed probability to the conclusion of an inductive argument, Peirce regarded inductive probabilities as truth-frequencies that may be displayed in repeated applications of a (non-demonstrative) mode of argument.

These two ideas have played a prominent role in the subsequent probabilistic theories of induction. The modern variants of Peirce's idea of truth-frequencies include the Reichenbach-Salmon theory of the 'weights' of individual events and the Neyman-Pearson theory of statistical inference. The idea of applying Bayes' Theorem to evaluate the posterior probability of a hypothesis relative to supporting evidence has been developed in two main directions. Frank P. Ramsey, Bruno de Finetti and L. J. Savage have laid the foundations of modern Bayesian statistics and decision theory, based upon a subjectivistic or 'personalist' theory of probability as a degree of rational belief. The so-called Cambridge School, including C. D. Broad, J. M. Keynes, W. E. Johnson, and Jean Nicod, created modern confirmation theory in 1920' s, and Rudolf Carnap constructed in the 1940's his system of inductive logic, based upon a logical interpretation of probability as a unique, 'purely' logical relation between a hypothesis and evidence.

The most important test for a theory of induction is provided by its ability to give a reasonable account of non-demonstrative scientific inference. Several influential philosophers of our age have believed that modern probabilistic inductive logic has failed to pass this severe test. For example, Karl Popper has claimed that inductive logic is impossible, and that attempts towards its development are based upon a misguided 'inductivist' view of science.[2] Imre Lakatos has argued that inductive logic is a 'degenerating', rather than 'progressive', research programme.[3] Wolfgang Stegmüller—after arguing that Carnap's system of inductive logic must be interpreted as a normative theory of rational inductive behaviour rather than as a theory of rational scientific inference—has suggested that Carnap was mistaken in his original understanding of what he was aiming at when he started to develop his system.[4]

The view that inductive logic, as a theory of scientific inference, is impossible or futile relies upon certain accidental features of the extant systems of inductive logic (especially the 'old system' of Carnap), and is not supported by the recent developments within the logical theory of induction. This thesis I try to illustrate in this paper. I shall pay my tribute to the 100th anniversary of Jevons's *Principles of Science*—a book "so superficial in argument yet suggesting so much truth", as Keynes once remarked[5]—by studying a problem which has been largely neglected by his followers, viz. the problem of evaluating the inductive probabilities of theories relative to observational evidence.

The branch of formal methodology that I call inductive logic is not a *logic* in the strict sense of the word. The logical interpretation of probability as a degree of 'partial entailment' is untenable, as several authors have convincingly shown.[6] Inductive logic, as I see it, is basically a study of the question of how the inductive probability of a hypothesis h on evidence e may depend upon various factors and assumptions. These factors include the logical form of hypothesis h, the factual information contained in evidence e, the conceptual systems that are used to express h and e, and extra-logical features characteristic to the relevant inductive situation. Inductive probabilities are thus assumed to have more structure than the subjective probabilities of the Bayesians. The relativity of inductive probabilities to conceptual systems is the power, not the weakness, of inductive logic: it enables one to study the dynamic interplay between conceptual change and induction.[7] The influence of 'pragmatic' factors on probabilities is expressed through a number of free extra-logical parameters. For example, such parameters may depend upon the universe of discourse conceptualized in a certain way, and they may reflect the regularity, or the scientist's estimate of the regularity, of this universe. Different systems of inductive logic assume different forms of these dependencies, so that they specify different kinds of probability models for the distribution of probabilities of statements belonging to a conceptual system. Every application of such a model requires the specification of the values of its parameters which, therefore, express factual or 'synthetic' presuppositions of induction.

Inductive logic proper is not based upon any particular view of the uses that inductive probabilities are put to. For example, inductive logic as such is free from the 'inductivist' prejudice that posterior probabilities measure degrees of support or confirmation. On the other hand, inductive probabilities may serve as arguments in various definitions of degrees of confirmation, corroboration, acceptability, information content and systematic power.[8] Inductive logic is thus a tool which can be used, together with semantic information theory and decision theory, to conceptualize the inductive aspects of scientific inference.[9]

The success of this programme for inductive logic has at least two necessary preconditions, however.

(1) Scientific inference deals with *general laws* that are expressed by genuinely universal sentences. Therefore, a reasonable treatment of inductive generalization should be possible in inductive logic.

(2) Scientific laws are not only general in their form, but also typically theoretical in their character. Theories have also an important role in science both as starting-points of inference and as background frameworks that are used to interpret observations. Therefore, it should be possible to give an account of the role of *theories* and *theoretical concepts* within scientific inference.

Jaakko Hintikka has given a solution to the problem (1) by developing, for monadic first-order languages, a system of inductive logic in which genuinely universal statements may receive non-zero probabilities.[10] Risto Hilpinen has taken the basic steps in extending this system to cover first-order languages with relational concepts.[11] A partial answer to some of the problems connected with condition (2) has been given in the monograph *Theoretical Concepts and Hypothetico-Inductive Inference* (1973), by Ilkka Niiniluoto and Raimo Tuomela, where

Hintikka's system of inductive logic is applied to situations in which new, or theoretical, concepts are introduced into the original language.[12] The resulting framework gives outlines to a systematic theory of hypothetico-inductive inference—including inductive systematization, inductive explanation and prediction, as well as corroboration by theories or relative to theories. This framework requires the use of inductive probabilities of the following form:

(1) $P(g/e \wedge T)$,

where e is a singular evidence statement in a language $\mathscr{L}_0$, g is a generalization in $\mathscr{L}_0$, and T is universal theory in a language $\mathscr{L}$ that contains $\mathscr{L}_0$ as its sublanguage. Thus, (1) may interpreted as the inductive probability of an observational generalization g relative to observational evidence e and to theoretical evidence or background assumption T.

The converse problem, viz. determination of the inductive probabilities of theories relative to observational evidence, is not systematically studied in the monograph referred to above—nor in other studies in inductive logic. This problem I undertake to examine in this paper by studying within Hintikka's system the behaviour of probabilities of the form

(2) $P(T/e)$,

where e is in $\mathscr{L}_0$ and T is a strong generalization in $\mathscr{L}$.

II. INDUCTIVE PROBABILITIES OF OBSERVATIONAL CONSTITUENTS

In this section, a brief summary of Hintikka's two-dimensional continuum of inductive methods is presented. We operate here within one language, $\mathscr{L}_0$, and show how the inductive probabilities of generalizations of $\mathscr{L}_0$ depend upon evidence which is complete with respect to $\mathscr{L}_0$. The behaviour of these probabilities is to be contrasted with the probabilities of generalizations relative to incomplete evidence (cf. Section 4 below).

Let $\mathscr{L}_0$ be an applied monadic first-order language with the set $\lambda = \{O_1, \ldots, O_k\}$ of logically independent primitive predicates. Conjunctions of the form

$$(3) \qquad \bigwedge_{i=1}^{k} (\pm)O_i(x) = (\pm)O_1(x) \wedge \ldots \wedge (\pm)O_k(x).$$

where the symbol '(+)' may be replaced by the negation sign or nothing, will be denoted by $Ct_j(x), j = 1, \ldots, K$, where $K = 2^k$. These expressions define the *Ct-predicates* $Ct_j, j = 1, \ldots, K$, of language $\mathscr{L}_0$. It is assumed that $\mathscr{L}_0$ is interpreted in an infinite universe U. The *Ct-predicates* Ct_j specify all different kinds of individuals that can be described in the vocabulary λ, and thereby partition the universe U into K *cells*. *Constituents* of $\mathscr{L}_0$ are statements which specify which cells of U are empty and which are instantiated. If $\mathbf{CT}_i$ is the set of Ct-predicates of $\mathscr{L}_0$ which are instantiated according to constituent C_i, then C_i is equivalent to the sentence

(4) $$\bigwedge_{Ct_j \in \mathbf{CT}_i} (\mathrm{E}x)Ct_j(x) \wedge (x)\left[\bigvee_{Ct_j \in \mathbf{CT}_i} Ct_j(x)\right].$$

The number

(5) $$w_i = \mathrm{card}(\mathbf{CT}_i)$$

is called the *width* of constituent C_i. The number of constituents C_i of $\mathscr{L}_0$ is 2^K. Constituents are maximally strong quantificational sentences of $\mathscr{L}_0$ (without singular terms), and they are mutually incompatible in the sense that $\vdash \neg(C_i \wedge C_j)$ for all $i, j \leqslant 2^K, i \neq j$. Each general sentence of $\mathscr{L}_0$ is equivalent to a finite disjunction of constituents; this equivalence defines its *distributive normal form*. General sentences with more than one constituent in their normal form are called *weak generalizations*; general sentences equivalent to a constituent are *strong generalizations*.

Let P_0 denote a probability measure for language $\mathscr{L}_0$. Within Hintikka's system of inductive logic, this measure is determined up to two extra-logical parameters, α and λ. Since parameter λ is primarily relevant to singular inductive inference, the normalizing assumption $\lambda(w) = w$ is adopted in this paper.

Let e be a singular evidence statement in $\mathscr{L}_0$ which describes a sample of n individuals from universe U. Evidence e is assumed to be *complete with respect to* $\mathscr{L}_0$ in the sense that e expresses for each individual $a_i, i = 1, \ldots, n$, in the sample the Ct-predicate of $\mathscr{L}_0$ that a_i exemplifies. For $j = 1, \ldots, K$, let n_j be the number of individuals in e exemplifying Ct-predicate Ct_i of $\mathscr{L}_0$ (here $n_j \geqslant 0$ and $n_1 + \ldots + n_K = n$), and let c be the number of different kinds of individuals in e. The representative function of Hintikka's system, with $\lambda(w) = w$, is defined as follows:

(6) *If* $Ct_j \in \mathbf{CT}_i$ *and* e *is compatible with* C_i, *then*

$$P_0(Ct_j(a_{n+1})/e \wedge C_i) = \frac{n_j+1}{n+w_i}.$$

It follows from (6) that

$$(7) \qquad P_0(e/C_i) = \frac{(w_i-1)!}{(n+w_i-1)!} \prod_{j=1}^{K} (n_j!).$$

The prior probabilities of constituents C_i of $\mathscr{L}_0$ are defined as a function of non-negative real valued parameter α:

$$(8) \qquad P_0(C_i) = \frac{\dfrac{(\alpha+w_i-1)!}{(w_i-1)!}}{\sum\limits_{j=0}^{K} \binom{K}{j} \dfrac{(\alpha+j-1)!}{(j-1)!}}.$$

From (7) and (8) we get:

$$(9) \qquad P_0(e) = \frac{\sum\limits_{j=0}^{K-c} \binom{K-c}{j} \dfrac{(\alpha+c+j-1)!}{(n+c+j-1)!} \prod\limits_{j-1}^{K} (n_j!)}{\sum\limits_{j=0}^{K} \binom{K}{j} \cdot \dfrac{(\alpha+j-1)!}{(j-1)!}}.$$

An application of Bayes' Theorem gives now a formula for the posterior probability of constituent C_i relative to e:

$$(10) \qquad P_0(C_i/e) = \frac{\dfrac{(\alpha+w_i-1)!}{(n+w_i-1)!}}{\sum\limits_{j=0}^{K-c} \binom{K-c}{j} \dfrac{(\alpha+c+j-1)!}{(n+c+j-1)!}}$$

The posterior probability of any weak generalization g of $\mathscr{L}_0$ can now be computed from

$$(11) \qquad P_0(g/e) = \sum_i P(C_i/e),$$

where i varies over all values for which C_i belongs to the normal form of g in $\mathscr{L}_0$.

It is seen from formula (10) that the probability $P_0(C_i/e)$ depends on constituent C_i through its width w_i, on evidence e through its size

n and through its 'variety' c, on the conceptual system (language $\mathscr{L}_0$) through the number K of Ct-predicates, and on the inductive situation through parameter α. The following consequences of formula (10) are of particular interest for us:

(H1) $P_0(C_i/e)$ depends upon the size n of sample e, and upon the number c of exemplified Ct-predicates, but, for fixed n and c, is independent of the numbers $n_1, \ldots, n_k$ of individuals in the cells of U (as long as C_i is compatible with e).

(H2) When $\alpha \neq \infty$ and $n \to \infty$, probability $P_0(C_i/e)$approaches in the limit either the value one or zero. The unique constituent of $\mathscr{L}_0$ with the asymptotic probability one is C_i such that

$$\mathbf{CT}_i = \{Ct_j | n_j > 0\}.$$

this constituent is here denoted by C_*; it states that there exists in universe U only such kinds of individuals that are exemplified in sample e. The width of C_* is c.

(H3) For fixed $n \neq \infty$, $P_0(C_i/e)$ is inversely proportional to parameter α. In particular, when $\alpha \to \infty$, $P_0(C_i/e)$ approaches the value zero. Parameter α thus serves as an index of caution with respect to inductive generalization: it may be taken to reflect the amount disorder or irregularity that obtains in universe U, or an estimate of such irregularity.

III. INTRODUCTION OF THEORETICAL CONCEPTS

Suppose that a number of new predicates are introduced into language $\mathscr{L}_0$. As the qualitative results are similar irrespective of the number of the new predicates, consideration may be confined to the case in which only one new predicate is introduced into $\mathscr{L}_0$.[13] Thus, let $\mathscr{L}$ be an extension of language $\mathscr{L}_0$ which is obtained by joining a new monadic predicate M to $\mathscr{L}_0$. Language $\mathscr{L}$ has then $k+1$ primitive predicates, viz. $O_1, \ldots, O_k, M$, and hence $2^{k+1} = 2K$ Ct-predicates.

Each Ct-predicate Ct_j of $\mathscr{L}_0$ is now equivalent to the disjunction of two Ct-predicates of $\mathscr{L}$, that is,

$$\vdash Ct_j(x) \equiv Ct^r_{j1}(x) \vee Ct^r_{j2}(x),$$

where

$$(12)\qquad \begin{aligned} Ct^r_{j1} &= Ct_j \wedge M \\ Ct^r_{j2} &= Ct_j \wedge \neg M. \end{aligned}$$

In other words, the cells of universe U determined by the Ct-predicates of $\mathscr{L}_0$ are split into two subcells by means of the new predicate M. In this way, the richer language $\mathscr{L}$ allows us to make a finer partition of U than the poorer one.

If a Ct-predicate Ct_j of $\mathscr{L}_0$ is instantiated in U, then there are three possibilities in regard of the subcells of Ct_j:

1° *Ct^r_{j1} is empty, Ct^r_{j2} is non-empty,*

2° *Ct^r_{j1} is non-empty, Ct^r_{j2} is empty,*

3° *both Ct^r_{j1} and Ct^r_{j2} are non-empty.*

Consequently, each constituent C_i of $\mathscr{L}_0$ is equivalent to the disjunction of 3^{w_i} constituents of $\mathscr{L}$. The constituents of $\mathscr{L}$ are denoted by C^r_i; their total number in $\mathscr{L}$ is $2^{2K} = 4^K$. Notations $\mathbf{CT}^r_i$ and w^r_i, corresponding to constituent C^r_i, are defined analogously to $\mathbf{CT}_i$ and w_i. Define further, for each constituent C^r_i of $\mathscr{L}$.

$$(13)\qquad \begin{aligned} \mathbf{D}^1_i &= \{j \in \{1, \ldots, K\} \mid Ct^r_{j1} \notin \mathbf{CT}^r_i, Ct^r_{j2} \in \mathbf{CT}^r_i\} \\ \mathbf{D}^2_i &= \{j \in \{1, \ldots, K\} \mid Ct^r_{j1} \in \mathbf{CT}^r_i, Ct^r_{j2} \notin \mathbf{CT}^r_i\} \\ \mathbf{D}^3_i &= \{j \in \{1, \ldots, K\} \mid Ct^r_{j1} \in \mathbf{CT}^r_i, Ct^r_{j2} \in \mathbf{CT}^r_i\}. \end{aligned}$$

Let P_1 denote a probability measure for language $\mathscr{L}$, defined by Hintikka's system. The probability $P_1(C^r_i/e)$ of a constituent C^r_i of $\mathscr{L}$ relative to evidence e stated in $\mathscr{L}_0$ cannot be evaluated without further assumptions of the nature of the new predicate M. Predicate M is called an *evidential theoretical predicate*, if it can be used in reporting and hence redescribing evidence originally stated in $\mathscr{L}_0$. Predicate M is *non-evidential*, if it cannot be so used.[14]

The case of evidential theoretical predicates can be treated without any new technical difficulties. Here evidence e can be redescribed in $\mathscr{L}$ so that it is complete with respect to $\mathscr{L}$, i.e., it is known for each individual in e which Ct-predicate Ct^r_j of $\mathscr{L}$ it satisfies. If the number of those Ct-predicates of $\mathscr{L}$ that are exemplied in e is denoted by c', and the number of individuals of e in the cells determined by $\mathscr{L}$ are n'_j,

$j = 1, \ldots, 2K$, then the formulae (6)–(10) hold for P_1, instead of P_0, if K is replaced by $2K$, c by c', n_j by n'_j, C_i by C^r_i, and w_i by w^r_i.

As an illustration, suppose that e is as above, g is a weak generalization in $\mathscr{L}_0$ which claims that certain b cells of U, determined by $\mathscr{L}_0$, are empty, and T is a theory in $\mathscr{L}$ which states that certain r cells of U, determined by $\mathscr{L}$, are empty. Suppose further that e, g, and T compatible. If M is an evidential theoretical predicate, so that e and g can be redescribed in $\mathscr{L}$, then we have the result[15]:

$$(14) \qquad P_1(g/e \wedge T) = \frac{\displaystyle\sum_{i=0}^{2K-r-b'-c'} \binom{2K-r-b'-c'}{i} \frac{(\alpha+c'+i-1)!}{(n+c'+i-1)!}}{\displaystyle\sum_{i=0}^{2K-r-c'} \binom{2K-r-c'}{i} \frac{(\alpha+c'+i-1)!}{(n+c'+i-1)!}}$$

where b' is the number of Ct-predicates of $\mathscr{L}$ which are empty by g, but not by T. (Here $0 \leqslant b' \leqslant 2b$.) If T is a tautology in $\mathscr{L}$, so that $r = 0$ and $b' = 2b$, we get as a special case of formula (14):

$$(15) \qquad P_1(g/e) = \frac{\displaystyle\sum_{i=0}^{2K-2b-c'} \binom{2K-2b-c'}{i} \frac{(\alpha+c'+i-1)!}{(n+c'+i-1)!}}{\displaystyle\sum_{i=0}^{2K-c'} \binom{2K-c'}{i} \frac{(\alpha+c'+i-1)!}{(n+c'+i-1)!}}$$

If α and n are sufficiently large in relation to K, formulae (14) and (15) can be approximated by

$$(14') \qquad P_1(g/e \wedge T) \simeq \frac{1}{\left(1+\dfrac{\alpha+c'}{n+c'}\right)^{b'}}$$

$$(15') \qquad P_1(g/e) \simeq \frac{1}{\left(1+\dfrac{\alpha+c'}{n+c'}\right)^{2b}}$$

It follows from (14′) and (15′) that $P_1(g/e \wedge T) > P_1(g/e)$ if and only if $b' < 2b$. By the symmetry of positive relevance relation, we get the result

$$(16) \qquad P_1(T/e \wedge g) > P_1(T/e) \quad \textit{if and only if} \quad b' < 2b.$$

This result states that generalization g is positively relevant to theory T relative to evidence e if and only if T excludes some subcells of those cells of U which are claimed to be empty by g. In other words, (16) expresses a condition under which a theory T can be *confirmed* by an observational generalization g. Since approximations similar to (14′) and (15′) hold in the case where M is a non-evidential theoretical predicate, the qualitative result (16) holds for a theory T irrespective of the evidential or non-evidential nature of the theoretical predicate M employed by T.[16]

Result (16) was obtained without calculating the probabilities $P_1(T/e \wedge \wedge g)$ and $P_1(T/e)$. In the next section, the inductive probabilities of constituents C_i^r of $\mathscr{L}$ relative to evidence e in $\mathscr{L}_0$ are studied in the case where M is a non-evidential predicate. This gives rise to a number of new problems, and to results that are qualitatively new as compared to (H1), (H2), and (H3).

IV. INDUCTIVE PROBABILITIES OF THEORETICAL CONSTITUENTS

Suppose that languages $\mathscr{L}_0$ and $\mathscr{L}$ are defined as above, and that M is a *non-evidential* theoretical predicate. As above, e is an evidence statement which is complete with respect to language $\mathscr{L}_0$. But since M is non-evidential, it is not known whether the individuals in e satisfy M or not. Evidence e is thus *incomplete with respect to* $\mathscr{L}$: of each individual in the sample it is known only that it satisfies a disjunction of two Ct-predicates of $\mathscr{L}$ (cf. (12)).

Formula (6) is now replaced by the following result:[17]

(17) *Suppose that e is compatible with constituent C_i^r of $\mathscr{L}$. Then*

$$P_1(Ct_j(a_{n+1})/e \wedge C_i^r) = \frac{n_j+1}{n+w_i^r}, \text{ if } j \in \mathbf{D}_i^1 \text{ or } j \in \mathbf{D}_i^2$$

$$= \frac{n_j+2}{n+w_i^r}, \text{ if } j \in \mathbf{D}_i^3.$$

(Cf. (13).) It follows from (17) that

$$(18) \qquad P_1(e/C_i^r) = \frac{(w_i^r-1)!}{(n+w_i^r-1)!} \prod_{j=1}^{K} (n_j!) \cdot \prod_{j \in \mathbf{D}_i^3} (n_j+1).$$

The prior probability $P_1(C_i^r)$ of C_i^r in $\mathscr{L}$ can be computed from formula (8), if w_i is replaced by w_i^r, and K by $2K$. Thus, we get the result

$$(19)\qquad P_1(C_i^r)P_1(e/C_i^r) = Q\frac{(\alpha+w_i^r-1)!}{(n+w_i^r-1)!}\prod_{j\in \mathbf{D}_i^3}(n_j+1),$$

where Q is a constant depending upon α, K, and n_j, $j = 1, \ldots, K$, but not upon i. By Bayes' Theorem, the probability $P_1(C_i^r/e)$ of the 'theoretical' constituent C_i^r relative to 'observational' evidence e in $\mathscr{L}_0$ can be computed from the formula

$$P_1(C_i^r/e) = \frac{P_1(C_i^r)P_1(e/C_i^r)}{\sum_h P_1(C_h^r)P_1(e/C_h^r)}$$

where h varies over the set, say $\mathbf{E}$, of indices j such that C_j^r is compatible with e. It follows now from formula (19) that

$$(20)\qquad P_1(C_i^r/e) = \frac{\dfrac{(\alpha+w_i^r-1)!}{(n+w_i^r-1)!}\prod_{j\in \mathbf{D}_i^3}(n_j+1)}{\sum_{h\in \mathbf{E}}\dfrac{(\alpha+w_h^r-1)!}{(n+w_h^r-1)!}\prod_{j\in \mathbf{D}_h^3}(n_j+1)}$$

The denominator of formula (20) is equal to $P_1(e)/Q$; the main technical difficulty in applying (20) is to compute this value. This is illustrated by two examples; the latter of them is representative of the general situation.

Example 1. Suppose that $c = 1$, i.e. that only one Ct-predicate of $\mathscr{L}_0$, say Ct_j, is exemplified in evidence e. Now, $n_j = n$ and $n_h = 0$ for all $h \neq j$. Among constituents C_h^r of $\mathscr{L}$ for which $j \in \mathbf{D}_h^1$ there are $\binom{2K-2}{m}$ ones such that $w_h^r = 1+m$, where $m = 0, \ldots, 2K-2$. The same holds for constituents C_h^r for which $j \in \mathbf{D}_h^2$. For all of these constituents

$$\prod_{i\in \mathbf{D}_h^3}(n_i+1) = 1.$$

Among constituents C_h^r of $\mathscr{L}$ for which $j \in \mathbf{D}_h^3$ there are $\binom{2K-2}{m}$ ones such that $w_h^r = 2+m$, where $m = 0, \ldots, 2K-2$. For all of these constituents

$$\prod_{i\in \mathbf{D}_h^3}(n_i+1) = n+1.$$

It follows that

$$\frac{1}{Q} P_1(e) = 2 \sum_{m=0}^{2K-2} \binom{2K-2}{m} \frac{(\alpha+m)!}{(n+m)!} + \sum_{m=0}^{2K-2} \binom{2K-2}{m} \times$$

$$\times \frac{(\alpha+m+1)!}{(n+m+1)!}(n+1) = \sum_{m=0}^{2K-2} \binom{2K-2}{m} \left\{ 2+(n+1) \times \right.$$

$$\left. \times \frac{(\alpha+m+1)}{(n+m+1)} \right\} \frac{(\alpha+m)!}{(n+m)!}.$$

If C_i^r is a constituent of $\mathscr{L}$ which is compatible with e, then we get by formula (20):

$$(21) \qquad P_1(C_i^r/e) = \frac{\dfrac{(\alpha+w_i^r-1)!}{(n+w_i^r-1)!} \prod_{j \in \mathbf{D}_i^3} (n_j+1)}{\sum_{m=0}^{2K-2} \binom{2K-2}{m} \left\{ 2+(n+1) \dfrac{(\alpha+m+1)}{(n+m+1)} \right\} \dfrac{(\alpha+m)!}{(n+m)!}}$$

Define the constituents C_1^r, C_2^r, and C_3^r of $\mathscr{L}$ by

$$\mathbf{CT}_1^r = \{Ct_{j1}^r\}$$
$$\mathbf{CT}_2^r = \{Ct_{j2}^r\}$$
$$\mathbf{CT}_3^r = \{Ct_{j1}^r, Ct_{j2}^r\}.$$

Then by formula (21)

$$P_1(C_1^r/e) = P_1(C_2^r/e)$$

$$P_1(C_3^r/e) = \frac{(\alpha+1)}{(n+1)}(n+1) P_1(C_1^r/e) = (\alpha+1) P_1(C_1^r/e).$$

If $\alpha \neq \infty$ and $n \to \infty$, then

$$P_1(C_1^r/e) = P_1(C_2^r/e) \to \frac{1}{\alpha+3}$$

$$P_1(C_3^r/e) \to \frac{\alpha+1}{\alpha+3}.$$

The asymptotic posterior probability of all other constituents of $\mathscr{L}$ which are compatible with e is zero.

Example 2. Suppose that $c = 2$, i.e., that two Ct-predicates of $\mathscr{L}_0$, say Ct_1 and Ct_2, are exemplified in evidence e. Now, $n_1 > 0$, $n_2 > 0$, $n_1+n_2 = n$, and $n_j = 0$ for $j > 2$. The product

$$\prod_{j \in \mathbf{D}_h^3} (n_j+1)$$

can now take four different values:

(1°) 1

(2°) n_1+1

(3°) n_2+1

(4°) $(n_1+1)(n_2+1)$

for a constituent C_h^r compatible with e. Among constituents corresponding to case 1°, there are $4\binom{2K-4}{m}$ ones for which $w_h^r = 2+m$. Corresponding to case 2°, there are $2\binom{2K-4}{m}$ constituents for which $w_h^r = 3+m$. The same holds for case 3°. Finally, corresponding to case 4°, there are $\binom{2K-4}{m}$ constituents for which $w_h^r = 4+m$. Here $m = 0, \ldots, 2K-4$. It follows that

$$\begin{aligned}\frac{1}{Q} P_1(e) = 4 \sum_{m=0}^{2K-4} \binom{2K-4}{m} \frac{(\alpha+m+1)!}{(n+m+1)!} \\ + 2(n+2) \sum_{m=0}^{2K-4} \binom{2K-4}{m} \frac{(\alpha+m+2)!}{(n+m+2)!} \\ + (n_1+1)(n_2+1) \sum_{m=0}^{2K-4} \binom{2K-4}{m} \frac{(\alpha+m+3)!}{(n+m+3)!}.\end{aligned}$$

Let C_* be the constituent of $\mathscr{L}_0$ which claims that Ct-predicates Ct_1 and Ct_2 of $\mathscr{L}_0$, and only them, are instantiated in universe U. C_* is equivalent to the disjunction of $3^2 = 9$ constituents of $\mathscr{L}$. These constituents $C_1^r, \ldots, C_9^r$ are defined by

$$\mathbf{CT}_1^r = \{Ct_{11}^r, Ct_{21}^r\}$$
$$\mathbf{CT}_2^r = \{Ct_{12}^r, Ct_{21}^r\}$$
$$\mathbf{CT}_3^r = \{Ct_{11}^r, Ct_{22}^r\}$$
$$\mathbf{CT}_4^r = \{Ct_{12}^r, Ct_{22}^r\}$$
$$\mathbf{CT}_5^r = \{Ct_{11}^r, Ct_{12}^r, Ct_{21}^r\}$$
$$\mathbf{CT}_6^r = \{Ct_{11}^r, Ct_{12}^r, Ct_{22}^r\}$$
$$\mathbf{CT}_7^r = \{Ct_{11}^r, Ct_{21}^r, Ct_{22}^r\}$$
$$\mathbf{CT}_8^r = \{Ct_{12}^r, Ct_{21}^r, Ct_{22}^r\}$$
$$\mathbf{CT}_9^r = \{Ct_{11}^r, Ct_{12}^r, Ct_{21}^r, Ct_{22}^r\}.$$

Thus, $w_1^r = w_2^r = w_3^r = w_4^r = 2$, $w_5^r = w_6^r = w_7^r = w_8^r = 3$, and $w_9^r = 4$. The posterior probabilities of these constituents are

$$P_1(C_1^r/e) = P_1(C_2^r/e) = P_1(C_3^r/e) = P_1(C_4^r/e)$$
$$= \frac{\dfrac{(\alpha+1)!}{(n+1)!}}{\dfrac{1}{Q}P_1(e)}$$

$$P_1(C_5^r/e) = P_1(C_6^r/e) = \frac{(\alpha+2)}{(n+2)}(n_1+1)P_1(C_1^r/e)$$

$$P_1(C_7^r/e) = P_1(C_8^r/e) = \frac{(\alpha+2)}{(n+2)}(n_2+1)P_1(C_1^r/e)$$

$$P_1(C_9^r/e) = \frac{(\alpha+2)(\alpha+3)}{(n+2)(n+3)}(n_1+1)(n_2+1)P_1(C_1^r/e).$$

Suppose that $\alpha \neq \infty$ and $n \to \infty$. The asymptotic posterior probabilities of C_i^r, $i = 1, \ldots, 9$, depend now upon the asymptotic behaviour of numbers n_1 and n_2 *vis-á-vis* n. Let

$$n_1/n \to N_1 \quad \text{and} \quad n_2/n \to N_2$$

when $n \to \infty$. Since $n_1+n_2 = n$, we have $N_1+N_2 = 1$. Then asymptotically, with $n \to \infty$,

$$P_1(C_1^r/e) = \frac{1}{4+2(\alpha+2)+N_1N_2(\alpha+2)(\alpha+3)}$$

$$P_1(C_5^r/e) = \frac{N_1(\alpha+2)}{4+2(\alpha+2)+N_1N_2(\alpha+2)(\alpha+3)}$$

$$P_1(C_7^r/e) = \frac{N_2(\alpha+2)}{4+2(\alpha+2)+N_1N_2(\alpha+2)(\alpha+3)}$$

$$P_1(C_9^r/e) = \frac{N_1N_2(\alpha+2)(\alpha+3)}{4+2(\alpha+2)+N_1N_2(\alpha+2)(\alpha+3)}.$$

Since the sum of the asymptotic posterior probabilities of C_i^r, $i = 1, \ldots$ $\ldots, 9$, is one, the asymptotic probability of all other constituents of $\mathscr{L}$ compatible with e is zero. Three special cases of the above formulas are separated below.

(A) $N_1 = 1$, $N_2 = 0$. In this case, $n_1/n \to 1$ and $n_2/n \to 0$, i.e., the number n_1 of individuals found in cell Ct_1 grows without limit, when $n \to \infty$, but the number n_2 of individuals found in cell Ct_2 is bounded by some finite number. By the above formulae, we get

$$\begin{aligned} P_1(C_i^r/e) &= \frac{1}{4+2(\alpha+2)}, \quad \textit{if } i = 1, \ldots, 4; \\ &= \frac{(\alpha+2)}{4+2(\alpha+2)}, \quad \textit{if } i = 5, 6; \\ &= 0, \quad \textit{if } i = 7, 8, 9. \end{aligned}$$

The most probable constituents in this asymptotic case are C_5^r and C_6^r; when α grows from 0 to ∞, their probability grows from 1/4 to 1/2.

(B) $N_1 = 0$, $N_2 = 1$, i.e., $n_1/n \to 0$ and $n_2/n \to 1$. In this case,

$$\begin{aligned} P_1(C_i^r/e) &= \frac{1}{4+2(\alpha+2)}, \quad \textit{if } i = 1, \ldots, 4; \\ &= \frac{(\alpha+2)}{4+2(\alpha+2)}, \quad \textit{if } i = 7, 8; \\ &= 0, \quad \textit{if } i = 5, 6, 9. \end{aligned}$$

Now the most probable constituents are C_7^r and C_8^r.

(C) $N_1 > 0$, $N_2 > 0$. In this case, an infinite number of individuals is asymptotically found in both cells Ct_1 and Ct_2. Now all constituents C_i^r, $i = 1, \ldots, 9$, have a non-zero asymptotic probability. The most probable among them are

$$C_5^r \textit{ and } C_6^r, \textit{ if } N_1 < \frac{1}{\alpha+3}$$

$$C_9^r, \textit{ if } \frac{1}{\alpha+3} < N_1 < \frac{\alpha+2}{\alpha+3}$$

$$C_7^r \textit{ and } C_8^r, \text{ if } N_1 > \frac{\alpha+2}{\alpha+3}.$$

Thus, C_9^r is the most probable constituent, if N_1 and N_2 do not differ too much from 1/2, i.e., if N_1 belongs to the interval

$$\left[\frac{1}{\alpha+3}, \frac{\alpha+2}{\alpha+3}\right].$$

If $\alpha = 0$, this interval is [1/3, 2/3], and when α grows from 0 to ∞, it approaches the interval [0, 1]. Note also that when $\alpha \to \infty$, we have

$$P_1(C_i^r/e) \to 0, \textit{ for } i = 1, \ldots, 8;$$
$$P_1(C_9^r/e) \to 1.$$

The most interesting results about the probabilities $P_1(C_i^r/e)$ of the constituents C_i^r of $\mathscr{L}$ (compatible with e) can now be summarized as follows:

(T1) $P_1(C_i^r/e)$ depends upon the size n of sample e, and upon the number c of different kinds of individuals in e, and also, as a rule, upon the numbers $n_1, \ldots, n_k$ of individuals of e in cells $Ct_1, \ldots, Ct_K$ of universe U.

(T2) When $\alpha \neq \infty$ and $n \to \infty$, there is a unique constituent C_* of $\mathscr{L}_0$ which has one as its asymptotic posterior probability. Evidence e in $\mathscr{L}_0$ is not decisive, in this sense, with respect to constituents of $\mathscr{L}$. There are always several constituents of $\mathscr{L}$ which asymptotically receive non-zero probabilities. They are all included among those constituents of $\mathscr{L}$ that logically entail the constituent C_* of $\mathscr{L}_0$. In other words, precisely those constituents of $\mathscr{L}$ which are compatible with C_* may receive non-zero asymptotic probabilities.

(T3) The asymptotic probabilities of the constituents of $\mathscr{L}$ are independent of the number $2K$ of the primitive predicates of $\mathscr{L}$, but depend on the number c of exemplified Ct-predicates of $\mathscr{L}_0$, on parameter α, and on the asymptotic behaviour of the numbers $n_1, \ldots, n_K$. The asymptotically most probable constituent of $\mathscr{L}$ is the one which claims that at least one subcell of each exemplified cell of $\mathscr{L}_0$ is instantiated in universe U, and that (i) both subcells of Ct_j are instantiated in U, if $n_j/n \to 1$ for some j, or (ii) both subcells of all exemplified cells Ct_j with $n_j/n \to N_j > 0$ are instantiated in universe U, whenever the dispersion or spread of these values N_j is not too great.

(T4) $P_1(C_i^\tau/e)$ may be asymptotically very high, that is, arbitrarily close to one.

(T5) The probability of the constituent which claims that all (or 'many') subcells of the exemplified cells of $\mathscr{L}_0$ are instantiated in universe U is directly proportional to parameter α, and grows towards the value one when $\alpha \to \infty$. Again, a great value of α reflects a great amount of disorder or irregularity of universe U.

Results (T1) and (T3), when compared to (H1), illustrate the fact that the dependence of the posterior probabilities of theoretical constituents on observational evidence is much more complex than that of observational constituents. Result (T5), when compared to (H3), shows that an increase in the value of parameter α decreases the probability of an observational constituent, but may increase that of a theoretical constituent. But this is precisely what one would expect, if α has the intuitive interpretation that Hintikka has given to it. Thus, (T5) serves to vindicate Hintikka's view of the role of α in inductive generalization.

Result (T2) illustrates the weakness of evidence—even if asymptotic—which is conceptually incomplete. It also suggests that the much debated principle of *Converse Consequence*

(CC) *If e confirms (supports) h and g logically entails h, then e confirms g, too,*

requires considerable modifications.[18]

One further difference between the probabilities of theoretical and observational constituents is worth mentioning here. Constituent C_* is the *boldest*, or least probable *a priori*, among those constituents of $\mathscr{L}_0$ which are compatible with evidence e. In other words, the most *informative* of the non-falsified strong generalizations of $\mathscr{L}_0$ receives the highest posterior probability relative to sufficiently large evidence. Among the theoretical constituents of $\mathscr{L}$ which are compatible with evidence e the situation is somewhat different, however. In Example 2, it was seen that the boldest of these constituents of $\mathscr{L}$ are C_i^r, $i = 1, \ldots$ $\ldots, 4$, but the constituents that evidence e favours most are among C_i^r, $i = 5, \ldots, 9$. Still, these latter constituents are bolder than any statement of $\mathscr{L}$ which is equivalent to an observational constituent or generalization of $\mathscr{L}_0$ (and is compatible with e).

When the posterior probabilities of constituents of $\mathscr{L}$ have been determined, the posterior probability any theory (weak generalization) in $\mathscr{L}$ can be computed by transforming it to distributive normal form and by taking the sum of the posterior probabilities of all constituents occurring in this normal form. In other words,

$$(22) \qquad P_1(T/e) = \sum_{i \in \mathrm{I}} P_1(C_i^r/e)$$

holds if

$$\vdash T \equiv \bigvee_{i \in \mathrm{I}} C_i^r.$$

Theoretical weak generalizations are not discussed in this paper; results (T1)–(T5) are sufficient to show that their confirmation by observational evidence is possible.

Another problem which is not discussed in this paper is to study what happens when the assumption of the logical independence of the primitive predicates $O_1, \ldots, O_k$, M of $\mathscr{L}$ is relaxed. Between these predicates there may obtain analytic, or definitory, connections which are expressed by a conjunction T of generalizations of $\mathscr{L}$. Statement T may also express factual background assumptions relative to which the confirmation of generalizations of $\mathscr{L}$ is considered.[19] In both cases, a study should be made of probabilities of the form

$$(23) \qquad P_1(C_i^r/e \wedge T).$$

V. CONCLUSION

The general result established in Section 4 can be summarized as follows: Within Hintikka's system of inductive logic, confirmation—e.g., in the positive relevance sense—of informative theoretical generalizations by incomplete, observational evidence is possible. This result as such is interesting for the prospects of inductive logic. Moreover, it strongly suggests that Stephen Barker was mistaken, when he claimed that a 'transcendent hypothesis of the second kind' (i.e., a statement involving non-evidential theoretical predicates) cannot be confirmed by induction.[20]

REFERENCES

[1] W. S. Jevons, *Principles of Science*, Macmillan, London, 1874 (2nd edition, 1877). For discussions about Jevons's theory of induction, see E. H. Madden's chapter 'W. S. Jevons on Induction and Probability', in R. M. Blake, C. J. Ducasse, and E. H. Madden, *Theories of Scientific Method: The Renaissance through the Nineteenth Century*, University of Washington Press, Seattle, 1960; and L. L. Laudan, 'Induction and Probability in the Nineteenth Century', in P. Suppes *et al.* (eds.), *Logic, Methodology and Philosophy of Science IV*, North-Holland, Amsterdam, 1973, pp. 429–438.

[2] K. R. Popper, *The Logic of Scientific Discovery*, Hutchinson Co., London, 1959 (revised edition, 1968), and *Conjectures and Refutations*, Routledge and Kegan Paul, London, 1963 (3rd revised edition, 1969). See also I. Niiniluoto, 'Review of Alex C. Michalos: The Popper-Carnap Controversy', *Synthese* **25** (1973) 417–436, and I. Niiniluoto and R. Tuomela, *Theoretical Concepts and Hypothetico-Inductive Inference*, D. Reidel, Dordrecht and Boston, 1973.

[3] I. Lakatos, 'Changes in the Problem of Inductive Logic', in I. Lakatos (ed.), *The Problem of Inductive Logic*, North-Holland, Amsterdam, 1966, pp. 315–417. For attempts to reconstruct some of Lakatos's ideas within inductive logic, see L. J. Cohen, 'The Inductive Logic of Progressive Problem-Shifts', *Revue Internationale de Philosophie* **95/96** (1971) 62–77; and I. Niiniluoto and R. Tuomela, *op. cit.*, p. 136 (see Note 2).

[4] W. Stegmüller, 'Das Problem der Induktion: Humes Herausforderung und moderne Antworten', in H. Lenk (ed.). *Neue Aspekte der Wissenschaftstheorie*, Vieweg, Braunschweig, 1971, pp. 13–74; *Personelle und Statistische Wahrscheinlichkeit. Probleme und Resultate der Wissenschaftstheorie und Analytischen Philosophie, Band IV*, Springer-Verlag, Berlin, Heidelberg, New York, 1973; 'Carnap's Normative Theory of Inductive Probability', in P. Suppes et al. (eds.), *Logic, Methodology and Philosophy of Science IV*, North-Holland, Amsterdam, 1973, pp. 501–513.

[5] J. M. Keynes, *A Treatise on Probability*, Macmillan, London, 1921 (reprinted: Harper Torchbooks, New York, 1962), p. 273.

[6] Already in 1926, Ramsey remarked against Keynes that 'there really do not seem to be any such things as the logical probability relations he describes'; see F. P. Ramsey, 'Truth and Probability', in *The Foundations of Mathematics* (ed. by R. B. Braithwaite), Routledge and Kegan Paul, London, 1931, pp. 156–198. For other criticisms of the logical interpretation of probability, see W. C. Salmon, 'Partial Entailment

as a Basis for Inductive Logic', in N. Rescher *et al.*, *Essays in Honor of Carl G. Hempel: A Tribute on the Occasion of his Sixty-Fifth Birthday*, D. Reidel, Dordrecht, 1969, pp. 47–82; and J. Hintikka, 'On Semantic Information', in J. Hintikka and P. Suppes (eds.), *Information and Inference*, D. Reidel, Dordrecht, 1970, pp. 3–27.

[7] This is a basic idea in Niiniluoto and Tuomela, *op. cit.* See especially Chapters 1.3 and 10.1.

[8] See, for example, Chapters 6, 7 and 8 in Niiniluoto and Tuomela, *op. cit.*

[9] I do not wish to suggest that inductive logic is 'global' in the sense that it alone is 'the analysis of all learning from experience'; cf. I. Hacking, 'Propensities, Statistics, and Inductive Logic', in P. Suppes *et al.*, *Logic, Methodology and Philosophy of Science IV*, North-Holland, Amsterdam, 1973, pp. 485–500. Statistical inference is partly based upon physical probabilities (propensities). But the main interest in inductive logic lies in the fact that it deals with problems—such as confirmation of universal laws—to which, or to their analogues in the statistical problems, current statistical methodology provides no answer. Moreover, applications of inductive logic are always relative to certain 'pragmatic' boundary conditions, which in a sense makes the results of inductive logic 'local'.

[10] J. Hintikka, 'Towards a Theory of Inductive Generalization', in Y. Bar-Hillel (ed.), *Proceedings of the 1964 International Congress for Logic, Methodology and Philosophy of Science*, North-Holland, Amsterdam, 1965, pp. 274–288; 'A Two-Dimensional Continuum of Inductive Methods', in J. Hintikka and P. Suppes (eds.), *Aspects of Inductive Logic*, North-Holland, Amsterdam, 1966, pp. 113–132.

[11] R. Hilpinen, 'Relational Hypotheses and Inductive Inference', *Synthese* **23** (1971) 266–286.

[12] *op. cit.* (see Note 2).

[13] Cf. Niiniluoto and Tuomela, *op. cit.*, pp. 41–43.

[14] Cf. Niiniluoto and Tuomela, *op. cit.*, pp. 7, 21, 33.

[15] Cf. Niiniluoto and Tuomela, *op. cit.*, p. 36.

[16] Cf. Niiniluoto and Tuomela, *op. cit.*, pp. 46–47.

[17] For the derivation of result (17), see Niiniluoto and Tuomela, *op. cit.*, pp. 27–29.

[18] For a brief discussion of CC, see Niiniluoto and Tuomela, *op. cit.*, pp. 224–228. Note that in Example 2 above evidence e (asymptotically) supports C_*, and all constituents C_i^r, $i = 1, \ldots, 9$, logically entail C_*. Which one (ones) of them is supported by e depends partly on evidence e (through N_1 and N_2) and partly on the inductive situation (through α).

[19] Confirmation (corroboration) of observational generalizations in $\mathscr{L}_0$ relative to background theories in $\mathscr{L}$ has been studied in Niiniluoto and Tuomela, *op. cit.*, 118–140. See also I. Niiniluoto, 'Two Measures of Theoretical Support', forthcoming in *Proceedings of the XVth World Congress in Philosophy, Varna, 1973*, Sofia, 1975.

[20] S. F. Barker, *Induction and Hypothesis: A Study in the Logic of Confirmation*, Cornell University Press, Ithaca, 1957, p. 97.

ROBIN GILES

A PRAGMATIC APPROACH TO THE FORMALIZATION OF EMPIRICAL THEORIES[1]

SUMMARY A concrete interpretation of a formal language can be given in terms of *documents*, each document describing an 'action' associated with the corresponding formal expression. Through this 'documentary interpretation' the correlative of a prime proposition turns out to be (a document describing) an *elementary experiment*: i.e. an experimental procedure which leads to a definite outcome, yes or no. If repetition of an elementary experiment always yields the same outcome it is possible to assign a truth value, 'true' or 'false', to the corresponding proposition. If this is not the case—and the inescapability of such *dispersive* experiments has been emphasized by quantum mechanics—there is no sense in assigning a truth value, and classical logic becomes ineffective.

To reach an alternative the following principle is adopted: the meaning of each proposition is to be given not in terms of conditions for its truth but *in terms of a definite commitment that is assumed by him who asserts it*. For instance, the assertion of a prime proposition A is taken to constitute a commitment to pay $1 should a trial of the corresponding elementary experiment yield 'no'. Of course, a 'tangible meaning', in this sense, is also assigned to each compound proposition: for example, the proposition $A \to B$ (A and B being any propositions) represents an offer to assert B if the other speaker will assert A.

A form of subjective truth can now be defined: a proposition A is *true relative to a speaker* **P** if **P** is willing to assert A; it is *false* if he is willing to assert $\neg A$. Most propositions are neither true nor false. However, if (e.g.) $A \to B$ is true one may say B is *at least as true as A*.

In this way a new formal language is obtained which is syntactically identical to but semantically richer than the classical first-order predicate calculus. In the new language it is possible to express *degrees of belief*. In fact, with certain simplifying assumptions, a unique quantitative 'truth value' can be assigned to each proposition. This leads to a logic that corresponds closely to the infinite-valued logic $Ł_\infty$ of Łukasiewicz. If these simplifying assumptions are not made a more general logic is obtained which is currently under investigation.

This report is a brief account of a longer work which is currently in preparation. An early version [1] has been published and a more up to date and extended form [2] is available on request.

I. INTRODUCTION

I shall take it as axiomatic that the aim of a physical[2] theory is to make predictions of the results of experiments, or more generally of phenomena. Other aims, such as 'to explain' or 'to provide understanding', may be regarded as being fulfilled when a convenient and effective system for making predictions has been obtained.

When we consider particular physical theories we notice at once that the language employed does not have a sufficient variety of terms to admit the enunciation of particular predictions: a formulation of Newton's theory of gravitation, for instance, does not contain terms for *particular* physical objects, yet these must be referred to when the theory is applied. Before it can be used, therefore, a physical theory must be embedded in a more extensive *physical language* which contains terms for all the objects to which (in a given application) the theory is intended to apply. Clearly, the effectiveness of the theory can only be assessed in the context of this language and it depends absolutely on the precision with which the physical language is formulated. In this paper I shall be concerned primarily with the formulation and interpretation of such a physical language.

The principle problems in setting up a physical language $\mathscr{L}$ are semantic: it is necessary to explain exactly how each sentence is to be interpreted in experimental terms. It is practically impossible to do this in other than a piecemeal fashion, which moreover depends on the use of some common language (e.g. English) and is thus severely limited in precision, *unless* $\mathscr{L}$ is a formal language. In this case, however, a procedure becomes available which may be graphically described as the 'documentary interpretation'.

II. THE DOCUMENTARY INTERPRETATION

Let us first consider a very simple example. Although we are concerned with physical theories it will simplify the description greatly, without introducing any differences in principle, if we take an example from 'experimental mathematics'.

2.1. *Example*: *the natural numbers*, N. Let us represent numbers by means of a single character, the stroke |: 1 will be represented by the expression '|', 2 by '||', 3 by '|||', and so on. Let us call these expressions *numerals* and let the *sum* of two numerals be given the obvious meaning. Suppose we have carried out a series of experiments on the addition of numerals and wish to convey a precise understanding of our results.

To this end we introduce a formal language $\mathscr{L}$(N) with four non-logical constants 1, S, A and E, to denote respectively the numeral |, the successor function (which assigns to each numeral the next one), the operation of addition, and the relation of equality. With this language we can record an observation that (for instance) $|+|| = |||$ by means of the prime proposition $E(A(1, S(1)), S(S(1)))$. Assuming the concept of a first order language is understood, the *syntax* of the particular language $\mathscr{L}$(N) is given by announcing that the symbols 1, S, A, E are respectively a constant, a unary function, a binary function, and a binary predicate. It remains to explain what the symbols *mean*; this must be done in such a way that if any prime proposition (for instance the one above) is taken as a prediction it is perfectly clear how it should be verified. To this end it is convenient to imagine four *documents*, one labelled by each of the primitive symbols. These bear inscriptions as follows:

1: "Draw a 'stroke'."

1 is a *constant document*: i.e. it represents a particular numeral.

S: "Add a stroke to the expression – – –."

Here '– – –' represents a blank space on the *function document* S. When any constant document, labelled by a constant term t, say, is pasted into this space the whole, now labelled by the constant term $S(t)$, becomes a constant document itself (representing the successor of the numeral represented by t). Similarly, the documents labelled A and E have two blank spaces each and the following inscriptions:

A: 'Place the expression (produced by the instruction) – – – immediately adjacent to the expression – – – so as to form a single expression.'

E: 'Construct the expressions – – – and – – –. Delete a stroke from the first expression and simultaneously from the second, and repeat this operation as often as possible. If in the end no strokes remain then the outcome is 'yes'; otherwise 'no'.'

It is easy to see how a straightforward pasting operation constructs, from (copies of) the four primitive documents, the document labelled by the prime proposition $E(A(1, S(1)), S(S(1)))$, which then describes in complete detail the procedure required for an experimental verification of the relation $|+|| = |||$ mentioned above. Clearly, the same applies to any prime proposition. I shall call the language $\mathscr{L}(\mathbf{N})$, accompanied by the above interpretations of the four primitive symbols, an *interpreted language*—in this case a *mathematical* language.

It is clear that simply by admitting, on the primitive documents, instructions for physical (rather than mathematical) operations and observations we can obtain a 'documentary interpretation' for a physical language. The following points, which are all illustrated by the above example, should be noted:

(a) For a full understanding of the language it is necessary and sufficient that the interpretation (given on the primitive documents) of each of the primitive symbols be understood. Although, in the above example, these interpretations were described in common language, it is not necessary that this be so. Provided only that the concepts involved are sufficiently 'direct' (i.e. close to immediate experience) they need not be explained but may simply be *demonstrated*—i.e. the definitions of the primitive terms become *ostensive definitions*. In fact, such a procedure is quite plausible in the above case of $\mathscr{L}(\mathbf{N})$.

(b) Since each primitive symbol necessitates a corresponding document (or ostensive definition) only a finite number of primitive symbols can be admitted. Since in virtually all cases an infinite number of 'objects' (e.g. numerals, in the above example) have to be referred to at least one function symbol must be employed.

(c) In general, though not in the above example, an interpreted language may involve objects of several types [3]. Appropriate restrictions then appear in the rules for the construction of terms but the documentary interpretation is otherwise unaffected. There is no difficulty, either, in adapting the documentary interpretation to deal with cases where there are symbols for functions which act on functions or predicates and produce functions or predicates, and so on. Such cases arise even with very simple physical examples and in mathematics already with the real numbers [4]. Naturally such examples usually involve a higher order language.

(d) In every case the text on a document labelled by a prime proposition A is a set of instructions for what I shall call an *elementary experiment*: namely, an experimental procedure that culminates in the occurrence of one of two possible *outcomes*, 'yes' or 'no'. A particular execution of an experimental procedure in accordance with these instructions will be called an (admissible) *trial* of this elementary experiment. (Strictly speaking, 'the elementary experiment determined by the prime proposition A' should be regarded as a generic term, referring to all (admissible) trials).

(e) The test as to whether all users of a given interpreted language understand it in the same sense is a pragmatic one: it is necessary that whenever a purported trial of any elementary experiment (corresponding to some prime proposition in the language) is carried out all observers agree as to (a) whether the trial is admissible, and (b) if so, what the outcome is. If all users are confident in so agreeing then the language has been satisfactorily defined. (Of course, should a disagreement nevertheless arise clarification of the language is at once called for).

The concept of an elementary experiment as the referent of a prime proposition in an interpreted language is fundamental for this study. I shall take the view that no other kind of experiment need be considered in formalizing an empirical theory. It is, of course, certainly the case that in conventional theories other experiments are referred to: indeed, a typical experiment (it is often claimed) yields a real number as its outcome. However, if we look at any practical case we find that the number of distinct outcomes that the experimenter is prepared to record is finite: his instrument has a finite range and is of limited accuracy. Moreover, looking more closely, we realize that in choosing which of these outcomes to record on each particular occasion he makes in fact a series of 'binary decisions' (e.g. 'Is the pointer to the right of this division'?). Thus in fact he is carrying out, albeit simultaneously, a series of elementary experiments.

III. DISPERSION

In discussing the language $\mathscr{L}(\mathrm{N})$ above no mention was made of the logical symbols. The reason is that a serious difficulty arises when one attempts to apply the classical interpretation of compound propositions. This difficulty owes its genesis to a feature which characterizes physical or empirical, in contrast to mathematical, interpreted languages.

In the case of a mathematical interpreted language (e.g. $\mathscr{L}$(N)) every elementary experiment (corresponding to some prime proposition) is *dispersion-free*: i.e. any two admissible trials give the same outcome. (We shall then say that the *language* is dispersion-free). However, in the case of a physical (or empirical) interpreted language this is not so: an elementary experiment in general shows dispersion—if several trials are carried out the outcome is not always the same.

Three causes of dispersion in an elementary experiment may be recognized. First, it may arise from *ambiguity* in the corresponding instructions. Here are two examples: "Drop an egg; if it breaks the outcome is 'yes'." (Clearly the conditions can be satisfied in various ways, some of which are more likely to yield a 'yes' outcome than others). "Measure the position of the bob of a swinging simple pendulum; if it is to the right of the equilibrium position the outcome is 'yes'." (There is no indication of *what* simple pendulum should be used, *how* it should be set moving, or *when* the observation should be carried out. This example illustrates the fact that to avoid ambiguity the instructions for an elementary experiment in physics must describe not only the observation in which the outcome is obtained, but must also include a sufficient description of the construction and/or preparation of the various systems involved. Note, in particular, that the customary use in 'quantum logic' of 'proposition' as a term to denote a two-valued observable is quite incompatible with our use of 'prime proposition' for a sentence in an interpreted language).

Dispersion in an elementary experiment E due to ambiguity can be reduced by 'clarification' of the instructions: i.e. by adding further conditions to the instructions for E. We then obtain what I call a *refinement* of E. When this process is carried to the limit we obtain a *perfect* elementary experiment: i.e. (roughly) one for which further refinement "makes no difference".

Imperfect elementary experiments can arise even in a mathematical language. (Example: "Think of a number. If it is odd the outcome is 'yes'.") However, it is universally agreed that science (and mathematics) is not really concerned with imperfect experiments. Henceforth, unless otherwise stated, we shall assume all elementary experiments associated with prime propositions in an interpreted language are perfect.

A second cause of dispersion in an elementary experiment in physics is the presence of thermal or other (so-called) *statistical fluctuations*. If, for example, the outcome of an elementary experiment is determined by the random motion (Brownian motion) of a small particle in a liquid then the experiment will show dispersion and this dispersion cannot (in general) be removed by refinement of the experiment. Indeed, Brownian motion can only be reduced by cooling, but—even if the temperature were not originally prescribed—this would eventually result in the liquid's freezing, probably violating a condition laid down in the original experiment.

One school of thought claims that nevertheless the dispersion can "in principle" be avoided by observing accurately the initial position and velocity of every particle in the liquid and then predicting the outcome by the laws of mechanics. Rather than discuss this argument in detail (see, e.g., [2]) let it suffice here to remark that the claim is in any case irrelevant from the point of view of any theory which is concerned with making actual predictions in the practical world.

The third source of dispersion arises from the phenomena of quantum mechanics. A very simple example will suffice. Let a radioactive source be placed before a Geiger counter and suppose the source is so weak that on average one particle is detected per minute. The experiment consists in switching on the counter for one minute, the outcome being 'yes' if a count is recorded. With suitable refinement, the elementary experiment so obtained is perfect, but (owing to the random nature of radioactive decay) it still shows a large degree of dispersion. This experiment is in no way exceptional: almost any elementary experiment whose theoretical treatment involves quantum mechanics shows dispersion which cannot be removed by a suitable refinement.

Fundamentally, there is no clear line of demarkation separating the thermal and quantum causes of dispersion. Other elementary experiments in everyday life whose outcome is determined by 'chance' (Example: "Spin a coin; if 'heads' the outcome is 'yes'.") may also be classed under one of these heads.

IV. THE MEANING OF A PROPOSITION

In classical logic it is assumed that every prime proposition acquires, through its interpretation, a *truth value*, 'true' or 'false'. This truth value assignment is then extended in a familiar way to every compound proposition. In fact, the whole procedure is formalized in *model theory*, in which the term 'interpretation' is assigned essentially to such a truth value assignment.

In classical logic the process by which a prime proposition acquires its truth value is rarely discussed. In the documentary interpretation we have a concrete picture of this process, and it is clear that the classical method of assigning truth values can be carried through only in the case of a dispersion-free language:[3] in any other case there is no obvious way in which to assign truth values even to the prime propositions. We must therefore either reject the documentary interpretation, in which each prime proposition is inseparably associated with an elementary experiment, or we must replace classical logic with a more appropriate structure. We choose the latter course, abandoning the doctrine that every proposition is either true or false.

Classically, the *meaning* of any proposition is determined by stipulating the conditions under which it is true, or in pragmatic terms by *laying down the practical consequences which follow from its being true.* With the demise of the concept of truth this no longer makes sense. However, bearing in mind that the practical function of a proposition is *to be asserted* we are led to replace this last notion of meaning by the following:

(1) *The meaning of any proposition is determined by laying down the practical consequences that follow from its having been asserted.*

At first sight this seems to be nonsense: the mere fact that a proposition has been asserted has usually no practical consequences whatever—after all, the speaker might be an idiot. However, there is a kind of statement, well known in legal circles, that does have practical consequences *at least for the speaker*: namely, an assertion that embodies some kind of commitment. If, then, we associate with each proposition an *obligation* which is incurred, whenever the proposition is asserted, by him who asserts it we shall be able to implement the principle (1).

There are also good practical reasons for this procedure. The assertion of a salesman that an automobile will not break down may be discounted. But if he commits himself to refund the cost price should it break down within a year his statement suddenly becomes significant!

Even in this simple example one can recognize too that in some sense the full meaning of the assertion is carried by the associated commitment. Indeed, if another purportedly different assertion were associated with exactly the same commitment then any practical action which would be justified by the first assertion would be equally justified by the second: i.e. for all practical purposes the assertions would be indistinguishable, Let us therefore adopt the following fundamental principle:

4.1. *Principle of tangibility*. The meaning of any proposition is to be given in a *tangible form*: i.e. in terms of an obligation that is incurred by- him who asserts it.

To illustrate this principle let us consider a few particular propositions. First, suppose A is a prime proposition occurring in some interpreted language. An assertion of A should express the belief that a trial of the corresponding elementary experiment would yield the outcome 'yes'. We therefore require an obligation that results in a penalty should a trial of A yield 'no'.[4] For simplicity let us choose a penalty independent of A. Any unit of 'goods' could be adopted; $ 1 is notationally convenient. This gives:

4.2. *Definition*. Let A be a prime proposition. He who asserts A agrees to pay $ 1 should a trial of A yield the outcome 'no'.

It is to be understood that (a) the trial may be set up by the 'opponent' (i.e. he (or one of those) to whom the assertion is addressed), (b) the opponent must announce, before commencing a trial, whether or not it is to be used in fulfillment of the obligation incurred under 4.2, and (c) when one such trial has been carried out (and the resulting debt, if any, paid) then the obligation incurred under 4.2 has been fully discharged.

Two features of the forthcoming logic are already illustrated by Definition 4.2. Firstly, an unwise assertion of a prime proposition can result in a loss of no more than $ 1, whatever happens. We shall see later that this 'principle of limited liability' applies also in the case of an arbitrary compound proposition. Secondly, to assert a proposition is not the same thing as to assert it twice. This represents a major difference from classical logic.

It might seem from Definition 4.2 that a prime proposition would rarely be willingly asserted. This is in fact the case. (But the same does not apply to a compound proposition, as we shall see later). Indeed, a speaker **P** will willingly assert a prime proposition A only if he feels *sure* that a trial of A will yield the outcome 'yes'. In this case it will be convenient to say that A is *true relative to* (or *for*) **P** and write $\mathbf{P} \models A$, In this way the notion of truth re-enters the theory, but only in a relative or subjective form. [However, it will sometimes be convenient to describe a proposition as *true* (without qualification) if it is true for every speaker that we propose to consider.] If **P** is sure that A will yield the outcome 'no' we shall say *A is false for* **P**. Naturally, in general a prime proposition A will be neither true nor false relative to a speaker **P**.

As a second illustration of the principle of tangibility consider the proposition 'E is dispersion-free', where E is an elementary experiment. To assign a tangible meaning to this statement we must lay down some offer which might reasonably be expected of one who believed E to be dispersion-free. Let us adopt the following:

4.3. *Definition.* The assertion "E is dispersion-free" shall be understood to mean: "Let two trials of E be carried out. If the outcomes differ I will pay you \$ 1."

(It is understood that the opponent may arrange the trials of E). In a similar way we may reach the following tangible meaning for the proposition "E is perfect".

4.4. *Definition.* The assertion "E is perfect" shall be understood to mean: "Let two trials t_1 and t_2 of E be carried out. I agree to pay you \$ 1 if t_1 yields 'yes' and t_2 yields 'no' provided you agree to pay me \$ 1 if t_2 yields 'yes' and t_1 yields 'no'."

For brevity I omit a justification of this definition (see [2]). It is important to recognize, however, that once 4.3 and 4.4 have been adopted they provide the definitions of 'dispersion-free' and 'perfect' and the previous explanations of these terms become—except for heuristic purposes—obselete. All precise reasoning is now based on the tangible meanings. As an example of such an argument notice how we now show that any dispersion-free elementary experiment is perfect: What has to be shown is that if I believe E is dispersion-free then I should believe E is perfect: i.e. if I am prepared to make the offer in 4.3 then I should also be willing to make the offer in 4.4. But this is clear, for if an opponent wins from me through 4.4 by means of trials t_1 and t_2 then he would certainly also have won if these same trials had been construed as a response to 4.3.

V. THE MEANING OF THE LOGICAL SYMBOLS

We can now return to the problem, described at the beginning of § 4, of developing a replacement for classical logic which can be applied, in connection with the documentary interpretation, in the case of an arbitrary (not necessarily dispersion-free) interpreted language. As in the case of classical logic, the logical structure will follow once the meanings of the compound propositions have been fixed. We are thus faced with the problem of assigning a tangible meaning (in the sense of 4.1) to each compound proposition.

As in the classical case, it is natural to use an inductive procedure, explaining the obligation associated with a compound proposition $A \rightarrow B$ (for instance) in terms of those associated with A and B. I now put forward a series of definitions, each of which may be regarded as assigning a tangible meaning to one of the logical symbols. It is important to recognize that these definitions are not determined a priori. The aim is to choose a set of definitions that will be of practical value, has convenient mathematical properties, and can reasonably be regarded as an explication of the intuitive notions associated with the logical terms. In addition, the definitions should, when applied in the case of a dispersion-free language, be compatible, in a suitable sense, with the classical definitions. It would, of course, be quite in order to use new connectives also, provided only that each is introduced by a tangible definition. However, for the present we shall use the same syntax as in the classical case.

The term 'opponent', mentioned briefly in § 4 (see 4.2) will now be used more extensively. It is to be understood in terms of the following 'rules of discourse', which are proposed primarily in the interests of simplicity. When a speaker, who may in this connection be called the *proponent*, makes an assertion it is to be regarded in the first instance as an *offer* to assume the indicated commitment. Any other speaker may respond to this offer; he says, perhaps, "I accept". Thereupon the commitment becomes effective and a 'debate' ensues between this *opponent* and the proponent, in which no other speaker may intervene.

In the following, A and B denote arbitrary propositions. For $\neg$ we adopt:

5.1. *Definition*. He who asserts $\neg A$ offers to pay \$ 1 to his opponent if he will assert A.

That this definition is intuitively reasonable is clear, at least in the case when A is a prime proposition. For the proponent will then recover his money from the opponent iff (= if and only if) the trial of A yields 'no'. The justification of the definition of $\vee$ (or) is even simpler:

5.2. *Definition*. He who asserts $A \vee B$ undertakes to assert either A or B at his own choice.

We now come to the connective $\rightarrow$ (implies). We could adopt the classical procedure, introducing $A \rightarrow B$ as an abbreviation for $B \vee \neg A$; this leads to a structure equivalent to classical logic. However, we obtain a much richer language—and, in particular, one in which it is possible to express *degrees of belief*—and at the same time we accord at least as closely with the intuitive meaning of 'implies' with the following definition:

5.3. *Definition*. He who asserts $A \rightarrow B$ offers to assert B if his opponent will assert A.

In clarification of this definition I add the following remark: If, during any debate, either speaker asserts a proposition of the form $A \rightarrow B$ the other speaker may choose to *admit* this assertion, in which case it is simply annulled. Alternatively, he may *challenge* it by asserting A; the original speaker must then fulfill his obligation by asserting B, whereupon the original assertion of $A \rightarrow B$ is annulled: an assertion may be challenged only once.

The definition of $\rightarrow$ is a crucial feature of the new logic. A similar definition was introduced by Lorenzen [5] in giving a *dialogue interpretation* of intuitionistic logic.[5] The resulting logics are quite different, however, mainly because Lorenzen did not employ Definition 4.2 (there are other differences as well).

Let us we add to the formal language the symbol F to stand for the particular prime proposition whose tangible meaning is given by:

5.4. *Definition*. He who asserts F agrees to pay his opponent \$ 1.

Then $\neg A$ may be identified with $A \rightarrow F$: these assertions incur the same commitment. The device of replacing $\neg$ by F is technically convenient. We therefore ignore 5.1 and introduce instead:

5.5. *Definition*. $\neg A$ is an abbreviation for $A \rightarrow F$.

In selecting a definition for $\wedge$ (and) the obvious choice would be to let the assertion of $A \wedge B$ be equivalent to the assertion of both A and B. However, there are several reasons why this is not suitable here. Firstly, $A \wedge A$ would not be equivalent to A. Secondly, the analogous definition of $\forall x A(x)$ (see 5.7 below) is clearly not viable. Finally, such a choice would mean abandoning the principle of limited liability. So, instead, we adopt the following obvious analogue of 5.2:

5.6. *Definition.* He who asserts $A \wedge B$ undertakes to assert either A or B at his opponent's choice.

Observe that, although the speaker is not obliged to assert *both* A and B, he must be prepared to assert either, since he cannot tell which his opponent may choose.

The definitions of the quantifiers $\exists$ and $\forall$ are natural generalizations of 5.2 and 5.6:

5.7. *Definition.* He who asserts $\exists x A(x)$ undertakes to assert $A(t)$ for some constant term t of his own choice. He who asserts $\forall x A(x)$ undertakes to assert $A(t)$ for some constant term t of his opponent's choice.

Here $A(x)$ denotes a formula in which no variable other than x occurs free and $A(t)$ denotes the result of replacing x by t at every free occurrence of x in $A(x)$. [In the less well known, but much more convenient notation of Smullyan [6] one would write A for $A(x)$ and A_t^x for $A(t)$.]

We have now given definitions for all the logical symbols in such a way that, by recursion, a *tangible meaning* is assigned to every proposition: he who asserts any proposition incurs, in so doing, a *definite obligation.* The general nature of this obligation is clear. Suppose that one speaker asserts a compound proposition, say $A \rightarrow B$. His opponent may, if he wishes, challenge it by asserting A whereupon the first speaker must assert B, his original assertion then being annulled. If A and B are compound propositions the debate will continue, in accordance with the rules, with new propositions being asserted and previous assertions being annulled. At any time during the debate a *position* will obtain in which each speaker is committed to a finite collection of assertions which we shall call his *tenet* (repetitions of assertions may occur in a tenet). As the debate continues the position changes, the propositions asserted becoming simpler, until eventually a *final position* is reached in which all propositions asserted are prime. The appropriate trials are then carried out and each speaker pays the other in accordance with 4.2 and 5.4.

We assume that every assertion must in due course be "dealt with" in accordance with the definitions. We make no ruling here as to the *order* in which the assertions are dealt with, since it turns out, for our immediate purposes, to make no difference. However, one point needs clarification. We allow the *opponent* complete freedom in arranging the trials of the prime propositions occurring in the final position. He may use what procedures he likes; all that is requires is that these procedures are in fact admissible trials, as specified in the corresponding documents. The reason for this apparent favoritism is easy to see: it was the proponent who, in his initial assertion, first mentioned these propositions. He had at that time, in the process of defining them, the chance of laying down any conditions he wished. The opponent's strategy during the debate has been determined by the form of these conditions, and it would not be fair to impose further conditions *a postiori*.

To illustrate the above definitions let us consider a simple example. For convenience in the discussion the proponent and opponent will be referred to as 'I' and 'you' respectively, and any position will be denoted by an expression consisting of my tenet on the right and yours on the left of a vertical stroke.

Consider the propositions $A \to B$ and $B \vee \neg A$, where for simplicity we shall assume that the propositions A and B are prime. In classical logic these propositions 'have the same meaning': each is true iff B is true or A is false. However, in the new logic they cannot be identified since their assertions represent different commitments. Indeed, if I assert $A \to B$ we shall (according to whether you challenge or not) reach one of the final positions $\emptyset|\emptyset$ or $A|B$.[6] Thus in asserting $A \to B$ I commit myself to be placed in one (at your choice) of these final positions. On the other hand, if I assert $B \vee \neg A$ I am obliged either to assert B, thus reaching the final position $\emptyset|B$, or $\emptyset|\neg A$ which, if you challenge, yields the final position $A|F$: i.e. I commit myself to be placed in one of the final positions $\emptyset|B$ or $A|F$, at *my* choice.

It is clear that, formally, these commitments are quite distinct. To see that they are actually so, suppose that I believe more strongly that an outcome 'yes' will result from a trial of B than from a trial of A.

Then I may reason that, in the position $A|B$, I am more likely to gain than to lose and so be willing to assert $A \rightarrow B$ whereas, unless I am *sure* of the outcome of at least one of A and B, I would expect to lose on average from the assertion of $B \vee \neg A$. Indeed, the argument shows that the commitment involved in asserting $A \rightarrow B$ is strictly weaker than that for $B \vee \neg A$.

This example brings out two points: firstly, the new definitions of the connectives yield a syntactically identical but semantically richer language than that of classical logic; secondly, there is in this language a possibility of expressing relative *degrees of belief*. In fact, it can be shown [2] that an arbitrary quantitative declaration of (relative or absolute) degree of belief can be expressed with arbitrary accuracy by a suitable proposition in the language.

The definition of (relative) truth introduced for prime propositions at the end of § 4 can easily be extended to arbitrary propositions. We say that a proposition A is *true for an observer* **P** and write $\mathbf{P} \models A$ if **P** will willingly[7] assert A: i.e. if **P** fears no loss in asserting A, and we say A is *false for* **P** if $\mathbf{P} \models \neg A$. (This is not to be confused with $\mathbf{P} \not\models A$ which means 'it is not the case that $\mathbf{P} \models A$'.) It is easy to see that this agrees in the case of a prime proposition with the earlier definitions.

Consider, now, the case of a dispersion-free language and suppose that (a) every prime proposition A is either true or false for the observer **P**. (This is not necessarily the case: it might be that **P**, in spite of being sure that A is dispersion-free, does not know whether A is true or false and so will not willingly assert either A or $\neg A$). Assume also (b) that **P** is able to scan any set of propositions and tell whether any (or all) of the propositions in the set are true (or false) for him. Then it is easy to deduce from 5.2–5.7 that the familiar classical inductive definitions of the logical connectives are satisfied: $\mathbf{P} \models (A \wedge B)$ iff $\mathbf{P} \models A$ and $\mathbf{P} \models B$, $\mathbf{P} \models (A \rightarrow B)$ iff $\mathbf{P} \models B$ or $\mathbf{P} \models \neg A$, and so on. It then follows by a simple recursion that the propositions which are true (false) in the new logic coincide with those which are classically true (false); in particular, every proposition is either true or false. This shows that our logic does indeed reduce in a satisfactory way to classical logic in the case where the latter is applicable.

I shall call **P** an *omniscient observer* and refer to the assumptions (a) and (b) as the *principle of omniscience* (a term introduced by Bishop [7] in a very similar sense). This principle, although a fundamental assumption of both classical logic and classical mathematics, is obviously unrealistic.

VI. STRUCTURES

Let us consider briefly how the methods of classical logic are applied in the case of a dispersion-free interpreted language. Assume, then, an omniscient observer **P**; for him every prime proposition is either true or false. In this way the interpretation determines a (classical) *prime valuation*: i.e. a map that assigns to each prime proposition a *truth value*, 'true' or 'false'. It is a well known result of classical logic (see e.g. [6]) that every prime valuation can be extended uniquely to a *valuation* (which assigns a truth value to *every* proposition). It follows in particular that the valuation which represents the truth values assigned by **P** is determined by its restriction to the prime propositions: in other words, we know all there is to know about **P** if we know what he thinks of the prime propositions.

If the observer **P** is not omniscient then for him not every prime proposition is assigned a truth value. We might then speak of an *indefinite* (or partial) *prime valuation*, meaning that only some prime propositions are assigned a truth value. Such an indefinite prime valuation admits, in an obvious way, an extension to an 'indefinite valuation', but this extension need not correctly reflect the state of belief of **P**. It is quite possible, for example, that **P** should believe that $A \to B$ is true (i.e. $\mathbf{P} \models A \to B$) without having knowledge of the truth values of A or B. Examples abound in mathematics. Thus the indefinite valuation which represents the beliefs of **P** is not determined by its restriction to the prime propositions. The methods of classical logic fail, even in the dispersion-free case, when the principle of omniscience is abandoned.

Now consider the situation when the new logic is used, the language involved not necessarily being dispersion-free. In this case little information will be conveyed by giving the set of true prime propositions and the set of false prime propositions (relative to **P**): indeed, both sets may very well be empty. However, many compound propositions may be true. For instance, we saw that $A \to B$ is true for **P** (A and B being prime) iff the final position $A|B$ is acceptable to **P**. Indeed, it is clear that in order to decide, in general, whether an arbitrary compound proposition can be safely asserted **P** must be prepared to say whether an arbitrary given final position is acceptable or not. In some sense, then, **P**'s state of belief is characterized by the set of final positions he considers acceptable. This set, which I shall refer to (tentatively) as *the structure for* **P** or simply as *the structure* **P**, plays the same role for the new logic as a prime valuation does in classical logic.

Obviously (for a rational observer—and this we shall always assume) not every set of final positions can be a structure—for instance, some final positions are surely unacceptable. I shall now state an axiom which purports to give necessary and sufficient conditions on a set of final positions that it be admissible as a possible structure for a (rational) observer. For this purpose some notation is necessary.

Let us write any tenet as a formal *sum* of the propositions in it and denote the position in which the proponent and opponent are committed to the tenets β and α respectively by the expression $\alpha|\beta$ (proponent on the right). Let $\mathbf{P}\models\alpha|\beta$ denote that the position $\alpha|\beta$ is acceptable for **P** (i.e. $\alpha|\beta$ belongs to the structure **P**). A little thought will show that, in view of the tangible definition (4.4) of 'perfect', our standing assumption that every elementary experiment is perfect implies that $\mathbf{P}\models A|A$ (or equivalently $\mathbf{P}\models A \to A$), for every prime proposition A and every observer **P**.

I claim that for any (rational) observer **P** the following axiom is satisfied (for brevity I write $\models$ for $\mathbf{P}\models$):

6.1. *Axiom.* Let $\alpha, \beta, \gamma, \delta$ be prime tenets and A be a prime proposition. Then, for any structure **P**:

(a) *If* $\models\alpha|\beta$ *and* $\models\gamma|\delta$ *then* $\models\alpha+\gamma|\beta+\delta$.

(b) $\models F|A$.

(c) $\not\models F$.

(d) $\models A|\emptyset$.

(e) If $\models n\alpha+F|n\beta$ *holds for arbitrarily large positive integers n then* $\models\alpha|\beta$.

(f) $\models\alpha+A|\beta+A$ *if and only if* $\models\alpha|\beta$.

The justification of this claim is given in detail in [8]. Very briefly: (b), (c), and (d) are evident; for (a) simply notice that if **P** is placed in the last position mentioned he is, to all intents and purposes, simultaneously in the other two positions; (f) follows from the assumption that A is perfect; for (e) observe that if **P** loses (on average) in the position $\alpha|\beta$ he would lose n times as much in the position $n\alpha|n\beta$ and thus, for sufficiently large n, he would also lose in the position $n\alpha+F|n\beta$.

In addition to this axiom there is a further property which plays the same role in the new logic as the condition (a) of the principle of omniscience does in the classical case:

6.2. *Definition.* A structure (or observer) **P** is *definite* or *probability-definite* (this last term is justified by the consequences) if, for every pair of tenets α and β, either $\mathbf{P}\models\alpha|\beta$ or $\mathbf{P}\models\beta|\alpha$.

Indeed, if **P** considers both $\alpha|\beta$ and $\beta|\alpha$ unacceptable he has evidently not yet been able to decide what (on average) will happen when the necessary trials are carried out: if he really expected to lose in $\alpha|\beta$ then he would expect to gain in $\beta|\alpha$. Thus to this extent he is not omniscient.

The consequences of Axiom 6.1 are worked out in [1] and [8] (see also [2]). I summarize the results first in the probability-definite case. Let us make the following definition.

6.3. *Definition.* A *prime valuation* $\langle\ \rangle$ is a map which assigns to each prime proposition A a number $\langle A\rangle$ and to each final position $\alpha|\beta$ a number $\langle\alpha|\beta\rangle$ such that:

(a) $0 \leqslant \langle A\rangle \leqslant 1$ *for all* A.

(b) $\langle F\rangle = 1$.

(c) $$\langle A_1+\ \dots\ +A_m|B_1+\ \dots\ +B_n\rangle = \sum_{j=1}^{n}\langle B_j\rangle - \sum_{i=1}^{m}\langle A_i\rangle,$$

for any prime propositions $A_1, \dots, A_m, B_1, \dots, B_n$.

We then have:

6.4. *Theorem.* Given any probability-definite structure **P** there is a unique prime valuation $\langle\ \rangle_{\mathbf{P}}$ such that

(d) for every final position $\alpha|\beta$, $\mathbf{P}\models\alpha|\beta$ iff $\langle\alpha|\beta\rangle_{\mathbf{P}} \leqslant 0$. Conversely, given any prime valuation $\langle\ \rangle$, there is a unique probability-definite structure **P** such that $\langle\ \rangle_{\mathbf{P}} = \langle\ \rangle$.

Thus each probability-definite structure is uniquely characterized by a prime valuation. The prime valuation corresponding to a given probability-definite structure (or observer) **P** has a very simple interpretation: $\langle A\rangle_{\mathbf{P}}$ may be described as **P**'s estimate of the probability that a trial of A will yield the outcome "no"; it is the expected loss or *risk value* of an assertion of A. Similarly, $\langle\alpha|\beta\rangle_{\mathbf{P}}$ is the risk value of the final position $\alpha|\beta$ and (d) simply asserts that a final position is acceptable if its computed risk value is not positive. Theorem 6.4 simply shows that, whatever considerations may have passed through the mind of an observer **P**, in the probability-definite case he *behaves as if* he had assigned subjective probabilities to every prime proposition and computed the acceptable final positions on this basis.

$\langle A\rangle$ may be thought of as the 'degree of falsehood' of A and $p(A) = 1-\langle A\rangle$ as its 'truth value'. $p(A)$ may take any value between 0 (falsehood) and 1 (truth).

Let us apply this result to the case of a dispersion-free language. In this case it is easy to show that the only possible risk values for a prime proposition A are $\langle A\rangle_{\mathbf{P}} = 0$ or $\langle A\rangle_{\mathbf{P}} = 1$, implying, (via (d) and the definition of $\urcorner$) that A is either true or false. Thus the prime valuation $\langle\ \rangle_{\mathbf{P}}$ reduces to a prime valuation in the classical sense given at the beginning of this section.

These considerations suggest that in the general (dispersive) case we are concerned with a logic having a continuous range of truth values ranging from 1 (true) to 0 (false). The suggestion bears up under closer study. Indeed, defining 'valuation' by simply omitting the word 'prime' from 6.3, we obtain

6.5. *Theorem.* Every prime valuation $\langle\ \rangle_{\mathbf{P}}$ has a unique extension to a valuation (also denoted $\langle\ \rangle_{\mathbf{P}}$) with the following property. Let A be any compound proposition. Then, for any $\varepsilon > 0$, **P** has a strategy (of debate) that will, from the initial position $\emptyset|A$, guarantee a final position $\alpha|\beta$ with $\langle\alpha|\beta\rangle < \langle A\rangle+\varepsilon$ but no strategy that will ensure $\langle\alpha|\beta\rangle < \langle A\rangle$.

We assume as always that **P** is rational. It follows that $\mathbf{P}\not\models A$ if $\langle A\rangle > 0$, for in this case he cannot (after asserting A) be sure of reaching an acceptable final position. On the other hand, in view of the "arbitrarily small fee" mentioned in the definition of truth, $\mathbf{P}\models A$ if $\langle A\rangle = 0$. This gives, since $\langle A\rangle \geqslant 0$ always:

6.6. *Corollary.* For every proposition A, $\mathbf{P}\models A$ iff $\langle A\rangle = 0$: in other words, *the true propositions are those whose risk value is zero.*

It can also be shown that the structure **P** is uniquely determined by the set of all true propositions.

In the course of the proof of 6.5 the following result is obtained. It arises rather directly from the pragmatic definitions 5.2–5.7 of the logical symbols:

6.7. *Theorem.* Let $\langle\ \rangle$ be any valuation. Then, for any propositions A and B,

$$\text{(a)} \quad \begin{aligned} \langle A \rightarrow B\rangle &= \sup\{0, \langle B\rangle - \langle A\rangle\}, \\ \langle A \vee B\rangle &= \inf\{\langle A\rangle, \langle B\rangle\}, \\ \langle A \wedge B\rangle &= \sup\{\langle A\rangle, \langle B\rangle\}, \\ \langle \neg A\rangle &= 1 - \langle A\rangle. \end{aligned}$$

Further, for any formula $A(x)$ in which no variable other than x is free

$$\text{(b)} \quad \begin{aligned} \langle \exists x A(x)\rangle &= \inf\{\langle A(t)\rangle\colon t \textit{ a constant term}\}, \\ \langle \forall x A(x)\rangle &= \sup\{\langle A(t)\rangle\colon t \textit{ a constant term}\}. \end{aligned}$$

With the trivial change, $p(A) = 1 - \langle A\rangle$, the relations 6.7(a) become the truth tables of a famous series of many-valued logics $Ł_n$, $n = 2, 3, \ldots$ and ∞, introduced by Łukasiewicz in the early 1920's. The subscript n indicates the number of truth (or risk) values. Indeed, suppose that the prime propositions take only risk values in the set $\{-r/(n-1)\colon r = 0, 1, \ldots, n-1\}$. Then it follows from 6.7 that this holds for all propositions. This gives $Ł_n$; $Ł_\infty$ is obtained by admitting all risk values in the interval [0, 1]. In the case of $Ł_2$, where the only risk values are 0 and 1 (corresponding to truth and falsehood respectively), the equations 6.7(a) reduce to the usual truth tables of the classical propositional calculus. This is the form that arises in the case of a dispersion-free language, since each prime proposition then takes one of the risk values 0, 1.

In the general case of a dispersive physical language the subjective probabilities assigned to the prime propositions can take any values. Thus *the logic $Ł_\infty$ plays the same role in the dispersive case as the classical propositional calculus does in the case of a dispersion-free language* (which includes, of course, any mathematical language).

$Ł_\infty$ is more practical than classical logic in that it is not restricted to dispersion-free interpreted languages. However, it is still unrealistic in that appeal must be made to the principle of omniscience. I have already remarked that 6.2 is the equivalent of part (a) of that principle and it is necessary to appeal to part (b) in dealing with the quantifiers during the proof of 6.5.

I now summarize the consequences of Axiom 6.1 (in so far as they have been worked out, see [8] and [2]) in the general case in which the structure (or observer) is not assumed to be probability-definite (Definition 6.2).

Forget, for the moment, about Axiom 6.1 and imagine an observer **P** who believes, as many physicists do, that associated with each elementary experiment (or prime proposition) A is an objective probability, $v(A)$, of outcome 'no', these probabilities being, however, largely unknown to him. We need not enquire what his interpretation of $v(A)$ is; it may, for instance, be the frequency interpretation or some sort of propensity interpretation. We are concerned only with the role which his belief plays in governing his behaviour: i.e. in determining which positions he considers acceptable. For every prime proposition A we will have, of course, $0 \leqslant v(A) \leqslant 1$. Moreover, since the elementary experiment F is sure, by definition, to produce the outcome 'no', $v(F) = 1$.

Let $\alpha = \Sigma m_i A_i$ and $\beta = \Sigma n_j B_j$ be any prime tenets, and set $v(\alpha|\beta) = \Sigma n_j v(B_j) - \Sigma m_i v(A_i)$; then, by 6.3, v is a prime valuation. Now **P**, as a believer in objective probability, will argue as follows. For each prime proposition A, $v(A)$ is, through the meaning of probability, the average loss, in dollars per trial, that he would experience in an infinite series of trials of A. Consequently, $v(\alpha|\beta)$ represents the average total loss in dollars that he would experience in the trials which result when he is placed in the position $\alpha|\beta$. Now, the objective probabilities $v(A)$ are not fully known to him. Let us suppose, however, that he is able to express his state of belief regarding the values of these probabilities by laying down a set $\mathscr{V}$ of valuations to which he believes v to belong. Then his mean loss per occasion, when placed in the position $\alpha|\beta$, will lie between $\inf\{v(\alpha|\beta)\colon v \in \mathscr{V}\}$ and $\sup\{v(\alpha|\beta)\colon v \in \mathscr{V}\}$. Suppose **P** considers a final position acceptable iff he is *sure* that, should he be placed in that position a large number of times, he would not lose, on the average. Then he will consider $\alpha|\beta$ acceptable, which we may write

(1) $\mathbf{P} \models \alpha|\beta$, *iff* $\sup\{v(\alpha|\beta)\colon v \in \mathscr{V}\} \leqslant 0$.

Now, it turns out [8] that, for any set $\mathscr{V}$, the relation defined by (1) satisfies Axiom 6.1: i.e. **P** is a rational observer in the sense embodied in that axiom. Conversely, given any structure **P** (satisfying Axiom 6.1) there is a set $\mathscr{V}$ of prime valuations such that (1) holds.[8] We can thus say that any observer who is rational, in the sense that 6.1 holds, *behaves as if* he believed in objective probabilities but was not fully aware of their values.

The quantity $\langle\alpha|\beta\rangle = \sup\{v(\alpha|\beta): v \in \mathscr{V}\}$, whose importance is shown by (1), may be called the *risk value* of the position $\alpha|\beta$. The *prime risk function* $\langle\ \rangle$, which assigns to each prime position its risk value, has properties somewhat weaker than those of a prime valuation. [Roughly, a prime risk function is to a prime valuation as a sublinear function (on a vector space) is to a linear function.]

In order to determine what sort of logic should rationally be used by a non-probability-definite observer **P**, it is natural first to ask whether Theorem 6.5 can be generalized to yield the existence of a unique *risk function*, an extension of the prime risk function $\langle\ \rangle$, which will determine the propositions which are true for **P** (i.e. those which can be safely asserted) via Corollary 6.6. One easily finds that this question cannot immediately be answered, owing to an ambiguity in the rules of debate introduced in § 5. The point was mentioned in § 5: At any time during a debate the two participants are each committed to a number of assertions. Our rules determine how each assertion should be dealt with, but they give no indication as to the *order* in which they should be treated: in game-theoretic terms, we need 'rules of order' at least to determine *whose move it is* in a given position. It can be shown (see [2]) that in the probability-definite case *and only in that case* the rules of order are irrelevant.

At the present time (March 1974) a particular set of rules of order has been selected and its consequences are under investigation. However, the situation is substantially more complicated than in the probability-definite case and it would be premature to discuss it further.

BIBLIOGRAPHY

[1] Giles, R., 'A non-classical logic for physics', *Studia Logica* **33** (1974), 4, pp. 397–415.

[2] Giles, R., 'Physics and logic': lecture notes (1973).

[3] Kreisel, G. and Krivine, J. L., *Elements of Mathematical Logic*, North-Holland 1971.

[4] Goodman, N. and Myhill, J., 'The formalization of Bishop's constructive mathematics', in *Toposes, Algebraic Geometry, and logic*, Springer Lecture Notes in Mathematics, 274, 1972.
[5] Lorenzen, P., *Metamathematik*, Bibliographisches Institut, Mannheim.
[6] Smullyan, R. M., *First-Order Logic*, Springer-Verlag, 1968.
[7] Bishop, E., *Foundations of Constructive Analysis*, McGraw-Hill, 1967.
[8] Giles, R., 'A logic for subjective belief', Harper and Hooker (eds.), *Foundations of Probability Theory, Statistical Inference, and Statistical Theories of Science*, vol. I, pp. 41—72, Reidel, 1976.

REFERENCES

[1] Research supported by the National Research Council of Canada.

[2] In spite of the title I speak mainly of physical theories. It will be seen, however, that the considerations involved in this paper are so general as to apply without change to arbitrary empirical theories.

[3] One other assumption, the 'principle of omniscience', is also needed. See below.

[4] For simplicity I henceforth denote a prime proposition and the corresponding elementary experiment by the same letter.

[5] In fact, the definitions of all the logical symbols resemble, and were indeed suggested by, those introduced by Lorenzen.

[6] Ø denotes the *empty tenet*, in which nothing is asserted.

[7] More precisely, if **P** is willing to assert A for an arbitrarily small fee. This modification makes practically no difference and is technically convenient.

[8] Indeed, if we add the condition that the set $\mathscr{V}$ should be convex—i.e. closed under the formation of weighted means ($v = \lambda v_1 + (1-\lambda) v_2$ means $v(A) = \lambda v_1(A) + (1-\lambda) v_2(A)$ for every A)—and 'closed in the weak topology', then **P** determines $\mathscr{V}$ *uniquely*.

B. L. LICHTENFELD AND A. I. RAKITOV

UNCERTAINTY, PROBABILITY AND EMPIRICAL KNOWLEDGE

An empirical knowledge arises as an effect of at least five factors. They are:

(1) the nature and state of the object;

(2) the characteristics of the experimental apparatus and tools of observation;

(3) actions of the experimentator and researching procedures;

(4) the paradigm of theoretical constructions that regulate our vision of things;

(5) the decisions upon each step of collecting empirical data and upon planning of carrying out and completing the experiment.

In each of these factors it is possible to find some *uncertainty*. To describe and evaluate it we need to use some concepts and technique of the theory of probability.

H. E. Kyburg ('Probability and Inductive Logic', London, 1970) mentions six main conceptions of probability. As known, they do not influence essentially calculation of probabilities. Each of these concepts is oriented for describing and explaining only one kind of uncertainty and requires more or less stronger abstraction and assumption. The strongest is the assumption concerning situations, actions and decisions that are necessary for building empirical knowledge.

The calculations based on such assumptions often lead to *paradoxes*, like the so-called 'Petersburg paradox of Bernoulli'. The procedure of building the empirical knowledge is finite by its nature. Sooner or later, we have to stop our experiment or observation, even in case we are not completely satisfied with the results. The reasons for this may be:

(a) the experiment is too prolonged;

(b) to continue the experiment is too expensive;

(c) the expected information is supposed to be unable to change the results obtained before;

(d) the amortisation of the apparatus and tools of observation, or at last the change of the aims of the researching groups.

Arising of paradoxes shows that the given concept of probability is unadequate to the empirical process. The experimentator has to take into consideration not only the probability of the situation, but also the probability of decisions and actions. In a scientific experiment certain events are often the products of definite decisions and actions of the experimentator. In turn these actions and decisions may be implicated by previous events. The probabilistic description of the real process of building the empirical knowledge should obviously take into account the interconnection of all these factors and *types of uncertainty*, corresponding to them. The suitable axiomatics must be based on less strong abstraction. This task is rather complicated and it is not clear whether there is a possibility to solve it completely. However if we take only two main factors—the set of *events* and the corresponding set of *actions* of the experimentator—then it will be possible to establish some modified system of probabilistic axioms. The functions included into these axioms are defined on two sets: the set of actions and the set of events. By the way, in such systems of axioms it is easy to avoid paradoxes like 'Petersburg paradox'.

We shall not differentiate actions and decisions, and besides we shall speak only of *objective events.* To simplify the task we shall regard only the case of finite number of points in the universal sets. Let $S_0 = \{s_1, s_2, \ldots, s_n\}$ be a set of *elementary events*; $D_0 = \{d_1, d_2, \ldots, d_k\}$ – a set of *elementary actions*; $\mathscr{F}$—some set of subsets from S_0 and $\mathscr{L}$—some set of subsets from D_0. We shall call the elements of $\mathscr{F}$—*random events*; the elements of $\mathscr{L}$—*possible actions.* Any pair $(S_i:D_j)$, where $S_i \in \mathscr{F}$, $D_j \in \mathscr{L}$ we call—a *random event* S_i, given D_j, and $(D_j:S_i)$—*possible action of* D_j, given S_i.

Axioms

1[a] *$\mathscr{F}$ is an algebra*	1[b] *$\mathscr{L}$ is an algebra*
2[a] *With any pair $(S_i:D_j)$ the number $p(S_i:D_j)$ is associated called the probability of S_i, given D_j.*	2[b] *With any pair $(D_j:S_i)$ the number $p(D_j:S_i)$ is associated, called the probability of D_j, given S_i.*

3^a $p(S_0: D_j) = 1$ *for any* j.	3^b $p(D_0: S_i) = 1$ *for any* i.
4^a *If* $S_p \cap S_m = \emptyset$, *then* $p(S_p \cup S_m: D_j) = p(S_p: D_j) + p(S_m: D_j)$ *for any* j *in case when the left-hand side and the right-hand side are defined simultaneously.*	4^b *If* $D_q \cap D_l = \emptyset$, *then* $p(D_q \cup D_l: S_i) = p(D_q: S_i) + p(D_l: S_i)$ *for any* i *in case when the left-hand side and the right-hand side are defined simultaneously.*
5^a *If* $D_q \cap D_l = \emptyset$, *then* $p(S_i: D_q \cup D_l) = \frac{p(S_i: D_q)p(D_q: S_0) + p(S_i: D_l)p(D_l: S_0)}{p(D_q \cup D_l: S_0)}$ *for any* i *in case when the left-hand side and the right-hand side are defined simultaneously.*	5^b *If* $S_p \cap S_m = \emptyset$, *then* $p(D_j: S_p \cup S_m) = \frac{p(D_j: S_p)p(S_p: D_0) + p(D_j: S_m)p(S_m: D_0)}{p(S_p \cup S_m: D_0)}$ *for any* j *in case when the left-hand side and the right-hand side are defined simultaneously.*

Beside doubled *Kolmogorov's axioms* there are two new axioms 5^a and 5^b in this system. They are necessary for composing the sets which make the condition of probability.[1]

If we take an exhaustive collection of mutually exclusive actions and events, for example the sets D_0 and S_0, then we shall get (by analogy with the Theorem of complete probability) the following formulas from axioms 5^a, 3^b and 5^b, 3^a:

$$
(1) \quad
\begin{aligned}
p(s_i: D_0) &= \sum_{j=1}^{k} p(s_i: d_j)p(d_j: S_0) \text{ for any } i \\
p(d_j: S_0) &= \sum_{i=1}^{n} p(d_j: s_i)p(s_i: D_0) \text{ for any } j
\end{aligned}
$$

$p(s_i: D_0)$ and $p(d_j: S_0)$ might naturally be called 'unconditional' probabilities of the events s_i and actions d_j.
We can regard them as the average value of conditional probabilities according to given collections of pairs.[2] This system can help us to

find any 'unconditional' probabilities on the base of given conditional probabilities.

Stochastically independent test sequence may be considered as a homogeneous *Markov process* with rectangular matrices $\{p_{ij}^{s}\}$ and $\{p_{ji}^{d}\}$. Every two multiplications result in an ordinary square matrix representing the probability of *event-to-event* transition. That is, an action was omitted between every two successive events which created an illusion of the process being absolutely isolated from the conditions of its realization and observation. It was possible to leave the action out because the whole set $\{d_j\}$ was actually considered as the only element d_f—the necessary conditions available for the experiment to be realized. This is evidently the case when for any i and $j \neq f$: $p(d_j:s_i) = 0$, $p(s_i:d_j) = 0$; $p(d_f:s_i) = 1$. This results in equality of an 'unconditional' probability of each event and its conditional probability in the action d_f:$p(s_i:D_0) = p(s_i:d_f)$. Thus, our model conforms to the conventional one.

In the primitive *Bernoulli test* D_0 is supposed to be composed at least of two points d_1 and $d_2 = \neg d_1$. In a complicated model, for example, drawing balls out of a number of boxes D_0 means: to take a ball out of the first box, out of the second box, ..., not to take the balls at all. Each space of elementary states is to include the point 'nothing has happened'.

Hence, a new definition of probability is included in the system of axioms. It does not contain any assumption concerning actually equal performance of the same action in all possible tests. Now we can interpret the probability not only as the measure of the set, but as the *measure of reflection* of the given action to the given event and vice versa. This position completely corresponds to our intuition. Advantages of the method are given below.

The Petersburg paradox was set forth by N. Bernoulli. A game is proposed to a book-maker. A coin is being tossed till the heads show for the first time. If the heads are made by the first toss, the player gets two roubles, by the second toss—four roubles, by the n-th toss—2^n roubles. What payment should be made before the game, so that it could be fair? Evidently, the fair payment is equal to the *expectation* (in the mathematical sense) of gain. But the expectation for the case is given by the diverging series:

$$\bar{B} = 2 \cdot 1/2 + 2^2(1/2)^2 + \dots + 2^n(1/2)^n = \infty$$

Thus any payment is too small, i.e. any finite sum for sufficiently large number of games would ruin the book-maker.

Where does the paradox come? There is no paradox at all in terms of the theory of probability. In this case the expectation of a random quantity is ∞. On the other hand, the game seems to be fair and it is not clear why it could not be played fairly. All the solutions of the paradox offered so far have been founded on the assumption that the book-maker's capital is limited or that *utility function* in this game is not strictly proportional to the money gained. But both methods are highly artificial. The limitation of the book-maker's capital might appear as an estimation of a real man's opportunities. In this model, however, the book-maker stands rather for a certain machine called 'gambling house'. Nothing leads us to the statement that the abilities of this machine are limited. Even more artificial is building up a suitable utility function. That involves absolute arbitrariness. Instead of the logarithmic function offered by D. Bernoulli (the nephew of the author of the paradox), one can take any function, provided the respective series converges. In any case these decisions are not derived from the game itself. It is required in the game that the test sequence of the length unknown beforehand is realized, and the theory of probability is not valid without supplementary concepts for the game situation.

Thus after each test the gambler has at least two actions to choose: d_1—'to go on gambling' or d_2—'to stop gambling'. Hence there are two rectangular matrices of two actions and three events: s_1—'heads', s_2—'tails', s_3—'nothing has happened'. The probability of elementary events and actions is shown in the tables.

I

$p(s_i : d_j)$	d_1	d_2
s_i	1/2	0
s_2	1/2	0
s_3	0	1

II

$p(d_j : s_i)$	s_1	s_2	s_3
d_1	0	p^d_{12}	1
d_2	1	$1 - p^d_{12}$	0

The third column of table II presents the start of the game, and action d_2 means the end of the game, i.e. payment. But the most important in them is p^d_{12}—the gambler's intention to go on gambling after 'tails', which he is supposed to inform the book-maker before the first game. For instance, at $p^d_{12} = 4/5$ the gambler must stop the game after each fifth 'tails' on the average and lose his stake. Now the probability of gain after n independent tossings of the coin is

$$p(d_2:s_1:d_1:s_2:d_1: \dots :d_1:s_3) = (1/2)^n(p^d_{12})^{n-1}$$

and gain expectation is

$$\bar{B} = \sum_{n=1}^{\infty} 2^n(1/2)^n(p^d_{12})^{n-1} = 1/1-p^d_{12}$$

In the example given above ($p^d_{12} = 4/5$) five roubles could be fair payment for the game. By the way, if the gambler is going to play only one game, then according to the 4/5 he is to stop the game precisely at the fifth 'tails'.

In a more comprehensive experiment the test is done either once or twice, etc. In the former axiomatics the number of the tests had to be strictly observed, but now it is determined by the experimentator's activities with certain probability.

The situation like 'Petersburg game' is valid for the scientific experiment planning. The study in the field of elementary particles physics is known to have risen in price and as a result of it expenditures for one experiment amount to millions of roubles. It is important to envisage the expenses that will be incurred.

After the experimental installation is ready the experiments begin. The positive outcome of the experiments gives the following: information, reliability, potential ground for future studies etc. It is natural to suggest that the utility should be inversely proportional to the a priori probability of its gaining. Hence, if the positive result is got at the n-th stage, then the utility $B_n = a/p_n$, where a shows usefulness of predetermined result. Let the probability p_n be approximately the same at each stage and equal to p^s_{11}, the stages being independent. In the axiomatics offered it is easy to make the expectation of the utility finite:

$$\bar{B} = \sum_{n=1}^{\infty} a/(p^s_{11})^n \cdot (p^s_{11})^n \cdot (p^d_{12})^{n-1} = a/1-p^d_{12}$$

The equation supports and determines qualitative considerations: a priori expected utility increases with the usefulness of the predetermined results and the a priori determination to go on with the experiments after failures. The determination can always be valued beforehand if the time of similar investigations is known with due account for the individual peculiarities of the work concerned.

Proceeding from the system of axioms which was proposed above one can say that the probability function is defined on the basis of binary relation of random reflection between the two sets. The function can be defined on the basis of a ternary relation as well if another set of conditions is taken into account. This process complicates still more the primary formulae of the calculus. We think, however, that it is possible to put a natural limit to it. The maximal number of interpretations to be obtained beyond this limit will describe more or less approximately all kinds of uncertainties in reality. An exhaustive description of these situations may yield either a generalization of concepts of probability or an exhaustive analysis of polysemy of probability.

REFERENCES

[1] The axioms 5^a, 5^b are independent of the others. To prove this one can construct the following measure (e.g. for 5^a):

$$p(S_i{:}D_j) = \begin{cases} N(S_i)/N(S_0), & \text{if } d_k \notin D_j; \\ 1, & \text{if } d_k \notin D_j \text{ and } s_n \in S_i; \\ 0, & \text{if } d_k \in D_j \text{ and } s_n \notin S_i; \end{cases}$$

$$p(D_j{:}S_i) = \frac{N(D_j)}{N(D_0)},$$ where $N(Q)$ is a number of points in Q.

[2] The definition of ordinary conditional probability is not changing, of course:

$$p(S_p/S_m{:}D_j) = \frac{p(S_p \cap S_m{:}D_j)}{p(S_m{:}D_j)}$$

EVANDRO AGAZZI

THE CONCEPT OF EMPIRICAL DATA

PROPOSALS FOR AN ITENSIONAL SEMANTICS OF EMPIRICAL THEORIES

I

The fact of necessarily including 'data' may be considered as the fundamental difference between empirical theories and formal ones. From a methodological viewpoint, this fact may be conceived as a rather radical difference in the way the two different classes of theories fulfil the essential condition of possessing some 'immediate truth'. *Formal theories* are characterized by the fact that the immediate truth of some of their sentences is 'stated' by the theory itself, whereas in the case of *empirical theories* such a truth is considered as something which is 'found', which comes from outside the theory; moreover, the theory is thought to construct its *internal truth* on the essential condition of keeping faithful to this *external truth* and of becoming able, in a way, to include it. In other words: every scientific theory has the problem of ascertaining the truth of its accepted sentences and this may be done rather often by generating it out of the truth of previously accepted statements, but this in turn is only possible if there are sentences which possess their own truth intrinsically: formal theories may be qualified as those which simply 'single out' some of their sentences as being endowed with such a truth, while empirical theories must learn from outside which are their immediately true sentences.

This way of considering 'data' is slightly different from the more usual one, which conceives them as *events*, as *facts*, as structures or properties of the so called 'external world', while we have directly connected them to the problem of the *immediate truth*. Nevertheless, it will be seen how this choice makes possible a rather precise treatment of some questions, which would hardly have been possible in the usual context: for the concept of truth applies properly to sentences and by that the reference to scientific theories becomes straightforward, as they are but particular systems of sentences. In such a way the problem of characterizing the concept of empirical data looses its common sense vagueness and becomes the more definite one of characterizing the concept of an immediately true sentence in an empirical theory.

We must now make more precise the above sketched distinction between formal and empirical theories, which was stated by saying that the first ones find their immediate truth 'inside' themselves, while the second ones find it 'outside'. If we look at this problem from a modestly formal point of view we must admit, first of all, that both kinds of theories, if properly conceived, turn out to be some sets of sentences which cannot posses as such any truth value, but must always receive it 'from the outside', after being interpreted on some suitable domain of individuals, in which the language of the theory may receive a *model*. This is true and obvious, but the different behaviour of the empirical and the formal theories comes out when one looks at the way they react towards the models of their languages. In case of a formal theory $\mathscr{F}\mathscr{T}$, which has some set of postulates (let us call it P), if it happens that a certain model of its language $\mathscr{L}$ is not a model of P, we do not worry and simply look for another model of the language, hoping to find out, at last, a model of $\mathscr{L}$ which would also be a model of P; but even if after a certain time our efforts do not come to a positive end, we do not discard our formal theory for that. It is in this sense that we can say that the truth of P, although it cannot be concretely shown unless a model of it is found, is nevertheless considered not essential in order to 'keep' P, which can also be expressed by saing that P is considered as true, to say so, 'inside' the theory. Quite different is the way we behave in the case of an empirical theory: if a model of its language $\mathscr{L}$ turns out not to be a model of a certain sentence S of it, this sentence is immediately discarded and the same happens for every set of sentences which cannot have as its model the model of the language. The difference appears now quite patently: in the case of the formal theories we have a stability of the sentences of the theory, which determines the choice of the *acceptable models* of the language; in the case of the empirical theories, we have a stability of the model of the language, which determines the selection of the acceptable sentences of the theory. If better understood, this fact must be expressed by saying that, while in the case of formal theories the language is commonly supposed to have many *possible models*, in the case of empirical theories it is supposed to have (at least theoretically, if not practically) one single model, i.e. its 'intended' model. There is surely nothing new in this remark, but its consequences do not seem to have been fully investigated yet. As a matter of fact, such a uniqueness of the model is in such a deep contrast with the current way of thinking in model theory, that one must expect, in a way, to find serious difficulties in applying its tools to the semantics of empirical theories. To put it more explicitly, one must be prepared to find difficulties when, after having applied the usual devices of the logico-mathematical semantics (which are all conceived in order to relate a formal language to 'arbitrary' universes) he will be faced with the problem of insuring the uniqueness of the interpretation that makes the formal sentences true of the intended model. Let us also notice that this model theoretic problem is strictly bound to the exact characterization of the concept of empirical data, because there could be no data without the possiblity of insuring to a language its *intended* meaning.

The purely theoretical reflections made above about the difficulty of taking full advantage of the model theoretical tools in the semantics of empirical theories are promptly confirmed by the actual efforts made to provide such an application of methods. It is well known, e.g., that the uniqueness of the model cannot be guaranteed by linguistical devices, like the one, e.g., of adding to the set of sentences of an empiric theory other particular sentences of the kind of the 'meaning postulates': model theory teaches us that, also in this case, we should not be able to distinguish the universe of our objects from other universes isomorphic to it (*isomorphism theorem*) and that, moreover, this would already be an exceptionally lucky case, for it is usually not possible to distinguish it even from non isomorphic universes; this always happens, in particular, in the case of infinite universes (results on categoricity).

As a consequence of these well known facts, the *semantic determinateness* of the language of an empirical theory has been investigated along some 'non verbal' paths, e.g. by resorting to the so called *ostensive definitions*. But these too revealed some weak points of their own: in fact, to define a predicate ostensively was meant to 'point out', one after the other, a certain amount of concrete objects for which the predicate is stated to hold and a certain other amount of objects for which it does not hold. This implies, obviously, that only a finite and yet a rather small set of 'positive standards' and of 'negative standards' can be concretely put forth and at this point semantic ambiguity appears inevitable. For, taking an object$x$ which belongs neither to the first, nor to the second set (and that must necessarily happen as they are both finite sets) we should not know wheter our predicate holds or does not hold of x. A possible way out could seem to be offered by saying that, after having ostensively provided a certain amount of positive and negative standards, this very fact should put everyone in the position of considering them as 'instantiations' of a certain predicate P, as 'examples' taken out of the 'class' which constitutes the proper denotation of P and which will contain every future object of which P hold true. But this solution too is weak for at least two reasons: first of all, because the objects which were 'pointed out' and sampled together as positive standards may well have more than one feature in common, which means that they could appear as instantiations of more than one predicate, and this would immediately imply a *semantic ambiguity* for P, that would correctly denote all these different classes. Secondly, even in the case that the selected positive standards should have in common just one single feature, and so uniqualy determine a class, the problem of accepting in this class any new object x should be solved every time on the basis of a certain 'criterion', which can only be a 'similarity' criterion. But this acceptance procedure is necessarily based on a 'judgment' and not any longer on an 'ostension'; moreover, in order for this judgment to be safe, it should be supported by the knowledge not of the positive standards, but of their unique common feature, which cannot be pointed out and, thus, goes beyond the ostensive procedure. On the other hand, without such a knowledge, the acceptance judgment would remain subjective and vague.

Both verbal and non verbal devices have also proved rather ineffective in overcoming semantic ambiguity. On the other hand, this overcoming appears of a decisive importance in order to speak of 'data' inside an empirical theory, for data are the touchstone to decide about the acceptability not only of single statements, but of the entire theory itself; they are, in a way, the only unshakeable part of it and they could not play this role unless they were free of any ambiguity.

If we turn our attention to the two types of failure we considered above, we should admit that one of them seems hardly avoidable: it is the one bound to the possibility of a 'verbal' characterization of the 'datum'. Against this possibility not only are very detailed and exact results obtained in model theory (like the isomorphism theorem and the categoricity results), but also a general epistemological consideration, i.e. the awareness that data are never, in science, the effect of a linguistic activity, but rather of some 'non verbal' activities, like observation, manipulation of instruments, modification of concrete situations, etc. It seems therefore advisable to look for a better exploitation of the possibilities hidden in the non verbal devices, which would be able to avoid the limitations that appeared in the case of the ostensive definitions.

The main source of inadequacy for the ostensive definitions seems to be the fact that they are able to let us perceive what the objects look like, but they are not in the position of showing us what the predicates look like. In other words, if I want to direct a little child toward the knowledge of the notion of 'red', I might point out to it e.g. a red ball, the red hood of its doll, a red skittle, but I could not be sure that, in place of the concept of 'red', it is not starting to form in its mind the concept of 'toy', for example. Moreover, it is quite possible that, from a psychological point of view, the concept formation for the everyday terminology follows indeed the path just sketched for the little child but it seems rather obvious that this is not the case for scientific languages. For these languages the question is not of becoming gradually trained in the more or less correct use of some vocabulary, but rather that of becoming acquainted with its exact meaning; as a consequence, while certain ostensive procedures might prove useful as mental training for the employment of a language according to the accepted standards of the community of the speakers of that language, it has not the 'logical' force sufficient to provide this language with the definiteness of meaning that is required in science.

But this argument, though of a certain value from a very general viewpoint, does not enter the very reason of the inadequacy of ostensive definitions for scientific assignments of meaning. Such a reason may be briefly sketched as follows: ostensive definitions can only show us 'things' of everyday experience, but not 'objects' of any specific science at all. This statement must be clarified a little, especially because we too have used the term 'object', previously, when speaking of the kind of concrete things which are 'pointed out' by ostensive definitions. We employed such a term, then, because the stage of our discourse was still rather informal, but from now on we must sharply distinguish between a 'thing' and a scientific 'object'; the reason of this distinction is quite natural: no exact science is concerned with *generic* 'things', but always with 'things considered from a certain viewpoint' and what matters turns out to be actually such a 'viewpoint'. So, e.g., a sheet of paper on which a red drawing is traced can be considered from the viewpoint of its weight, and becomes an 'object' of physics, but it can also be considered from the viewpoint of the composition of the cellulose it is made out of, or from the viewpoint of the composition of the red ink with which the drawing was traced on it, and it becomes thus an 'object' of chemistry; and if it is considered from the viewpoint of those spatial properties of the drawing which remain invariant under certain deformations of the sheet itself, it becomes an 'object' of topology, and so on. In other words, our sheet of paper, though being one and a single 'thing', turns out to become a very large group of 'objects', according to the different sciences that may be concerned with it.

Now, how can one clarify this notion of 'viewpoint' which makes a scientific 'object' out of a common sense 'thing'? The answer can be given by considering what the different sciences do in order to treat 'things' from their 'viewpoint': they submit them to certain specific manipulations of an *operational* character, which put the scientist in the position of answering certain specific questions he can formulate about these things. Such operational procedures may be the use of a ruler, of a balance, of a dynamometer, in order to establish some physical characters of the 'thing' like its length, its weight or the strength of some force exerted on it; they may be the employment of some reagents to determine its chemical composition, etc.

At this point, the whole situation becomes a bit clearer: the true question, in case of those empirical predicates that may be called *observational*, is not to point out their 'empirical' or 'factual' denotation (which would mean, if properly understood, to point out by a complete enumeration all the members of the class denoted by the predicate—which is impossible—or to point out only a finite number of them—which would fall short of the goal, as already remarked), nor to point out their 'abstract' denotation (that is their 'intention', which cannot be pointed out because it is a mental entity); the true question is to be able to answer in a positive or negative way about the truth of certain *sentences*. As a consequence, if certain operational criteria are at hand, which prove sufficient for that purpose, we must say that these very criteria are in the position of 'operationally' defining our observational predicates (i.e. the predicates that enter in these sentences).

We shall now sketch briefly how in such a way the difficulties that were met in the case of the ostensive definitions disappear. Let us consider, e.g., the predicate 'inflammable' and assign to it, as operational criterion for testing whether it holds true of a given object or not, that of putting x on a flame: the answer will be positive if that flame spreads on it, otherwise it would be negative. Such operational procedure being a 'criterion', it may be applied a potentially infinite number of times and, in such a way, the class of the objects which whould be assigned to the denotation of this predicate will be also potentially infinite, thus eliminating a first weak point of the ostensive definition. Secondly, this criterion being univocally determined, it will be affected neither by some difference, nor by any casual resemblance between the objects which have been actually grouped together up to a certain moment (that means that if they were, by chance, all red things, there would be no risk of taking that quality for that of being inflammable, for it was not mentioned in the description of the operation and it does not concern the 'objects', as a consequence). The example we selected shows, further, how the criterion of operationality allows one to consider as observational also many predicates which are not such from the viewpoint of the ostensive definition: in fact, while it is thinkable to construct directly, by pointing out, a set of red objects, it is not thinkable to make the same with inflammable objects (or with object endowed with any other 'dispositional' property), on the basis of simply perceiving them.

In an operational definition there is, to be precise, an ostensive aspect (one must point out concretely the different 'instruments' for operating and also the way of employing them), but this aspect regards only a finite and actually small number of ostensions, which concern the *predicate* and not the objects it refers to, and which allow an unambiguous definition of the predicate together with an idefinite possibility of applying it to objects.

After this brief explanation, the concept of an *empirical datum* may be cleared in the following way: it is a sentence which proves true according to the direct and immediate application of some of the operational criteria accepted to define the *ground-predicates* of that particular empirical science. Which are such ground-predicates must still be better determined: we shall see that they are the ones that directly enter the definition of the *objects* of an empirical science, but in order to see that it proves useful to leave now this informal discourse and enter a more formal treatement of the subject.

II

According to the oversimplifications which are currently accepted in the literature, we can suppose an empirical theory $\mathcal{T}$ to be expressed in a first order language $\mathcal{L}$, which must contain, among its descriptive constants, some observational predicates $O_1 \ldots O_n$ as well as theoretical predicates $T_1 \ldots T_p$. What makes this theory an empirical one is the existence of a model $\mathfrak{M}$ of its language, which may be identified with a structure of the following kind:

$$\mathfrak{M} = < U, \bar{R}_1 \ldots \bar{R}_s >$$

where U is a non-empty set of 'individuals' and $\bar{R}_1 \ldots \bar{R}_s$ are relations on U, the total number of which should be $s = n+p$, so that every O-predicate and every T-predicate may be interpreted on one of these relations (or, to put it differently, may be considered as the 'name' in $\mathcal{L}$ of that relation). The set U provides the range of the individual variables of $\mathcal{L}$. Once the model of $\mathcal{L}$ is fixed, it is straighforward to define the model of every sentence α of $\mathcal{L}$: unary relations are identified with subsets of U, n-ary relations with sets of ordered n-tuples of elements of U and then, taken a sentence $\alpha = Px$ we say that α is *true* in $\mathfrak{M}$ (or that $\mathfrak{M}$ is a *model of* α) if the individual $\bar{x}$ of U 'named' by x belongs to the subset $\bar{P}$ of U 'named' by P; if $\alpha = Rx_1 \ldots x_n$ we say that $\mathfrak{M}$ is a *model of* α if the ordered n-tuple $\langle \bar{x}_1 \ldots \bar{x}_n \rangle$ of individuals of U named by $x_1 \ldots x_n$ belongs to the set $\bar{R}$ of ordered n-tuples 'named' by R. The way of defining when α is true in $\mathfrak{M}$ for every non atomic α is well known.

There are two more or less explicit assumptions lying at the basis of this discourse: (i) that the individuals of U and the relations on U must be conceived as 'given'; (ii) that the set U must be decidable (i.e., taken an individual $\bar{x}$, it is always possible to decide whether $\bar{x} \in U$ or $\bar{x} \notin U$), while the relations on U are not necessarily decidable (i.e., taken an n-tuple $\langle \bar{x}_1 \ldots \bar{x}_n \rangle$ it is not always decidable whether $\langle \bar{x}_1 \ldots \bar{x}_n \rangle \in R_i$ or $\langle \bar{x}_1 \ldots \bar{x}_n \rangle \notin R_i$). This essential indecidability of the relations on U is the reason of the semantic ambiguity we have spoken about in the preceding section of this paper for it means that, given a certain $\bar{x}$, we are sometimes unable e.g. to state whether it belongs to $\bar{P}$ or not, which amounts to saying that we cannot decide whether $\mathfrak{M}$ is a model of Px or not. The above mentioned failures of the verbal and non verbal devices to avoid semantic ambiguity express the impossibility of making the relations on U decidable by means of those devices. We shall explicitly remark that, owing to the ineffectiveness of the ostensive definitions, this ambiguity holds also if we restrict ourselves to what we could call the observational submodel $\mathfrak{M}^o$ of our language (i.e. the model of the O-predicates of it):

$$\mathfrak{M}^o = \langle U, P_1^o \ldots P_n^o \rangle$$

We shall now try to explain how the semantics of an empirical theory should look in order to fit the methodological approach of the operational criteria of definition for predicates we put forth in the preceding section. Given the language $\mathscr{L}$ of an empirical theory, we shall still distinguish, among its descriptive constants, the O-predicates $O_1 \ldots$ $\ldots O_n$ and the T-predicates $T_1 \ldots T_p$, but the O-predicates will be considered, this time, as *operational* and not as *observational* (remember that *dispositional predicates* may turn out to be operationally definable, though not being observational in a strict sense). Our first problem (and actually the only one which will be discussed in this paper) concerns the semantic definiteness of the operational predicates; we shall therefore confine our treatement to the operational submodel $\mathfrak{M}^o$ of $\mathscr{L}$ or, if we prefer, to the model $\mathfrak{M}^o$ of the operational sublanguage $\mathscr{L}^o$ of language $\mathscr{L}$. Our model will be something of the following kind:

$$\mathfrak{M}^o = \langle \Omega, O, R, \bar{P}_1^o \ldots \bar{P}_n^o \rangle$$

where Ω is a finite set of 'instruments', O is a finite set of 'operations', R is a finite set of 'results' (i.e. of observational outcomes of concrete operations), while every $\bar{P}_i^o$ is an element of $\Omega \times O \times R$. To make this statement clear through an example, let Ω contain a gold-leaf electroscope ω_1, let O contain the operation o_1: 'to put $\bar{x}$ in contact with the free plate of ω_1', let R contain the 'result' r_1: 'the gold-leaf of ω_1 is repelled'. In this case, $\bar{P}_1^o$ could be e.g. $\langle \omega_1, o_1, r \rangle$, i.e., intuitively, 'operation

o_1 is performed on $\bar{x}$ and the gold-leaf of the instrument ω_1 is repelled' which could be seen as an operational definition of the unary predicate of 'being electrically charged'.

The most peculiar feature of our definition of $\mathfrak{M}^o$ is that no explicit mention of any universe U is made in it, contrary to what happens in every 'extensional' semantics, while a clear 'intensional' character is expressed by the fact that relations are effectively 'given' by reference not to set theoretic entities, but to some meaningful 'conditions'. On the other hand, in our example we have spoken of an $\bar{x}$ to be put in contact with the electroscope. This could sound strange, but it is in agreement with our previous distinction between 'things' and 'objects': $\bar{x}$ is here an undefinite 'thing', which becomes an 'object' of the theory $\mathscr{T}$ only at the moment *all* the operational precedures accepted in $\mathscr{T}$ (i.e. explicitly codified in Ω and O) prove applicable to it. Then, in our semantics the individuals of the universe surely must come out, but they are not 'given': they are 'singled out' step by step by the application of the *operational criteria*. The set of individuals is thus 'constructed' and remains always 'open', exactly as every empirical science requires. As a matter of fact, a pipe is not 'usually' thought to be an object of the electric science, but this does not prevent if from being taken into consideration by this science, if somebody would be interested in studying its electric properties, and, from that moment on, it would become an object of the electric science indeed.

If one should find too embarassing to accept that objects are constructed by predicates, we could make the innocent admission that there is somewhere an 'overall universe of discourse' to which all the individual variables of every language may be referred, provided it remains understood that a theory $\mathscr{T}$ is solely concerned with that subset of the overall universe to which all the operational criteria explicitly stated by the semantics of $\mathscr{T}$ do actually apply.

This methodological choice, beside being rather close to the actual practice of scientific inquiry, has a lot of advantages. First of all, as already remarked, it leaves the universe of the objects of a theory 'open' and potentially infinite; secondly, it leaves similarly open and potentially infinite, for quite analogous reasons, the subset of objects (or the set of n-tuples of objects) which corresponds to every predicate. Moreover, every O-relation is decidable for, in order to be accepted as an 'object' of the theory, a certain $\bar{x}$ must have proved manipulable by all the prescribed operations, but every such manipulation always gives a result which is selected as a kind of defining 'clause' of a certain $\bar{P}_i^o$. This implies that, at the same time, an $\bar{x}$ enters as an object of $\mathscr{T}$ and is effectively decided upon as far as its belonging to every O-relation (alone, or inserted in some n-tuple with other objects) is concerned. This implies, of course, that no semantic ambiguity is here possible, as it is easily seen when we proceed to explain the concept of the model of a sentence α.

Let us consider, for brevity, only the simple case of an atomic sentence $O_i x_1, \ldots, x_n$. O_i is interpreted on a certain $\bar{P}_i^o$, which is supposed to hold true of an n-tuple of objects $\langle \bar{x}_1, \ldots, \bar{x}_n \rangle$ if and only if, submitting them to certain manipulations by means of some ω_i belonging to Ω, according to a given operation o_i belonging to O, there will be a certain result r_i stated in R. As a consequence, when an assignment is made, which maps the individual variables of a sentence on some generic 'things' of the 'overall universe', it must first of all come out whether these 'things' can also be admitted as belonging to the universe of $\mathscr{T}$, and in this case it is automatically decidable whether $\bar{P}_i^o$ holds true of them or not. Indicating by *Ver* ($\mathfrak{M}^o$) the set of atomic sentences which are true in $\mathfrak{M}^o$ (or the set of atomic sentences of which $\mathfrak{M}^o$ is a model), we shall say, for $\alpha \equiv O_i x_1, \ldots, x_n$:

$$\alpha \in Ver(\mathfrak{M}^o) \leftrightarrow \langle \omega_i, o_i \rangle, \textit{ applied to } \langle \bar{x}_1, \ldots, \bar{x}_n \rangle \textit{ gives as a result } r_i.$$

Let us remark how suitable it is to have operational criteria 'singling out' the objects instead of having these as 'given': suppose we have $\alpha \equiv Px$ and that we have interpreted x on the 'thing' $\bar{x}$ which should be a 'toothache', while P has been interpreted on our previously illustrated predicate meaning 'being electrically charged'. If we were in the traditional situation of considering the objects as 'given', we should conscientiously say that α is false in $\mathfrak{M}^o$, as the predicate of being electrically charged does not hold true of the toothache; but this conclusion would puzzle many people, who would rightly point out that Px turns out to be 'meaningless' more than 'false' in $\mathfrak{M}^o$. If we adopt, instead, the viewpoint of our intensional semantics, we should immediately see that the operational criterion attached to P (i.e. to apply the electroscope, etc.) cannot be used with such an $\bar{x}$ and by this simple fact $\bar{x}$ would not belong to our universe and sentence α could not pretend to be thought neither true nor false, but simply meaningless in our theory, exactly like every man on the street should maintain.

But what to say if, e.g., we take as $\bar{x}$ the moon? It surely does not sound meaningless to ask whether the moon is electrically charged but, on the other hand, it is surely not possible to test such a predicate on it by resorting to an electroscope, as is prescribed by our operational definition. Should we then discard the moon from the objects of our theory? The answer to this question requires some additional considerations. First of all, we must remember that our discourse was restricted to the operational predicates only and the fact that, in the common scientific practice, predicates which have been originally defined in an operational way are applied also to 'unaccessible' objects, already suggests that this might be possible thanks to the 'mediation' of the theory, i.e. thanks to the presence of some *T*-predicates in it. From this viewpoint we can say that the inclusion of something in the universe of the objects of a theory may happen either directly, as a consequence of the application of the operational criteria, or indirectly, through the employment of theoretical tools. But here we have a slightly different question: the problem is not so much that of having *T*-predicates which might refer to operationally unaccessible objects, as that of having an *O*-predicate (like that of being electrically charged), which seems to apply outside the domain of its definitory operations. This problem is actually not easy and I have tried to treat it somewhere else, by suggesting that an operational concept be defined not by a single operation, but by an 'equivalence class' of operations, two operations being called equivalent (i) if there is a certain set of objects to which both can apply and (ii) if they give on these objects the same 'result'.[1] This can only happen if at least some fragment of the theory is employed and the consequence is an enlargement of the universe: in fact, the objects of the theory are obliged to be possible arguments of *all* the predicates of the theory and this means that, if two operations o_1 and o_2 of the theory $\mathcal{T}$ can be applied to two different sets of objects, only the intersection of these sets is included in the universe of $\mathcal{T}$. But if we accept the predicates to be defined not by single operations, but by equivalence classes of operations, it follows in our example that, if operations o_1 and o_2 are equivalent, the union, and not the intersection of their sets of objects is included in the universe of $\mathscr{T}$. The theory allows thus a first enlargement of its universe by stating the 'equivalence' of some different operations, but it can also ensure a 'connection' between predicates which can allow the *inference* from the fact that a certain O-predicate holds true of a certain $\bar{x}$ to the fact that a certain other O-predicate also holds true of $\bar{x}$, this inference being testable by the actual performance of the operations involved. Once the validity of this inference is tested, it becomes the basis for admitting its validity also for those cases in which it cannot be directly tested, i.e. also when the first O-predicate can be operationally tested on a certain $\bar{y}$, while the second cannot be. In this case we are allowed to say that the second predicate too holds true of $\bar{y}$, though we cannot test it. In such a way we have actually a 'prolongation' of the model $\mathfrak{M}^o$, which comes to including objects that are still characterized by O-predicates without actually being manipulable by *all* the operations of the theory.

If we now take all the above sketched remarks together (their formal treatment does not imply any problem and we omit it for brevity) and conceive O-predicates, beside being defined through equivalence classes of operations, as 'prolongable' thanks to the theory, we can qualify such predicates as ground-predicates and require that every object of the theory be characterized with reference to *all* of them. The reason for privileging them in such a measure as to consider them as the 'makers of the objects' is strictly bound to what was said in the first section of this paper about the fact of clipping a scientific 'object' out of a common sense 'thing': we remarked then that an object comes out when a thing is questioned from certain viewpoints and there are tools for answering immediate questions about it. The operational criteria are such tools, they are the effective incarnation of such viewpoints and it is thus quite legitimated to assume the O-predicates related to them as basic predicates of the empirical theory concerned with the 'objects' so emerged. Remark, furthermore, that when an empirical theory needs to put its sentences to the test, this cannot be done unless one comes step by step down to these operational procedures, which receive a confirmation of their foundational character from this fact too.

After the above considerations it appears quite obvious to qualify as 'empirical data', or as 'data' of an empirical theory, all the atomic sentences which are true in $\mathfrak{M}^o$ and all the negations of the atomic sentences which are false in $\mathfrak{M}^o$, i.e. all the atomic sentences (possibly negated) which are built up exclusively by resorting to O-predicates.

Without entering the complex questions which arise when T-terms come into play, we shall briefly hint at some points which seem worth mentioning.

It is perhaps not useless to point out that the kind of 'intensional' semantics proposed in this paper is not, after all, so complicated and cumbersome as may be judged at first sight. It is in fact a rather naive belief to think that it would be easy to 'give' actually a universe U of individuals, like it is presupposed in the current extensional semantics: it must be found much easier, from a concrete point of view, to 'give' three finite sets of 'instruments', 'operations' and 'results', which can rather easily be described in the metalanguage and even be practically 'pointed out' if needed. When we pass to the interpretation of the predicates, the current extensional semantics assignes to them some particular set-theoretical entities, which are very easy to mention but are practically impossible to show and this is immediately reflected itself upon the concept of the model of a sentence: here again it is easily said that α is true in $\mathfrak{M}^o$ if the relation P_i^o holds true of the objects $\langle \bar{x}_1 \ldots \bar{x}_n \rangle$, but how such a fact may be ascertained remains a rather enigmatic affair. On the contrary, the operational definition of the predicates makes such a crucial step quite manageable, as it was shown before.

Another interesting remark is that no isomorphism theorem holds in our semantics. The reason is simple: in the extensional semantics, if two universes U and U' have the same cardinality and a certain relation $\bar{R}$ is 'given' on U, a correspondent relation $\bar{R}'$ can be easily 'induced' on U' by simply stating:

$$\langle f(\bar{x}_1) \ldots f(\bar{x}_n) \rangle \in \bar{R}' \quad \leftrightarrow \quad \langle \bar{x}_1 \ldots \bar{x} \rangle \in \bar{R}$$

$f(\bar{x}_1 \ldots\ldots f(\bar{x}_n)$ being the images of $\bar{x}_1 \ldots\ldots \bar{x}_n$ under the one-one correspondence f, which must exist in order to insure that the two universes have the same cardinality. In the case of our semantics, nothing of the kind is possible for nobody can be sure that, 'given' two operationally defined predicates $\bar{P}_1^o$ and $\bar{P}_2^o$, everytime $\bar{P}_1^o$ holds true of its objects, $\overline{P_2^o}$ holds true of some 'corresponding' objects of its domain, and that because such a 'correspondence' between objects cannot be established as a rule; on top of that, relations cannot be 'induced' from one model to another for, if they are obtained by 'copying' the operational definition of the first model, they simply turn out to coincide with those from which they were supposed to derive and the two models do thus coincide; if their are characterized by different operations, there is no a-priori warranty of they keeping 'parallel' in their behaviour. This fact holds even if $\overline{P_1^o}$ and $\overline{P_2^o}$ are referred to the same 'universe' (i.e. when they belong to the same theory $\mathscr{T}$). In fact, it is quite possible that, sometimes, we can prove something like

$$\forall x(P_1 x \leftrightarrow P_2 x)$$

but this simply means that we have found an *empirical law* connecting two different properties of our objects. If P_1 and P_2 are both O-predicates, we could take advantage of this law and declare the two operational criteria on which the predicates are founded as 'equivalent', putting them in the same 'equivalence class', but we are not compelled to do that (think of the predicate 'magnetic' defined by the operational criterion of attracting iron filings or by the criterion of inducing an electric current by motion near a circuit). In this a last case we would prefer to say that we have discovered a new empirical testable property of our objects. This can be generalized to the case of non operational predicates, and expresses the fact that in empirically sciences (but in mathematics as well) we frequently arrive at establishing the 'equivalence' of certain properties, without meaning by that, that they are one and the same property and this is perhaps one of the ways of giving an exact characterization of the fact that science always proceeds by 'synthetic' and 'synthetic a-priori' judgements or, if we prefer, that there cannot be properly scientific inquiry without 'data' (empirical and not).

REFERENCE

[1] This problem was first treated in the book by E. Agazzi, *Temi, e problemi di filosofia della fisica*, Milan, 1969 (p. 128–130) without any formal apparatus. In the paper by M. L. Dalla Chiara Scabia and G. Toraldo di Francia, 'A Logical Analysis of Physical Theories', in *Rivista del Nuovo Cimento*, **3** (1973) p. 1–20 it was further investigated in a formal way.

MARIAN PRZEŁĘCKI

INTERPRETATION OF THEORETICAL TERMS: IN DEFENCE OF AN EMPIRICIST DOGMA

It should be stated in advance that the present paper does not contain any new results in the formal methodology of science. It is not formal, but philosophical as to its nature. I shall discuss in it an account of the interpretation of theoretical terms known under the name of the partial interpretation view of the meaning of theoretical terms, and my main concern will be with some of its general assumptions and consequences. The view is based on certain formal as well as philosophical assumptions. The formal assumptions are those characteristic of the model theoretic approach. The philosophical assumptions amount to the so called thesis of semantical empiricism. These assumptions entail certain paradoxically sounding consequences concerning the interpretation of an important class of theoretical terms. The consequences have most clearly been pointed out by John. A. Winnie in his well-known paper 'The Implicit Definition of Theoretical Terms' (*Brit. J. Phil. Sci.* 1967). Because of their somewhat instrumentalistic character, they have been considered obviously inadequate by some scientifically minded philosophers and, hence, regarded as destructive for the assumptions from which they result. In what follows I try to question this opinion. The criticism directed against the partial interpretation view seems to me unjustified. As far as I can see, the view is in accord with the actual interpretation of theoretical terms within scientific theories, and the alternate, naively realistic, account of their interpretation is hardly tenable. This is the main claim argued for in the present paper.

Let us recall briefly the view to be discussed and some of its most controversial implications. Let $\mathscr{L}$ be a language of some empirical theory. As a formal framework for $\mathscr{L}$'s semantics, I shall here employ a model theoretic conceptual apparatus—known well enough to dispense with its detailed explanation. According to this approach, an interpretation of language $\mathscr{L}$ will be identified with a suitable set theoretic entity, called a *model for* $\mathscr{L}$. For a language $\mathscr{L}$ of the simplest kind, it will consist of a universe for the variables of $\mathscr{L}$ and of denotations for the non-logical predicates of $\mathscr{L}$. The universe is conceived of as some non-empty set, and the denotations—as sets of objects belonging to the universe (or as sets of n-tuples of such objects—in the case of n-place predicates). The objects themselves will be called *objects designated by the given predicate.* It is assumed known under what conditions a sentence α of language $\mathscr{L}$ is said to be true in a model $\mathfrak{M}$.

As mentioned before, the philosophical assumptions underlying the view to be discussed represent a kind of empiricist epistemology, and make up what is sometimes called the doctrine of *semantical empiricism.* Let us sketch its main tenets. According to some version of this standpoint, the empirical terms of a language $\mathscr{L}$ may be divided into three classes: observation terms, or *O*-terms, "mixed" terms, or *M*-terms, and theoretical terms, or *T*-terms. (I follow Winnie's terminology here; usually both *M*- and *T*-terms are called theoretical in a broader sense, and *T*-terms are distinguished from the others as theoretical terms *sensu stricto.*) *O*-terms refer to observational properties (and relations) and attribute them to observable objects only. All the remaining terms refer to some theoretical properties (and relations); but while *M*-terms attribute them to observable as well as unobservable objects, *T*-terms apply them to unobservables only. Terms like 'green', 'mass', 'electron' may serve as corresponding examples.

The distinctions are admittedly vague and ambiguous. Any attempt at making them more precise presents a task that cannot be undertaken in this paper. I must restrict myself to a short comment only, intended to remove some, at least, ambiguities. And so, the distinction between observable and unobservable things will be assumed to coincide roughly with that between macro- and micro-objects. The concept of an observational property (or relation) appears to be more enigmatic. One way of conveying its sense refers to the concept of ostension. So understood, the distinction between observational and theoretical properties will correspond to that between ostensive and non-ostensive properties (whatever these may mean). Yet for our further considerations it is not important where exactly these distinctions are to be drawn. What is decisive is the very fact of existence of unobservable things and non-observational properties. And this seems hardly questionable.

Now, the central problem arising with regard to a language $\mathscr{L}$ so characterized is the question of interpretation of its theoretical terms: how are the theoretical terms assigned their denotations? Or, to put it in Winnie's wording: how do these terms which designate unobservables come to do so? It should be emphasized that the question is to be understood as a logical (or methodological), and not as a factual (e.g. historical or psychological) one: as a question '*quid iuris*', not '*quid facti*'. An answer to it constitutes the fundamental tenet of semantical empiricism. It amounts to the following. There are only two ways of interpreting a term: by what we do, or by what we say. What a given term is to designate may be assigned to it directly—by pointing at the relevant objects, or indirectly—by describing the objects. Now, it is observable objects only that can be assigned to a term in a direct way; unobservable entities can be assigned to it in an indirect way only. In consequence, the only terms that may be interpreted directly are the O-terms. The M- and T-terms have to be interpreted indirectly. Any direct way of interpretation reduces to some kind of ostensive procedures. It is assumed that in this way the O-terms are provided with their proper interpretation. An indirect way of interpretation consists in accepting some sentences, called postulates, which connect the terms being interpreted with some terms already endowed with interpretation. A proper interpretation of the M- and T-terms is thus defined as such which makes the corresponding postulates true while retaining the proper interpretation of the O-terms.

A formalization of these ideas may be sketched as follows. Let $\mathscr{L}_o$ be the observational sublanguage of language $\mathscr{L}$, containing O-terms as its only descriptive terms. The *proper model for* $\mathscr{L}_o$, $\mathfrak{M}_o^*$, is assumed to be fixed uniquely by some ostensive procedures. Its universe, $U_{\mathfrak{M}_o^*}$, and denotations assigned by it to O-terms consist of observable objects solely. The *proper model for the whole language* $\mathscr{L}$, $\mathfrak{M}^*$, is characterized by the following conditions:

(i) the postulates for M- and T-terms are true in $\mathfrak{M}^*$;

(ii) the universe of $\mathfrak{M}^*$, $U_{\mathfrak{M}^*}$, includes the universe of $\mathfrak{M}_o^*$, $U_{\mathfrak{M}_o^*}$;

(iii) the denotations of O-terms in $\mathfrak{M}^*$ are identical with their denotations in $\mathfrak{M}_o^*$;

(iv) the denotations of T-terms in $\mathfrak{M}^*$ are restricted to the set

$U_{\mathfrak{M}^*} - U_{\mathfrak{M}_o^*}$, i.e. to the subset of unobservable objects of the universe of $\mathfrak{M}^*$.

It is evident that the above conditions do not characterize the model $\mathfrak{M}^*$ uniquely; they do not amount to a definition of $\mathfrak{M}^*$. What they do define is a class of models, $\mathcal{M}^*$—let us call it the class of proper models for $\mathcal{L}$.

Now, the Winnie's observation makes us realize how weak the conditions are, and how comprehensive, in effect, the class $\mathcal{M}^*$ is bound to be. If $\mathcal{M}^*$ contains a model $\mathfrak{M}$ whose universe includes some unobservable objects, then $\mathcal{M}^*$ will contain a model $\mathfrak{M}'$ whose universe includes, in place of those objects, some abstract entities, e.g. numbers. In other words, if for some model $\mathfrak{M} \in \mathcal{M}^*$, the set $U_{\mathfrak{M}} - U_{\mathfrak{M}_o^*}$ is non-empty, then there exists a model $\mathfrak{M}' \in \mathcal{M}^*$ such that the set $U_{\mathfrak{M}'} - U_{\mathfrak{M}_o^*}$ is identical with a set of numbers. So, the objects designated by T-terms may always be construed as numbers. Among the proper interpretations of a theoretical term there is always a numerical interpretation. A similar conclusion applies to the interpretation of M-terms, though to a lesser extent. *Some* of the objects designated by those terms, viz. the unobservable ones, also may be identified with numbers.

These consequences have been considered unacceptable—especially by those who declare for realism in the realism-instrumentalism controversy. Their argument runs as follows. Though unobservable, the objects designated by the theoretical terms in empirical theories must certainly be physical entities. Any interpretation that identifies them with numbers, or other abstract entities, is obviously an unintended interpretation. Yet such unintended interpretations are not excluded from the class of proper interpretations defined as above. So, "there must be something amiss somewhere"—Winnie concludes. In what follows, I want to question this conclusion. My contention is that the account of the interpretation of theoretical terms outlined above is essentially in keeping with their actual interpretation within scientific theories. The seemingly paradoxical consequences pointed out by Winnie are, despite appearances, compatible with that interpretation.

To fix our attention, let us consider the term 'electron'—a classic example of a theoretical term in physics. On the above account, its denotation turns out to be determined in an extremely loose and ambiguous way. It is not fixed uniquely; it is even not restricted to the sets of physical objects: some sets of abstract objects, e.g. numbers, are admitted as well. In consequence, the predicate 'electron' turns out to be completely vague within its appropriate domain, i.e. within the domain of unobservable objects: for every such object, there is always some proper model of the given language which assigns this object to the denotation of the predicate, and some other which does not. Is that what the actual interpretation of the term 'electron' is like, the way it is interpreted in physical theories? I am not able to give any competent and detailed analysis of the term's interpretation. Some general considerations, however, seem to suggest a positive answer to the question.

A common account of the interpretation provided for the term 'electron' by the relevant physical theories usually refers to such facts as the definability of the term 'electron' by means of such physical quantities as mass, electric charge, and the like, or the alleged 'observability' of electrons. Are these facts compatible with the partial interpretation view of the term 'electron'? Can we account for them within our present framework? Suppose that electrons may be identified as objects endowed with such-and-such a mass, such-and-such an electric charge, and so on; in other words, that the term 'electron' is definable by means of the terms: 'mass', 'electric charge', and the like. What does this imply as far as the interpretation of this term is concerned? It should be noticed here that the defining terms do not belong to observation terms: they are either theoretical terms, or mixed terms at best. Let us take as an example a possible consequence of such a definition, which states that electrons are objects with a mass equal k (say $9 \cdot 10^{-28}$ g):

(I) $E(x) \rightarrow m(x) = k.$

The term m is a typical mixed term. According to the usual assumption, its interpretation is determined by a set of postulates of the following kind:

(1) $R(x, y) \leftrightarrow m(x) \leqslant m(y),$

(2) $m(x \circ y) = m(x) + m(y),$

(3) $m(a) = 1,$

which connect the quantitative term m with the qualitative terms R and o. But, as I tried to show elsewhere ('Empirical Meaningfulness of

Quantitative Statements', *Synthese* 1974), the latter terms are not observation terms either. They are mixed terms themselves, and their connections with observation terms assume forms of certain reduction sentences:

(i) $O_1(x, y) \rightarrow (R(x, y) \leftrightarrow O_2(x, y))$,

(ii) $O_3(x, y, z) \rightarrow (x \, o \, y = z \leftrightarrow O_4(x, y, z))$.

Now, the point is that postulates of this type cannot provide the terms R and o with any fixed interpretation outside the domain of observable objects, i.e. outside the set $U_{\mathfrak{M}_o^*}$. In particular, they cannot exclude some numbers from membership in the relation R so interpreted. This characteristic is transmitted to the interpretation of the term m as effected by postulates (1)–(3). Among other things, it is numbers that constitute some arguments of the function m. In consequence, neither the statement (I), nor any adequate definition of the term E, from which it follows, can endow the term E with a fixed physical interpretation and exclude all unintended, especially numerical, ones.

As far as the alleged 'observability' of electrons is concerned it is evident that this characteristic cannot be taken literally. Electron is not observable in any straightforward sense. What is observable are certain macro-objects characterized with the help of this term, e.g. as objects 'containing free electrons', or the like. It is such objects only that are directly referred to in certain well-known observational criteria. In a highly simplified form, a criterion of this kind may be rendered as follows:

(II) $O_5(x) \rightarrow \exists y(P(y, x) \wedge E(y))$.

The observation predicate O_5 may here be taken to describe certain observational state of the Wilson cloud chamber, and predicate P—the part-of relation. In spite of its observational character, no criterion of the type (II) can endow the term E with a fixed physical interpretation. This is easily seen if we realize that the term P is not an observational, but a mixed one, and thus only indirectly and loosely connected with observation terms. The postulates for P will typically include two kinds of statements: theoretical postulates and correspondence rules. The first may be taken to consist of the axioms of mereology (containing P

as the only non-logical term), the second—of some reduction sentences of the type:

$$O_6(x, y) \rightarrow (P(x, y) \leftrightarrow O_7(x, y)).$$

So interpreted, the predicate P may well designate some numbers—as far as its unobservable designata are concerned. And such objects will, in effect, make up some proper denotations of the term E—in spite of its observational criteria.

It turns thus out that, under our assumptions, neither definability nor 'observability' of electrons can guarantee this term the kind of interpretation usually ascribed to it: to assign to it some uniquely fixed set of physical objects as its proper denotation. Does this fact discredit our assumptions? Or might we, in spite of it, consider such an account of the term's interpretation an adequate one? I think we might. The following facts seem to speak for its adequacy. Though the term 'electron' is completely vague (in the domain of unobservable objects), its interpretation is by no means completely arbitrary. It is not that any set of unobservable objects may be asisgned to it as its proper denotation. The class of such denotations is restricted to sets which have certain structural properties (such as non-emptiness), and which bear certain structural relations to the proper denotations of other terms. Owing to these restrictions, some sentences which involve the term 'electron' in an essential way turn out to be empirically decidable statements. This claim needs a few comments. A sentence α is said to involve the term E in an essential way if its truth value depends on the interpretation of E; i.e. if there is a proper interpretation of its remaining terms such that under one interpretation of E α turns out to be true, and under another—false. To explain the concept of empirical decidability mentioned above, we shall first define the concept of decidability. A sentence α will be called *decidable* if it is true under all its proper interpretations, or if it is false under all such interpretations. α will be said to be *empirically decidable* if it is decidable and neither α nor its negation are consequences of the language's postulates. Now, which exactly sentences involving the term E in an essential way will belong to the class of empirically decidable statements depends, of course, on what are the postulates for E. Let us suppose that they contain the postulate (II). Then, if only $O_5(a)$ is true for some a, the statements:

$\exists y(P(y, a) \wedge E(y))$ and $\exists y E(y)$ will prove to be empirically decidable. Under our intended interpretation, the hypothesis is certainly true, and the statements read as follows: 'this cloud chamber contains (free) electrons' and 'electrons exist'. On the other hand, a sentence of the form $E(a)$, where a refers to an unobservable object, may serve as an example of an empirically undecidable statement—under any plausible interpretation of the term E. But there is nothing strange in this. A statement 'a is an electron' is never employed in scientific practice. It is my contention that the class of decidable statements defined as above includes all statements about electrons actually asserted by the scientist. I cannot justify this claim in any reliable way. I put it forward as a hypothesis only.

Notwithstanding all the arguments here adduced, it must be admitted that the account of the interpretation of theoretical terms advanced in the present paper is clearly incompatible with a common sense view of the matter. But it is the latter that seems to be untenable. The belief in a fixed empirical interpretation of any theoretical term is, in my opinion, an illusion. I know of no way in which such a belief could be substantiated. There is no procedure of 'pinning down' just the set of electrons as the denotation of this term. What is more, there seems to be no procedure by which this denotation could be restricted to the physical objects only. We can, of course, (and we do) stipulate that the electrons be spatio-temporal objects, endowed with mass, and so on. But all these are words only. They determine the interpretation of the word 'electron' to such an extent only to which their interpretation has been determined. When I postulate that x be an (unobservable) objects endowed with a certain mass, what I, in fact, do amounts to the claim that x is an argument of a function which satisfies such-and-such axioms and which in a certain subdomain (the subdomain of observable things) takes such-and-such objects as its arguments. As seen above, this cannot preclude x from being an abstract object, say a number. This is a consequence of the empiricist assumptions. And I cannot see how these could be avoided. Our only contact with reality is at the macro-level. We have a direct access to macro-objects solely. Our access to micro-objects is through the medium of words only. The unobservable entities can only be spoken about. They cannot be 'grasped' in any non-verbal way. And they hardly need be—as far as the scientific aims are concerned. Their 'elusiveness' does not prevent them from playing an essential role in scientific theories.

The approach discussed by us is based not only on some philosophical, but also on some formal assumptions, viz. the model theoretic ones. And it is these assumptions that are questioned sometimes by those who find the approach unacceptable. As the model theoretic framework is usually considered extensional, its extensional character is blamed for the paradoxical consequences of the present approach, and a recourse to some intensional framework is postulated. The problem is too involved to be discussed in the present paper. Hence, a short comment only. The concept of intension is notoriously vague and ambiguous, and those explications which have succeeded in making it sufficiently precise do not seem to be of any help in providing a more satisfactory account of the interpretation of theoretical terms. This, in particular, is true of an explication couched in model theoretic terms. According to it, the intension of a term may be defined through reference to the class of all possible models of the given language ('all possible worlds')—e.g. as a function which to each possible model assigns the denotation of the term in that model. A concept of intension defined along these lines may be incorporated in our model theoretic account: it suffices to identify a possible model of a given language with a model of its postulates. I do not see how a recourse to such a concept could bring an essentially different and more satisfactory solution to the problem of interpretation of theoretical terms.

Finally, let us look at the account propounded in this paper from the point of view of the realism-instrumentalism issue. May it be called a realistic one? In what sense, if any? There is no straightforward answer to this question. The account seems to be realistic in that it provides an interpretation for all theoretical terms, and, in consequence, allows to define the concept of truth for all theoretical sentences. But it seems instrumentalistic in how it conceives that interpretation. The interpretation of a theoretical term is identified not with its single intended denotation, but with a whole class of denotations, containing some obviously unintended ones. There is assumed to be no way to specify the intended denotation, e.g. a physical one, and to distinguish it from all the unintended denotations, e.g. the numerical ones. Every theoretical term is provided with a multiplicity of interpretations. Their variety results in the complete vagueness of the term (in its proper field of application). In spite of this, I am inclined to consider the account as realistic in essence. But if one feels that it should be called instrumentalistic rather than realistic, I will not protest. I do not find the epithet abusive. And if I were to point out a philosophical standpoint that underlies the account here advanced, I would mention the view propounded by Quine in his famous essay on 'Ontological Relativity'. The unavoidable multiplicity and variety of interpretations of all theoretical terms reflects the ontological relativity of our conceptual framework, so convincingly argued for by Quine.

Let me close these remarks with a few words devoted to a current criticism raised against one of the fundamental assumptions underlying the view discussed in the present paper. It is an assumption involved in the observational-theoretical distinction, viz. that claiming the presence of observation terms in the language of empirical theories. These are to be observation terms in a rather strict and absolute sense—conceived of as terms interpreted by some ostensive procedures. Now, there are no such terms in the vocabulary of actual scientific theories—the objection reads; their vocabulary consists of theoretical terms only, if the latter are to be understood according to the above distinction. In reply to this objection I would like to make two comments.

Let me notice, first, that the questionable assumption does not play an essential role in the argumentation presented in the paper. The argumentation may easily be generalized so as to dispense with the concept of observation terms altogether. In its generalized form the argumentation hinges on the following, less problematic, claim:

There are empirical terms which are interpreted only indirectly, and which designate objects not being designated by any term interpreted in a direct way.

That is all what is needed in order to arrive at our main conclusions.

Now, identifying direct way of interpretation with some ostensive procedure and, in consequence, terms interpreted in that way with observation terms provides a kind of explanation for the above assumption. But, is this assumption so interpreted still acceptable? Do any observation terms belong to the vocabulary of actual scientific theories? The answer seems to depend on what kind of entity a scientific theory is supposed to be. On certain narrow conceptions of scientific theory, most empirical theories (in particular, all physical ones) do not contain anything like observation terms literally understood. But on some broader conceptions, every empirical theory is bound to include an observational vocabulary. Let us take classical particle mechanics as a typical example. What is its vocabulary like? According to its current conception, it is a theory comprising Newton's laws as its only physical axioms and the function symbols: 'position', 'mass' and 'force' as its only physical (i.e. non-logical and non-mathematical) terms. None of these terms, of course, is an observation term in the sense here considered. The same theory, however, may be looked upon in a different, more comprehensive, way. On its current conception, the theory does not comprehend any of the sub-theories which underlie it, in particular—any of the theories of measurement of its fundamental physical quantities. These underlying sub-theories are not part of the theory proper. A physical theory, such as classical particle mechanics, is here identified with the very top layer of the relevant conceptual structure, with its most theoretical level. Now, there certainly are a number of problems for which such a conception of physical theory seems to be an appropriate one. But it is equally obvious that there are some other problems for which it is much too restrictive. If, e.g., what we are interested in is the problem of empirical content of a given physical theory, we cannot, in its analysis, abstract from the theories of measurement underlying it. Only a reconstruction of measurement procedures characteristic of the relevant physical quantities can make explicit the empirical content of the theory's fundamental concepts. And so, when examining problems of this kind, we must conceive a given theory in a more comprehensive way, and include into it, in addition to its theoretical laws, the so called correspondence rules as well. As far as physical theories are concerned, the correspondence rules are typically statements embodying the methods of measurement for the relevant physical quantities. Now, if we conceive an empirical theory in such a comprehensive way (as we actually have done in the present paper) we can safely assume that its vocabulary includes some observation terms. When we carry an analysis of measurement procedures far enough, we are bound to come across some observation terms in the strict sense being here considered.

It is only fair to say what price we have to pay for this kind of approach. Including into a given theory, as its essential part, all theories of measurement underlying it, we replace a well defined, neat structure by a vague and unwieldy whole, whose analysis presents an extremely difficult and thankless task. But it is a task we cannot avoid if our concern is just with the content, or interpretation, of the theory's fundamental concepts. We cannot then treat their interpretation as something given, we must analyze the way in which it is being provided. It has to be admitted that all solutions to this problem offered thus far are hardly satisfactory: they are either indefinite, couched in loose metaphoric terms, or unrealistic, based on drastic simplifications and idealizations—or both. But it is my contention that, with regard to a genuine theoretical question, a wrong answer is better than none, because it presents a challenge to further inquiries and improvements.

VEIKKO RANTALA

DEFINABILITY PROBLEMS IN THE METHODOLOGY OF SCIENCE

I. PRELIMINARIES

In empirical sciences problem of definability and identifiability have frequently come up. Usually they have concerned definability of certain terms (theoretical terms) by means of certain other terms whose 'values' can be obtained, some way or other, by observations. From the strict logical point of view, however, the notion of definability has been used in a rather loose sense, which has been different in different sciences.

As soon as an empirical theory can in principle be axiomatized in some suitable system of formal logic, it is possible to remove the vagueness of the concept of definability on the basis of results and notions from the theory of definability in recent logic.

It is the aim of this paper to consider certain empirical theories which can be formalized in first-order logic, in order to find out what the methodologists mean when they speak about definability or identifiability in different cases. Thus we shall make an effort to relate certain definability problems in empirical sciences and certain exactly formulated notions of different kinds of definability in first-order logic. It will appear that these definability problems can be construed as special cases of more general definability questions in first-order logic. Since the aim of this paper is mainly logical, it may seem to empirical scientists that it does not increase the substantial understanding of these sciences. We nevertheless hope that there is some interest in placing the concepts of departmental sciences into a general logical framework.

We are not going to attempt an exhaustive exposition of definability and identifiability problems in empirical sciences but to consider only selected examples from econometrics and physics.

In the logical definability theory there are plenty of important results obtained recently. They look expecially promising for applications to empirical sciences in that they relate syntactical (proof-theoretical) and semantical (model-theoretical) notions to each other. Before presenting a summary of some of these results it is perhaps appropriate to present some standard notions of the model theory, needed later. It is possible to do it here only concisely. For an almost exhaustive treatment of the model theory, see Chang and Keisler (1973).

Let $L(\varkappa)$ be a first-order finitary language (with identity) whose set of non-logical constants (primitive terms) is $\varkappa$, and $T(\varkappa)$ a theory in $L(\varkappa)$ (a set of sentences closed under deduction in $L(\varkappa)$). A structure M for $L(\varkappa)$ is of the form $\langle D, \underline{\varkappa}\rangle$ where D is a non-empty set (the domain of the structure) and $\underline{\varkappa}$ is the set of interpretations of the members of $\varkappa$. If $c \in \varkappa$ is a k-place predicate symbol, then its interpretation $\underline{c}$ in D is a k-place relation on D, i.e., $\underline{c}$ is a subset of D^k. If c is k-place function symbol, then $\underline{c}$ is a function on D^k to D. If c is an individual constant, then $\underline{c}$ is an individual of M ($\underline{c} \in D$).

If $\varkappa = \lambda \cup \mu$, we shall write simply $L(\lambda, \mu)$ for $L(\lambda \cup \mu)$, $\langle D, \underline{\lambda}, \underline{\mu}\rangle$ for $\langle D, \underline{\lambda \cup \mu}\rangle$, etc. If $k = \lambda \cup \{c\}$, we shall write $L(\lambda, c)$ for $L(\lambda \cup \{c\})$, etc. We may also use superscripts to differentiate two interpretations of the same constant or set of constants, e.g., $\underline{c}'$ and $\underline{c}''$ are both interpretations of c.

The presence of a certain constant or set of constants in a formula will be indicated by writing them out as arguments in the following way: $F(\lambda, c_1, \ldots, c_h)$ is a formula in which the constants $c_1, \ldots, c_h$ are present as well as some constants of λ.

We write $M \models T(\varkappa)$ if M is a model of $T(\varkappa)$, i.e. if M is a structure for $L(\varkappa)$ in which all the axioms of $T(\varkappa)$ are true, and $T(\varkappa) \vdash F$ if F is a theorem of $T(\varkappa)$.

Suppose that $\varkappa = \lambda \cup \mu$. We denote by $T(\lambda)$ that subtheory of $T(\varkappa)$ which consists of all those formulas of $T(\varkappa)$ from which the members of μ are absent. Thus $T(\lambda)$ is the maximal subtheory of $T(\varkappa)$ in the language $L(\lambda)$. Let $\mu = \{c_1, \ldots, c_n\}$. If $T(\varkappa)$ is finitely axiomatizable, having $A(\lambda, c_1, \ldots, c_n)$ as its single axiom, then $T(\lambda)$ is also the set of first-order consequences of the (possibly second-order) formula $(\mathrm{E}X_1) \ldots$ $\ldots (\mathrm{E}X_n) A(\lambda, X_1, \ldots, X_n)$ where the X_j are first-or second-order variables corresponding to the c_i's (cf. Craig (1960) and Tuomela (1973)). If especially $c_1, \ldots, c_n$ are individual constants, then $(\mathrm{E}X_1) \ldots (\mathrm{E}X_n)$ $A(\lambda, X_1, \ldots, X_n)$ is the first-order sentence. Hence it is the single axiom of $T(\lambda)$.

Let $M = \langle D, \underline{\lambda}\rangle$ be a structure for $L(\lambda)$. A structure $M' = \langle D, \underline{\lambda}, \underline{\mu}\rangle$, which has the same domain and the same interpretation of the members of λ, is called an *expansion* of M to the language $L(\lambda, \mu)$. M is the *restriction* (or *reduct*) of M' to $L(\lambda)$. If M' is a model of $T(\lambda, \mu)$, then M is a model of $T(\lambda)$. On the other hand, there can be models of $T(\lambda)$ (a fortiori, structures for $L(\lambda)$) which cannot be expanded to a model of $T(\lambda, \mu)$.

In connection with empirical theories, the members of λ are often referred as *observational terms* and the members of μ as *theoretical terms*. Accordingly, $L(\lambda)$ is the *observational language*, and the maximal subtheory $T(\lambda)$ is the *observational subtheory* of $T(\lambda, \mu)$.

Let $M = \langle D, \underline{\varkappa}\rangle$ be a structure for $L(\varkappa)$. A structure $M' = \langle D', \underline{\varkappa}'\rangle$ for $L(\varkappa)$ is a *substructure* of M and M is an *extension* of M' iff $D' \subseteq D$ and $\underline{\varkappa}'$ is $\underline{\varkappa}$ restricted to D', that is, for every $c \in \varkappa$, $\underline{c} \in \underline{\varkappa}$, $\underline{c}' \in \underline{\varkappa}'$:

(i) $\underline{c} = \underline{c}'$ when c is an individual constant,

(ii) for every k-tuple $\langle d_1, \ldots, d_k\rangle$ of elements of D' $\underline{c}(d_1, \ldots, d_k) = \underline{c}'(d_1, \ldots. d_k)$ when c is a k-place function symbol,

(iii) for every k-tuple $\langle d_1, \ldots, d_k\rangle$ of elements of D', $\underline{c}(d_1, \ldots, d_k)$ holds in M iff $\underline{c}'(d_1, \ldots, d_k)$ holds in M' when c is a k-place predicate symbol.

II. RESULTS ON DEFINABILITY

We shall give a brief summary of some known results in the theory of definability in first-order logic. A more exhaustive summary can be found in Tuomela (1973).

We shall consider a theory $T = T(\lambda, \mu)$ and a constant $c \in \mu$. Let $\langle D, \underline{\lambda}\rangle$ be a structure for $L(\lambda)$ and $\mathscr{M}$ the set of all its expansions $\langle D, \underline{\lambda}, \underline{\mu}\rangle$ to $L(\lambda, \mu)$ such that $\langle D, \underline{\lambda}, \underline{\mu}\rangle$ is a model of T. We can ask, how many different interpretations of c at most there exist in the different models in $\mathscr{M}$, i.e., what is at most the cardinal number of the set of all $\underline{c}$'s such that $\langle D, \underline{\lambda}, \underline{c}\rangle$ can be expanded to a model $\langle D, \underline{\lambda}, \underline{\mu}\rangle$ of T. If this cardinal is finite (say n) for every $\langle D, \underline{\lambda}\rangle$ it is said that c is *finitely identifiable* (more exactly, *n-foldly identifiable*) in T. If it is $\overline{\overline{D}}$ for every $\langle D, \underline{\lambda}\rangle$ (where $\overline{\overline{D}}$ is infinite), c is *restrictedly identifiable* in T. If it is $\aleph_0$, c is *countably identifiable* in T (in terms of λ) (cf. Hintikka

(1972) and Tuomela (1973)). Expecially, if c is 1-foldly identifiable, c is *semantically definable* in T (in terms of λ). Thus c is semantically definable in T (in terms of λ) when, loosely speaking, the interpretation of λ in any model of T fixes uniquely the interpretation of c.

When we shall state the results below, we first suppose that c is a k-place predicate symbol P. In what follows, we shall often use the following abbreviations: $\bar{x}_n$ for a sequence $\langle x_1, \ldots, x_n \rangle$ of distinct variables, $(\bar{x}_n)$ and $(\mathrm{E}\bar{x}_n)$ for prefixes $(x_1) \ldots (x_n)$ and $(\mathrm{E}x_1) \ldots (\mathrm{E}x_n)$, respectively.

Theorem 1 (Beth (1953)). The following conditions are equivalent:

(i) P is semantically definable in T (in terms of λ).

(ii) There is a formula $F(\bar{x}_k)$ of $L(\lambda)$ such that

$$T \vdash (\bar{x}_k)(P(\bar{x}_k) \equiv F(\bar{x}_k)).$$

The condition (ii) says that P is *explicitly definable* in T (in terms of λ).

Theorem 2 (Svenonius (1959)). The following conditions are equivalent:

(i) P is definable (in terms of λ) in every model of T.

(ii) There are formulas $F_i(\bar{x}_k)$ $(i = 1, \ldots, n)$ of $L(\lambda)$ such that

$$T \vdash \bigvee_{i=1}^{i=n} (\bar{x}_k)(P(\bar{x}_k) \equiv F_i(\bar{x}_k)).$$

That P is *definable* in M means that an explicit definition of P holds in M. When (ii) holds, it is said that P is *piecewise definable* in T (in terms of λ). For the conditions under which a piecewise definition can be reduced to an explicit definition, see Hintikka and Tuomela (1970), and Tuomela (1973).

Theorem 3 (Kueker (1970)). The following conditions are equivalent:

(i) P is n-foldly identifiable in T (in terms of λ).

(ii) There are formulas $S(\bar{x}_h)$ and $F_i(\bar{x}_h, y_k)$ $(i = 1, \ldots, n$; the x_i are distinct from the y_j's) of $L(\lambda)$ such that

(a) $T \vdash (\mathrm{E}\bar{x}_h) S(\bar{x}_h)$

(b) $T \vdash (\bar{x}_h)(S(\bar{x}_h) \rightarrow \bigvee_{i=1}^{i=n} (\bar{y}_k)(P(\bar{y}_k) \equiv F_i(\bar{x}_h, \bar{y}_k)))$.

If $n = 1$ in (b), P is explicitly definable. Then the explicit definition obtained from (a) and (b) is:

$$T \vdash (\bar{y}_k)(P(\bar{y}_k) \equiv (\mathrm{E}\bar{x}_h)(S(\bar{x}_h) \wedge F_1(\bar{x}_h, \bar{y}_k))).$$

In general, (ii) cannot be converted to a piecewise definition. A piecewise definition results when one can get rid of the parameters $x_1, \ldots, x_h$. This is possible if for every formula $G(x)$ of $L(\lambda)$ there is a formula $H(x)$ of $L(\lambda)$ such that $T \vdash (\mathrm{E}x)G(x) \rightarrow (\mathrm{E}!\, x)(G(x) \wedge H(x))$ (cf. Kueker (1970)).

Theorem 4 (Chang (1964), and Makkai (1964)). The following conditions are equivalent:

(i) P is restrictedly identifiable in T.

(ii) There are formulas $F_i(\bar{x}_h, \bar{y}_k)$ $(i = 1, \ldots, n)$ of $L(\lambda)$ such that

$$T \vdash \bigvee_{i=1}^{i=n} (\mathrm{E}\bar{x}_h)(\bar{y}_k)(P(\bar{y}_k) \equiv F_i(\bar{x}_h, \bar{y}_k)).$$

All these theorems are equivalences which give syntactical counterparts to certain semantical notions of definability. Theorems 1,3–4 treat quantitatively different kinds of definability in that they give syntactical counterparts to restrictions which T can impose to the number of possible interpretations of P when the interpretation of λ is given. But definability can be considered also qualitatively by fixing one's attention to the question whether P is explicitly definable or not, rather than to the questions of cardinality restrictions as above. Then it appears that it is possible to develop a general theory of definition in which the results stated above find their place in a natural way. In this theory further conditions can be placed on a par with the conditions (i)–(ii) of each of Theorems 1–4 (cf. Hintikka (1972) and Rantala (1973)). Although this theory may have some methodological import for problems which are connected with definability and identifiability in empirical theories, it is not possible to discuss it here in any detail.

There is an important further kind of definability to be considered here. It is said that P is *conditionally definable* in T (in terms of λ) if there are formulas $G(\bar{x}_k)$ and $F(\bar{x}_k)$ of $L(\lambda)$ such that

$$T \vdash (\bar{x}_k)(G(\bar{x}_k) \rightarrow (P(\bar{x}_k) \equiv F(\bar{x}_k))).$$

This does not impose any quantitative restrictions on the interpretations of P, without further conditions. It is reduced to an explicit definition only if $T \vdash (\bar{x}_k)G(\bar{x}_k)$. Anyway, it has an important semantical meaning. Suppose that we are trying to expand $\langle D, \underline{\lambda}\rangle$ to a model of T. If $\langle d_1, \ldots, d_k\rangle \in D^k$ does not satisfy $G(\bar{x}_k)$, then we have a freedom to choice $\langle d_1, \ldots, d_k\rangle \in \underline{P}$ or $\langle d_1, \ldots, d_k\rangle \notin \underline{P}$. But if it satisfies $G(\bar{x}_k)$, we have not such a freedom. What is often important in applications of an empirical theory is to know whether P will be uniquely interpreted with respect to those $\langle d_1, \ldots, d_k\rangle$ which satisfy $G(\bar{x}_k)$ (cf. section 4, below).

If c is a function symbol or an individual constant. Theorems 1–4 can immediately be modified to apply to c. If c is a $(k-1)$-place function symbol f. P must be replaced by f and $P(\bar{x}_k)$, $P(\bar{y}_k)$ by the formulas $f(\bar{x}_{k-1}) = x_k, f(\bar{y}_{k-1}) = y_k$, respectively. If c is an individual constant a, P is replaced by a and $P(\bar{x}_k)$, $P(\bar{y}_k)$ by a formula $x = a$ (and the prefixes are modified accordingly). In the case of a conditional definition of f there is a further modification: $G(\bar{x}_k)$ must be replaced by $G(\bar{x}_{k-1})$.

III. IDENTIFIABILITY IN ECONOMETRICS

In econometrics, there occurs the problem of identification of an econometrical 'model' (a set of equations). That amounts to the uninue determination of the coefficients, and the problem of identification of parameters. It may happen that a 'model' is identifiable but all its parameters are not. The notion of identifiability is mixed up with the special problems of econometrics thus somewhat prohibiting the understanding of its logical character. We shall consider identification of parameters, but here it can be done only rather schematically, and the following treatment obviously needs more precision.

We shall try to get as close as possible to the notion of identifiability in econometrics by using the logical tools exhibited in the previous sections. Thus the following definition of identifiability is perhaps a slight generalization of the notion of identifiability in the econometrical sense.

Suppose that a set of equations, with variables $x_1, \ldots, x_h$, is stated to present a mathematical 'model' for an empirical phenomenon. Then it is specified to what entity each of the variables refers. For example, in econometrics the variables can refer to the prices or quantities of commodities, etc., in physics they can refer to time, particles, etc. In other words, it is asserted that this set of equations is satisfied by all those values of the variables which belong to certain *subsets* of the universe of discourse D. Thus we can define a relation $\underline{O}$ on D by stating $\langle x_1, \ldots, x_h \rangle \in \underline{O}$ iff $x_i \in \underline{P}_i$ (i $= 1, \ldots, h$) where the $\underline{P}_i$ are these specified subsets of D.

We shall consider an econometrical (non-stochastic) 'model' as a (consistent) formalized theory of the form $T = T(\varphi, O, a_1, \ldots, a_k)$ which includes 1) the theory of real numbers (with a set of constants φ), 2) the axioms S and $(\bar{x}_h)(O(\bar{x}_h) \to F(\varphi, a_1, \ldots, a_k, \bar{x}_h))$ where S is a sentence corresponding to the apriori restrictions for the parameters a_i, and $F(\varphi, a_1, \ldots, a_k, \bar{x}_h)$ is a formula corresponding to a set of algebraic equations. O is a h-place predicate symbol, and $a_1, \ldots, a_k$ are individual constants.

In intended (standard) models $\langle D, \underline{\varphi}, \underline{O}, \underline{a}_1, \ldots, \underline{a}_k \rangle$, D is considered as the set of real numbers. We shall call $\underline{O}$ the set of observations. $\underline{O}$ is required to be non-empty. For simplicity, by structures of the form $\langle D, \underline{\varphi}, \underline{O} \rangle$ (with possible superscripts) we mean in this section only those structures for $L(\varphi, O)$ which can be expanded to a model of T (i.e., in which the sets of observations are 'compatible' with T).

It seems that in econometrics identifiability of a parameter of a 'model' means potential definability of this parameter. A parameter is identifiable if it is possible to determine its value uniquely on the basis of every suitably chosen (large enough, in a sense) set of observations. More exactly: a_i is identifiable if for every $M = \langle D, \underline{\varphi}, \underline{O} \rangle$ there is a model $M' = \langle D, \underline{\varphi}, \underline{O}' \rangle$ such that $\underline{O} \subseteq \underline{O}'$ and every expansion of M' to a model of T has the same interpretation (value) of a_i.

A syntactical counterpart can be stated to this condition. Let us denote by $I_n(\bar{x}_h^1, \ldots, \bar{x}_h^n)$ the following formula with nh free variables $x_j^i (i = 1, \ldots, n; j = 1, \ldots, h)$:

$$\bigwedge_{i=1}^{i=n} (O(\bar{x}_h^i) \wedge (E\bar{z}_k^i)(\bigwedge_{\substack{j=1 \\ j \neq i}}^{j=n} F(\varphi \bar{z}_k^i, \bar{x}_h^j) \wedge \neg F(\varphi \bar{z}_k^i, \bar{x}_h^i))).$$

The observations (n-tuples of individuals in $\underline{O}$) $o_1, \ldots, o_n$ are *mutually independent* if $\langle o_1, \ldots, o_n \rangle$ satisfies $I_n(\bar{x}_h^1, \ldots, \bar{x}_h^n)$ in $\langle D, \underline{\varphi}, \underline{O} \rangle$. This notion of independence can be formulated in first-order logic since the parameters a_i are individual constants. Its obvious intuitive meaning is that, for every $i = 1, \ldots, n$, if $o_1, \ldots, o_{i-1}, o_{i+1}, \ldots, o_n$ satisfy any given set of equations of the form F, this does not 'force' o_i to satisfy it, $o_1, \ldots, o_n$ are then distinct from each other.

For every $n = 1, 2, \ldots$, let A_n be the sentence which says that in every set of observations there are at least n mutually independent observations:

$$(\mathrm{E}\bar{x}_h^1) \ldots (\mathrm{E}\bar{x}_h^n) I_n(\bar{x}_h^1, \ldots, \bar{x}_h^n).$$

Let A_0 be the formula $\neg A_1$.

If a model satisfies A_n (for a given n), then its set of observations $\underline{O}$ has at least n elements (but not always conversely). If $\underline{O}$ includes a subset of n mutually independent observations, then $\underline{O}$ includes subsets of $1, \ldots, n-1$ mutually independent observations. This holds since the sentence A_n implies logically the sentences $A_1, \ldots, A_{n-1}$.

For every T, there is a characteristic number n_T $(0 \leqslant n_T \leqslant k)$ which expresses the greatest possible number of mutually independent observations in the models of T (and in their restrictions to $L(\varphi, O)$). More exactly, there is a $n_T = 0, 1, 2, \ldots$ such that $T \cup \{A_{n_T}\}$ is consistent but $T \cup \{A_n\}$ is inconsistent for every $n > n_T$. If a model includes this maximal number of mutually independent observations, we call this model *observationally maximal.*

Now we state that a_i is identifiable (in the sense of econometrics) iff a_i is explicitly definable (in terms of $\varphi \cup \{O\}$) in the theory $T \cup \{A_{n_T}\}$. Thus indentifiability of a_i in an econometrical 'model' does not necessarily mean its explicit definability in the theory T which is the formalization of this 'model'. It means only that every observationally maximal model $\langle D, \varphi, \underline{O} \rangle$ can be expanded to a model of T uniquely with respect to a_i. Whether these considerations can be generalized for other kinds of theories, will be examined later.

We shall consider some examples. Although they are simple, we hope that they sufficiently illustrate connections of some of the different kinds of definability (in the sense of section II, above) with identifiability.

Example 1. T includes the following axioms (apart from the theory of real numbers):

$$a^2 = b \wedge a \neq 0 \wedge c = 1$$

$$(x)(y)(O(x, y) \rightarrow ax + by = c).$$

The following result shows that a is 2-foldly identifiable:

$$T \vdash (\mathrm{E}x)(\mathrm{E}y)O(x, y)$$
$$T \vdash (x)(y)(O(x, y) \rightarrow a = f_1(x, y) \vee a = f_2(x, y))$$

Here $f_1(x, y) \neq f_2(x, y)$ if $x^2+4y > 0$. Hence for any model $M = \langle D, \varphi, \underline{O} \rangle$, where $\underline{O}$ is of the form $\{(x, y)\}$ with $x^2+4y > 0$, there are two different interpretations of a when M is expanded to a model of T. But a is identifiable in the sense of econometrics: If $\overline{\overline{\underline{O}}} \geqslant 2$, then the interpretation of a is fixed uniquely, i.e., a is explicitly definable in $T \cup \{A_2\}$ $(n_T = 2)$. (For this reason, an econometrician is perhaps inclined to say that the 'model' defines a explicitly.) It should be observed that there are models $\langle D, \underline{\varphi}, \underline{O} \rangle$ of $T(\varphi, O)$ which are not observationally maximal $(\overline{\overline{\underline{O}}} = 1)$ but which nevertheless fix uniquely the value of a.

Example 2. $b = \sin a \wedge a \neq 0 \wedge b \neq 0 \wedge c = 1$

$$(x)(y)(O(x, y) \rightarrow ax+by = c).$$

Here a is countably identifiable in T, but explicitly definable in $T \cup \{A_2\}$, hence identifiable in the sense of econometrics.

Example 3. $a \neq 0 \wedge b \neq 0 \wedge c = 1$

$$(x)(y)(O(x, y) \rightarrow ax+by = c).$$

We obtain:

$$(x_1)(y_1)(x_2)(y_2)(O(x_1, y_1) \wedge O(x_2, y_2) \wedge (x_1 \neq x_2 \vee y_1 \neq y_2)$$
$$\rightarrow a = (y_2-y_1)/(x_1 y_2 - x_2 y_1)).$$

This does not amount to explicit definability (finite identifiability with one disjunct) of a since T does not imply the sentence

$$(\mathrm{E}x_1)(\mathrm{E}y_1)(\mathrm{E}x_2)(\mathrm{E}y_2)(O(x_1, y_1) \wedge O(x_2, y_2) \wedge (x_1 \neq x_2 \vee y_1 \neq y_2)).$$

a is restrictedly identifiable in T and again explicitly definable in $T \cup \{A_2\}$, hence identifiable.

Example 4. $a = d^2 \wedge b = ed \wedge c = 1$

$$(x)(y)(O(x, y) \rightarrow ax+by = c).$$

d and e are 2-foldly identifiable in T but they are not explicitly definable in $T \cup \{A_2\}$, hence not identifiable.

We do not try here to formulate the (rather vague) notion of overidentifiability, but it is possible that Examples 1 and 2 are cases of overidentification: The a priori restrictions in them are, in a sense, stronger than is necessary for identification of a. There are models $\langle D, \underline{\varphi}, \underline{O}\rangle$ of $T(\varphi, O)$ which fix uniquely the value of a although they are not obervationally maximal. According to this view, the 'model' of Example 3 would just identify a: Only observationally maximal models fix the value of a. The whole pattern of independent observations are needed to evaluate a uniquely.

It is not clear whether this standpoint at all corresponds to the intentions of econometricians when they speak about overidentification. Anyway, econometrical overidentification means some kind of extra strength of a priori restrictions for parameters. But Simon's concept of overidentifiability (see the axiomatization of Ohm's Law in Simon (1970)) is rather different in that it is associated with an extra number of observations. In the case of overidentification, he allows different subsets of a given set of observations $\underline{O}$ to produce different values for a parameter. But then $\underline{O}$ is not compatible with the theory T in question, and the corresponding structure for $L(\varphi, O)$ cannot be expanded to a model of T at all. Thus, if the theory is logically consistent (as Ohm's Law is), Simon's 'overidentification' does not differ from 'justidentification'. If measurements in an empirical experiment produce this kind of $\underline{O}$ (apart from the limits of approximation, of course), then T simply does not describe that phenomenon for which this theory has been introduced.

Also Simon's notion of underidentifiability obviously differs from its namesake in econometrics. The voltage v and the internal resistance b are identifiable (in the sense of econometrics) in the theory of Ohm's Law, i.e., potentially uniquely determinable on the basis of observations, since the axiom system does not state that $\overline{\overline{\underline{O}}} = 1$. Instead, Simon's 'general definability' (in Simon (1970)) seems to be near to 'identifiability' in econometrics. It should be noted, however, that the number of observations is not decisive for identification but the number of mutually independent observations.

There is perhaps reason to note that an axiom system for an empirical theory has also other models (in the model-theoretic sense) than the intended 'physical' or 'econometrical', etc., models (whose sets of observations can be thought to be obtained by measurements in the real world). If identification problems are treated as logical problems, one cannot consider only these intended models but all the models of a theory. Whether experiments can be arranged so that a sufficient number of mutually independent observations are available (if they are available theoretically), is not a logical question. It seems that much trouble in the questions of definability in empirical theories is due to the fact that the following questions have not been separated from each other: 1) Is a term explicitly definable in a theory? 2) Do the (empirical) observations in the real world determine this term uniquely and consistently with the theory?

IV. DEFINABILITY IN PARTICLE MECHANICS

Definability problems in physics are in general more difficult than in econometrics (so far considered), for physical theories are mathematically more complicated, and theoretical terms used in the existing axiomatizations of various parts of physics are often function symbols, not individual constants as above. Moreover, the questions of what should be taken explicitly for axioms when a theory is formalized, and what kind of language is needed, are not so straightforward as in econometrics. These problems must be resolved for a given theory before any definite meaning can be given to 'definable'.

We shall mainly restrict ourselves to certain examples. These examples are well-known axiomatizations of physical theories, and in connection with these theories there have been some discussion about definability of theoretical terms.

Mach's suggestion for the definition of mass in the classical particle mechanics (in Mach (1942), for example) gave rise to an interesting discussion of the subject, discussion which is mainly due to Pendse (in Pendse (1937), (1939), and (1940)) and Narlikar (1939). There definability of mass amounts to the problem whether the equations based on Newton's laws can be stated so that the masses, or the mass ratios, of the particles in question can be uniquely solved from these equations in terms of observables (time and the accelerations of the particles).

In the light of modern logical concepts of definability the discussion leaves the matter a little vague. This vagueness is due to the lack of exactness as to the *logical* basis of the theory in question. But this discussion is very important in the *physical* sense, for it immediately shows the circumstances in which the masses or the mass ratios can be uniquely calculated (in principle) from given equations and thus gives us the material for an exact treatment of the subject.

Without going into details of Mach's and Pendse's considerations, we shall turn to more recent formulations of particle mechanics. In McKinsey, Sugar, and Suppes (1953), Suppes (1957), and Simon (1970) the theory of particle mechanics is not completely formalized but defined as a set-theoretical predicate. This kind of formulation is simpler than a complete formalization but it is not the best possible for syntactical considerations. However, e.g., from Montague (1961) it is seen how a complete formalization can be done. Therefore we shall consider that question only when necessary.

Each of these papers pays some attention to definability of mass and force. However, their treatment of definability is not quite satisfactory, for it is not always quite clear what is meant by 'definability' in them.

The intended (physical) models of a theory $\mathscr{T}$ of particle mechanics include the set of real numbers in their domains. Hence the possible unique calculability of the masses or the forces of the particles would not necessarily mean explicit definability in $\mathscr{T}$. To see why it is so, let us consider as an example the axiomatization of particle mechanics in Suppes (1957), especially the following dynamical axiom (we can consider it as a shorthand notation for the formalized form):

$$(p)(t)(P(p)\wedge T(t) \rightarrow m(p)D^2 s_p(t) = \sum_{q\in P} f(p, q, t)+g(p, t)).$$

It is immediately obtained:

$$\mathscr{T} \vdash (p)(t)\,(P(p)\wedge T(t) \rightarrow g(p,t) = m(p)D^2 s_p(t) - \sum_{q\in P} f(p, q, t)).$$

This does not give an explicit definition of external force g (in terms of the other constants) but a conditional definition. It cannot be converted to an explicit definition in $\mathscr{T}$ since $\mathscr{T}$ does not fix the values of $\underline{g}$ for those individuals $\underline{p}$, $\underline{t}$ of a model for which $\langle \underline{p}, \underline{t}\rangle$ is not in $\underline{P}\times\underline{T}$. g is not even finitely identifiable since the models are of uncountable cardinality (and $\underline{P}$ is finite).

Thus we see that there is no chance to define explicitly mass or force on the basis of the axiomatizations mentioned above, without further axioms, since mass and force are function symbols in these axiomatizations. The reason was seen to be very simple, and there is no need to use Padoa's method to discover non-definability.

Of course, there are means to revise an axiomatization in order to convert a conditional definition to explicit one and still preserve the function symbols. In the case above, it suffices, e.g., to strengthen the axiom system by adding the axiom $(p)\,(t)\,(\neg P(p) \vee \neg T(t) \rightarrow g(p, t) = 0)$. This could be considered as an explication of the intuitive idea that at a given time forces are acting only on those individuals of a model which are considered as particles. However, this kind of revision of an axiom system has only theoretical import when an empirical theory is in question. As to applications of a theory, all we want is that we can uniquely calculate the values of a function in important cases, e.g., masses or forces for particles. But it is not harmful if we know what we are talking about when we in different cases use the word 'definable'.

Next we shall consider in some detail definability of mass in terms of observables. After Pendse, this subject has been mainly treated of by Simon (cf. Simon (1947) and (1970), also McKinsey, Sugar, and Suppes (1953)). But in his notion of definability there also occurs vagueness, considered from the strict logical point of view.

For precision, we set forth his axiomatization of Newtonian particle mechanics (to be found in Simon (1970)):

A system $\Gamma = \langle P, T, s(p, t), m(p) \rangle$ that satisfies Axioms $N1$–$N4$ is called a *Newtonian system of particle mechanics.*

*AXIOM N*1. P is a non-empty, finite set.

*AXIOM N*2. T is an interval of real numbers.

*AXIOM N*3. If p is in P and t is in T, then $s(p, t)$ is a three-dimensional vector such that $D^2 s(p, t)$ exists, where D denotes differentiation with respect to t.

*AXIOM N*4. If p is in P, then $m(p)$ is a real-valued function, $(p \in P)m \neq 0$, such that:

(4a) $(t \in T) \quad \sum_{p \in P} m(p) D^2 s(p, t) = 0,$

(4b) $(t \in T) \quad \sum_{p \in P} m(p) D^2 s(p, t) \times s(p, t) = 0.$

A Newtonian system of particle mechanics that satisfies Axiom *N*5 is called *holomorphic*.

*AXIOM N*5. If $P^* \subset P$, and if E_a and E_b are the statements obtained by replacing P by P^* in (4a) and (4b), respectively, then $\neg(E_a \wedge E_b)$.

In interpretation, P is a set of particles, T an interval of time, the $m(p)$ the mass function and the $s(p, t)$ the function designating the positions of the particles.

To see what it exactly means if it is said (in Simon (1970)) that the mass ratios for all pairs of particles are definable by means of the s's, we shall assume that this axiom system is formalized in first-order logic, without concepts of set theory. For an explicit treatment of this kind of formalization, see Montague (1961). Thus we suppose that that every formula and term we shall use has a counterpart in the formalized theory. Instead of s, we ought to take, e.g., s_1, s_2 and s_3 (scalar components) for primitive terms, but for brevity, we shall refer $\lambda = \varphi \cup \cup \{P, T, s\}$ as the set of observational terms. Here φ is the set of constants needed in the theory of real numbers. Then the models of the theory $\mathscr{T}$ of Newtonian particle mechanics are of the form $\Gamma = \langle D, \underline{\varphi}, \underline{P}, \underline{T}, \underline{s}, \underline{m} \rangle$ (with possible superscripts). We can also use the name 'system' for Γ. We shall replace $D^2 s(p, t)$ by the 3-tuple $\langle a_1(p, t), a_2(p, t), a_3(p, t) \rangle$ of its scalar components (the second derivatives of $s_1(p, t), s_2(p, t), s_3(p, t)$) and $D^2 s(p, t) \times s(p, t)$ by $\langle a_4(p, t), a_5(p, t), a_6(p, t) \rangle$.

With these modifications, Axiom *N*4 becomes (if we do not introduce individual constants for the names of the particles, i.e., the members of $\underline{P}$):

$$(p)(P(p) \rightarrow m(p) \neq 0)$$

$$(t)(\bar{p}_n)\Big(T(t) \wedge \bigwedge_{i=1}^{i=n} P(p_i) \wedge U(\bar{p}_n) \wedge (p)\big(P(p) \rightarrow \bigvee_{i=1}^{i=n} (p = p_i)\big) \rightarrow \bigwedge_{i=1}^{i=6} \Big(\sum_{j=1}^{j=n} a_i(p_j, t) m(p_j) = 0\Big)\Big),$$

where the last conjunction is the shorthand notation for the formula corresponding to the conjunction of the six scalar equations obtained from (4a) and (4b), and $U(\bar{p}_n)$ is the formula which says that $p_1, \ldots, p_n$ are distinct.

Axiom *N*5 is not quite correctly formulated to correspond to the intended meaning of a holomorphic system in Simon (1947). For this axiom says that for no system (model) $\Gamma = \langle D, \varphi, \underline{P}, \underline{T}, \underline{s}, \underline{m} \rangle$ there is a subsystem (submodel) $\Gamma^* = \langle D^*, \varphi^*, \underline{P}^*, \underline{T}^*, \underline{s}^*, \underline{m}^* \rangle$ such that $\underline{P}^* \subseteq \underline{P}$, $\underline{P}^* \neq \underline{P}$. If Γ^* is a subsystem of Γ, then it holds that $\underline{m}^*(d) = \underline{m}(d)$ for every $d \in \underline{D}^*$. Axiom *N*5 (in this form) is only necessary, not sufficient condition for the uniqueness of the mass ratios, since it does not exclude the existence of a model $\Gamma' = \langle D', \varphi', \underline{P}', \underline{T}', \underline{s}', \underline{m}' \rangle$ of T which has the following properties: $\overline{\overline{\underline{P}'}} < \overline{\overline{\underline{P}}}$, $\underline{T}' = \underline{T}$, there is a $\mathscr{T}$ bijective mapping f from $\underline{P}'$ into $\underline{P}$ such that $\underline{s}'(p, \underline{t}) = \underline{s}(f(p), \underline{t})$ for every $p \in \underline{P}'$, $\underline{t} \in \underline{T}$ (and possibly $\underline{m}'(p) \neq \underline{m}(f(p))$). ($f$ is not necessarily an isomorphic embedding of Γ' into Γ.) The axiom system does not exclude the possibility that two different particles may have same position at every instant of time. Whether this kind of situation can occur in the actual world (in intended models) is not a model-theoretical but an empirical question.

It is the spirit of the proof of Theorem I in Simon (1947) that not even models like Γ', above, exist for any Γ. Clearly this amounts to the algebraic condition on which (roughly speaking) one can find $n-1$ (homogeneous) equations (for n unknown) which determine uniquely the ratios of the unknown. Thus it is closely related to the independence condition A_{n_T} in section III, and we are close to the notion of identifiability in econometrics.

This condition can be expressed in first-order language as follows (we can dispense a function variable in favour of a number of individual variables):

*AXIOM N*51.

$$\neg(E\bar{p}_k)(Ep)(E\bar{z}_k)\Big(\bigwedge_{i=1}^{i=k} P(p_i) \wedge U(\bar{p}_k) \wedge P(p) \wedge \bigwedge_{i=1}^{i=k} (p \neq p_i)$$

$$\wedge \bigwedge_{i=1}^{i=k} (z_i \neq 0) \wedge (t)\Big(T(t) \rightarrow \bigwedge_{j=1}^{j=6} \Big(\sum_{i=1}^{i=k} a_j(p_i, t) z_i = 0\Big)\Big)\Big).$$

What does it mean syntactically that the mass ratios are 'definable'? Indeed, it can be shown that the theory $\mathcal{T}_0$ (in the form sketched above) of holomorphic Newtonian particle mechanics implies a sentence of the form

$$(x)(y)(z)\left(P(x) \wedge P(y) \wedge x \neq y \rightarrow \left(\frac{m(x)}{m(y)} = z \equiv F(x, y, z)\right)\right),$$

where m does not occur in F.

Whether this can be considered as a proper conditional definition is uncertain since it does not define conditionally any primitive symbol. Of course, it is possible to revise the axiom system by using (instead of m) as a primitive symbol a 2-place function symbol r to correspond to the mass ratios. Moreover, if it is postulated that $r(x, y) = 0$ when x or y is not a particle, and $r(x, x) = 1$ when x is a particle, an explicit definition of r is obtained. Anyway, we see that the interpretation of the observational terms fix uniquely the mass ratios of the *particles* in every model. One can, of course, take an attitude that for applications this is all that matters since one is interested only in the masses of the particles (that is, of the members of P).

Simon's result (that in a holomorphic system the mass ratios are uniquely determined by the observables, and only in such a system) seems philosophically significant and natural, for it implies that in an isolated universe there can be no 'absolute' unit of mass to which the masses of all the other particles could be uniquely compared if the universe includes independent subsystems. The masses of two particles which are located in different subsystems cannot be compared uniquely to each other. There is no reason to expect that two systems were dependent with respect to masses if they are independent with respect to forces.

There has been in the literature (e.g., in Pendse (1939) and in Simon (1947)) some discussion of the seemingly unsatisfactory situation that the values of the mass ratios of given particles are dependent on a selection of a reference system (coordinate system). From the purely model-theoretic point of view, there is nothing confusing in this situation. For the reference system is fixed by the interpretation of the observables (in fact, the interpretation of s) in D, and it is to be expected that two different structures (although they had the same domain and the same interpretation of P) for the observational language may introduce different interpretations for a theoretical term in their expansions to the richer language.

Also the question whether a given reference system can be transformed so that a given motion of particles becomes an isolated motion has the clear-cut model-theoretic meaning: Whether s can be reinterpreted so in a given structure for the observational language that the structure obtained can be expanded to a model of an isolated motion, i.e., to a Newtonian system of particle mechanics.

BIBLIOGRAPHY

Beth, Evert W., 'On Padua's Method in the Theory of Definition', *Indagationes Mathematicae*, vol. **15** (1953), pp. 330–339.

Chang, C. C., 'Some New Results in Definability', *Bulletin of the American Mathematical Society*, vol. **70** (1964), pp. 808–813.

Chang, C. C., and Keisler, H. J., *Model Theory*, North-Holland, Amsterdam (1973).

Craig, W., 'Bases for First-Order Theories and Subtheories', *Journal of Symbolic Logic*, vol. **25** (1960), pp. 97–142.

Hintikka, Jaakko, 'Constituents and Finite Identifiability', *Journal of Philosophical Logic*, vol. **1** (1972), pp. 45–52.

Hintikka, Jaakko, and Tuomela, Raimo, 'Towards a General Theory of Auxiliary Concepts and Definability in First-Order Theories', in Jaakko Hintikka and Patrick Suppes, editors, *Information and Inference*, D. Reidel, Dordrecht (1970), pp. 298–330.

Kueker, D. W., 'Generalized Interpolation and Definability', *Annals of Mathematical Logic*, vol. **1** (1970), pp. 423–468.

Mach, E., *The Science of Mechanics*, 5th American ed., La Salle I11 (1942), pp. 264-277.

Makkai, M., 'A Generalization of a Theorem of E. W. Beth', *Acta Math. Acad. Sci. Hungar.*, vol. **15** (1964), pp. 227–235.

McKinsey, J. C. C., Sugar, A. C., and Suppes, Patrick, 'Axiomatic Foundations of Classical Particle Mechanics', *Journal of Rational Mechanics and Analysis*, vol. **2** (1953).

Montague, R., 'Deterministic Theories', in Washburne, R., editor, *Decisions, Values and Groups II*, Pergamon Press, Oxford (1961), pp. 325–370.

Narlikar, V. V., 'The Concept and Definition of Mass in Newtonian Mechanics', *Phil. Mag., ser.* 7, **xxvii** (1939), pp. 33–36.

Pendse, C. G., 'A Note on the Definition and Determination of Mass in Newtonian Mechanics', *Phil. Mag., ser.* 7, **xxiv** (1937), pp. 1012–1022.

Pendse, C. G., 'A Further Note...', *ibid. ser.* 7, **xxvii** (1939), pp. 51–61.

Pendse, C. G., 'On Mass and Force in Newtonian Mechanics', *ibid., ser.* 7, **xxix** (1940), pp. 477–484.

Rantala, Veikko, 'On the Theory of Definability in First-Order Logic', Reports From the Institute of Philosophy, University of Helsinki, No. 2 (1973).

Simon, H. A., 'The Axioms of Newtonian Mechanics', *Phil. Mag.*, *ser.* 7, **xxxiii** (1947), pp. 888–905.

Simon, H. A., 'The Axiomatization of Physical Theories', *Philosophy of Science*, vol. **37** (1970), pp. 16–26.

Suppes, Patrick, *Introduction to Logic*, Van Nostrand, Princeton (1957).

Svenonius, Lars, 'A Theorem about Permutation in Models', *Theoria*, vol. **25** (1959), pp. 173–178.

Tuomela, Raimo, *Theoretical Concepts*, Springer-Verlag, Wien-New York (1973).

ROBERT L. CAUSEY

LAWS, IDENTITIES, AND REDUCTION*

I. INTRODUCTION

The language of a scientific theory is often assumed to consist of certain logical and mathematical symbols plus nonlogical constants.[1] In particular, the 'logical symbols' are often those used in set theory together with mathematical symbols definable in set theory. The nonlogical constants are considered to be predicate constants (including function symbols) which are interpreted as denoting sets, or sets of ordered *n*-tuples, of objects in the domain of the theory in question. Within such a language a universal law, of the simplest possible form, is represented by a sentence of the form

(1) $(x)(\alpha x \rightarrow \beta x)$

where '$\rightarrow$' is the material conditional, and α and β are predicates.

Of course, it is often argued that (1) does not distinguish nomological generalizations from accidentally true generalizations and hence that (1) is not a completely adequate representation of the form of a law.[2] In this article, I am concerned only with genuine laws, so the nomological/ /accidental distinction can be ignored. I will argue that the predicates in *law-sentences* cannot always be interpreted as merely denoting sets, and that instead they should usually be interpreted as denoting *kinds* of things and *attributes*. An example will help to motivate the later, more general discussion.

We introduce the following predicates and function symbols: '*sink x*' means that the object *x* has the disposition to sink in H_2O in a gravitational field, '*den*(*x*)' denotes the density of *x*, '*gold x*' means that *x* is a sample of gold. We then have the law-sentences

(2) $(x)(gold\ x \rightarrow sink\ x)$

(3) $(sink\ x \leftrightarrow den(x) > den(H_2O))$

where '$\leftrightarrow$' is the material biconditional, and '$>$' means 'greater than'.

Treating the predicates set-theoretically, i.e., as mere names for their extensions, (2) and (3) imply the law-sentence (4)

$$(4) \qquad (x)\,(\mathit{gold}\ x \rightarrow \mathit{den}(x) > \mathit{den}(H_2O)).$$

Treating the predicates as mere names for their extensions, (4) would state the same law as (2), but, intuitively, they state different laws. Furthermore, from information about the masses of their atoms and molecules, plus information about the crystalline and liquid structures of gold and H_2O, we could explain (4). Yet, in order to explain (3), and hence also (2), we would also use laws of the mechanics of fluids. If the explanations of two different law-sentences require obviously different sets of laws in their explanans, then the two law-sentences in question would seem to represent or state different laws. Finally, notice that (2) and (4) can be proved equivalent under the assumption of (3). But (3) represents a causal law which is explainable, and which provides a kind of causal connection between (2) and (4). It is natural to conclude that (2) and (4) represent different laws by virtue of the causal connection between them which is represented by (3).

Let us say that two law-sentences in a theoretical language are *nomologically-equivalent*, *n-equivalent*, or *n-eq*, if and only if they represent the same law, What is the criterion for *n*-equivalence?

Peter Achinstein says that two propositions expressing laws express the same law if they are either logically equivalent, or empirically equivalent, where two propositions are empirically equivalent if they can be derived from one another with the help of additional empirical assumptions which are logically independent of each of these propositions.[3] The above example, and others like it, show that Achinstein's condition is too broad.[4] Yet logical equivalence is too narrow, and there is instead a more subtle criterion for *n*-equivalence. In order to develop and apply it, I will give up a purely set-theoretical interpretation of theoretical languages.

Consider (3). Informally, we can say that the reason (3) states a law and is subject to a causal explanation is because '*sink x*' and '$\mathit{den}(x) > \mathit{den}(H_2O)$' are predicates which denote different attributes. The law-sentence (3) states that these attributes are co-extensional, and this fact is causally explainable. On the other hand, an identity between designating terms (other than definite descriptions) is not causally explainable. This is true when the identity is analytic (such as 'Bob = Bob', 'gold = gold') or synthetic (such as 'a smallest possible sample of water = an H_2O molecule'). An identity merely asserts that two terms denote the same object, and hence it is a sentence which is not subject to a causal explanation. By combining this observation with an ontology which includes *kinds* and *attributes*, it is possible to develop conditions for *n*-equivalence, the reduction of theories, and other methodological concepts.

II. NONCAUSAL SENTENCES AND N-EQUIVALENCE

Consider a language which describes kinds of objects in a specified domain of things, as well as attributes of these objects. This language consists of symbols used in set theory and mathematics plus a set $\mathscr{L}$ of nonlogical predicates and function symbols. $\mathscr{L}$ is the union of two disjoint subsets, $\mathscr{T}$ and $\mathscr{A}$, where $\mathscr{T}$ is the set of *thing-predicates* and $\mathscr{A}$ is the set of *attribute-predicates.* A thing-predicate is interpreted as a name for a *kind* of element of the domain, and an attribute-predicate is a name for an *attribute* (a property, relation, or quantity). In a previous publication [4] I have discussed the interpretation of such predicates in detail.

We are concerned with *law-sentences* of $\mathscr{L}$, which represent or state laws, and with *identity-sentences* (*identities*) of $\mathscr{L}$, which state identities. These can be *thing-identities* between two thing-predicates or *attribute--identities* between two attribute-predicates. Such identities imply that the identified predicates are co-extensional, but they also imply that this co-extensionality is not a law and not subject to causal explanation. Thus they are different from (3), which states a nomological co-extensionality.

A sentence that is not subject to causal explanation is said to be *noncausal.* Roughly speaking, it is a sentence without a cause. Analytic sentences and logical truths are necessarily noncausal. Thing-identities and attribute-identities are also noncausal, and they can be synthetic. We are concerned here only with true noncausal sentences, for obviously any false sentence is noncausal. Fortunately, a false sentence would not be contained in a true theory.

Within $\mathscr{L}$ the following appear to be the only possible kinds of true noncausal sentences: logically true sentences, analytically true sentences (which would follow logically from definitions of defined terms), true thing- and attribute-identities true assertions of nonidentity between predicates, and logical consequences of sets of true noncausal sentences. In particular, notice that '$(x)(Qx \rightarrow Qx)$' is noncausal, so if Q is identical with R, then '$(x)(Qx \rightarrow Rx)$' is noncausal, although it appears to be a law-sentence. If Q is nomologically co-extensional with R, then '$(x)(Qx \rightarrow Rx)$' is a law-sentence. Any law-sentence is in principle subject to a causal explanation, so no law-sentence can be noncausal, and no noncausal sentence can represent a law. Also notice that '$(x)(y)(Uxy \leftrightarrow Vyx)$' might be synthetic noncausal. For instance, it follows from the synthetic identity of V with U_*, and the definition of U_* as the converse of U. If necessary, 'U_*' can be added to $\mathscr{L}$ as a defined term. I now propose the following:

C_1 Let L_1, L_2 be law-sentences in $\mathscr{L}$.
Then L_1 *n*-eq L_2 if and only if there is a set **N** of true noncausal sentences of L such that $L_1 \leftrightarrow L_2$ is derivable from **N**.

I think that everyone would agree that L_1 and L_2 represent the same law if they are logically equivalent, which means here that they can be proved equivalent using the axioms of set theory. In this case an explanatory derivation of one of them could be changed into an explanatory derivation of the other without using any different causal premises. Similarly, if one of them is used as a premise in an explanatory derivation, then the other could replace it in this derivation. In short, we can say that there is no causal connection between L_1 and L_2 if they are logically equivalent.

C_1 is a generalization of this idea. If L_1 and L_2 can be proved equivalent from a set of noncausal premises, then there is no causal connection between them. However, in general they cannot arbitrarily be substituted for one another because they might contain some different predicates. In such a case they can be substituted into derivations along with the appropriate identities. Since identities are noncausal sentences, such substitutions plus additions of identities do not change the laws represented in derivations representing explanations.

In practice we will be most interested in cases of this sort: L_1 contains occurrences of predicate α, L_2 contains occurrences of predicate β, and L_1 and L_2 are uniform substitution instances of each other under substitution of β for α in L_1, and α for β in L_2, where α and β denote the same kind or attribute. Of course, it is necessary that the denotations of α and β be the same and not merely nomologically co-extensional. Since attribute-identities are more problematic than thing-identities, the remaining discussion will focus on attribute-identities.

Let the following law-sentences be formulated in $\mathscr{L}$:

(5) $(x)(Ax \rightarrow Bx)$

(6) $(x)(Ax \rightarrow Cx)$

(7) $(x)(Bx \leftrightarrow Cx)$.

On intuitive grounds, (5) and (6) represent different laws because they state different causal connections involving the different attributes B and C. If B and C were identical, then (5) would be equivalent to (6) solely by virtue of this noncausal identity of reference. However, actually (5) and (6) are more weakly equivalent by virtue of the causal law (7). This weak equivalence does not follow from the references of the predicates in $\mathscr{L}$, but rather from the causal law (7). (7) might have been false, in which case (5) and (6) could have been independent. This kind of reasoning, plus examples such as (2), (3) and (4), should help to defend C_1. Thus, if L_1 and L_2 are two law-sentences which can be proved equivalent *only* under the substitution of nomologically co-extensional predicates, then L_1 and L_2 represent different laws.

III. THEORIES AND EXPLANATIONS

Suppose that within $\mathscr{L}$ we have a theory $\mathbf{T} = \mathbf{F} \cup \mathbf{I} \cup \mathbf{D}$, where $\mathbf{F}$ is the set of fundamental law-sentences of $\mathbf{T}$, $\mathbf{I}$ is a set of true identities, and $\mathbf{D}$ is the set of derived law-sentences of $\mathbf{T}$. Explanations in $\mathbf{T}$ are represented by suitable derivations from sets of fundamental law-sentences and identities, and the explanans of any explanatory derivation must nontrivially contain at least one law-sentence that is not n-equivalent to the explanandum law-sentence. Each law-sentence in $\mathbf{D}$ is explainable in this manner from subsets of $\mathbf{F} \cup \mathbf{I}$. No law-sentence in $\mathbf{F}$ is so explainable within $\mathbf{T}$, although it is possible that the law-sentences in $\mathbf{F}$ might be explainable from the fundamental law-sentences of some other theory.

Suppose that **E** is the explanans of an explanatory derivation of derived law L_1 and suppose L_1 *n*-eq L_2. Then there is a set **N** of non-causal sentences from which $L_1 \leftrightarrow L_2$ is derivable. If $\mathbf{N} \subseteq \mathbf{T}$, then L_2 is derivable in **T** from $\mathbf{E} \cup \mathbf{N}$. In general, this new derivation should be considered to represent a different explanation because it will usually involve extra premises which are synthetic identities. However, it is only different in a weak sense because both derivations use the same set of causal premises, i.e., the same set of law-sentences.[5]

If two explanatory derivations represent the same explanation they are said to be *explanatorially-equivalent*, *e-equivalent*, or *e-eq*. It is difficult to give a complete criterion for e-equivalence: however, the following sufficient condition seems plausible [3]:

C_2 Let α, β be predicates, and let $\mathbf{D}_1$, $\mathbf{D}_2$ be derivations such that $\mathbf{D}_2$ is obtainable from $\mathbf{D}_1$ by uniform substitution of β for α, and such that $\mathbf{D}_1$ is obtainable from $\mathbf{D}_2$ by uniform substitution of α for β. Then, if α and β denote the same kind or attribute, then $\mathbf{D}_1$ e-eq $\mathbf{D}_2$.

Law-sentences and explanatory derivations are linguistic representations of certain types of causal relationships. C_1 and C_2 imply that suitable substitutions of identicals leave invariant the causal relationships that are represented. This is natural since identities are noncausal sentences. Of course, substitutions of synthetically identical predicates need not preserve analyticity.

In practice it could be difficult to decide whether two co-extensional attributes are identical or only nomologically co-extensional. If the co-extensionality is explainable as a derived law of **T**, then they, are not identical. If the co-extensionality is not explainable in **T**, then we might not be able to decide with confidence whether we have an identity or a fundamental law of **T**, unless we can later explain it by means of some other theory. On the other hand, if we have no reason to suppose that we do not have an identity, then we might hypothesize, on grounds of ontological, nomological, and explanatory simplicity, that we do have an identity. An extreme example will illustrate this point

Suppose that **T** is a theory of human behavior which contains three principal kinds of predicates: behavioral predicates, psychological predicates, and neurophysiological predicates. Also assume the following:

(i) the set of derived law-sentences of **T** is the set of purely behavioral law-sentences, and each behavioral law-sentence is subject to both a psychological and a neurophysiological explanation, (ii) there is a one-to-one correspondence between the psychological and the neurophysiological predicates under which corresponding predicates are co--extensional, (iii) this correspondence induces, by uniform replacement of corresponding predicates, a one-to-one correspondence between the psychological explanatory derivations of behavioral law-sentences and the neurophysiological explanatory derivations of behavioral law-sentences. In such a situation one could be a dualist and insist that the psychological attributes are distinct from the neurophysiological attributes. But this dualism yields a peculiar, nonunified theory with an extreme kind of causal overdetermination, for each behavioral law would have two different explanations corresponding to two kinds of parallel causes. On the other hand, by identifying the neurophysiological attributes with the psychological attributes, the ontology of the theory is simplified, the number of laws and explanations is decreased, and the theory becomes more unified.

This is clearly an oversimplified example to illustrate a possible situation in which identification has obvious advantages. I do not wish to suggest that nonidentical attributes can be identified by fiat merely because it seems more convenient to do so. The following discussion of reduction will show that this would be a serious mistake.

IV. REDUCTIONS

So far I have discussed single theories, and it is not likely that a single theory will contain many identities. Now let us consider two theories, $\mathbf{T}_1$, $\mathbf{T}_2$, in two languages, $\mathscr{L}_1 = \mathscr{T}_1 \cup \mathscr{A}_1$, $\mathscr{L}_2 = \mathscr{T}_2 \cup \mathscr{A}_2$. What is required for an adequate reduction of $\mathbf{T}_2$ to $\mathbf{T}_1$?

It is generally agreed that the laws of $\mathbf{T}_2$ must somehow be explained by the laws of $\mathbf{T}_1$, and that this will require adjoining a set **P** of connecting principles to $\mathbf{T}_1$ and then deriving the laws of $\mathbf{T}_2$ from $\mathbf{T}_1 \cup \mathbf{P}$. Also, it is frequently maintained that nontrivial connecting principles must be empirical laws, and in particular, that they must be nomological co-extensionalities between the predicates in $\mathscr{L}_2$ and predicates in, or defined in, $\mathscr{L}_1$.[6] I have recently argued that nomological co-

-extensionalities are inadequate connecting principles, and that instead **P** should consist only of thing-identities and attribute-identities [3]. With the help of the results above, I will very briefly summarize the reasons.

First of all, suppose that $\mathbf{T}_2$ is derived from $\mathbf{T}_1 \cup \mathbf{P}$, where **P** is a set of nomological co-extensionalities. Then we have added new laws to $\mathbf{T}_1$ and have not really explained the laws of $\mathbf{T}_2$ by the laws of $\mathbf{T}_1$, but rather by the new theory $\mathbf{T}_1 \cup \mathbf{P}$. Thus, the principal aim of reduction, to understand $\mathbf{T}_2$ in terms of $\mathbf{T}_1$, has not been achieved. Moreover, we have added to $\mathbf{T}_1$ a set **P** of mysterious new laws which need explanation. This is easily illustrated by the following example.

At one time chemists did not understand why some substances are optically active. They eventually discovered that a substance is optically active if and only if it has a dissymmetric molecular structure. This is not an attribute-identity, but rather a mysterious nomological co-extensionality which needs explanation. It would obviously not be a satisfactory connecting principle in the reduction of macroscopic properties to molecular properties. In fact, theories have been developed to explain it.[7] Thus, an alleged reduction to $\mathbf{T}_1$ by means of added nomological co-extensionalities between attributes is not a reduction to $\mathbf{T}_1$ at all, but rather a reduction to a new, somewhat nonunified theory $\mathbf{T}_1 \cup \mathbf{P}$.

Instead of this unsatisfactory result, I have proposed that **P** should contain only identities. This requirement prohibits using nomological co-extensionalities, such as that between optical activity and dissymmetric molecular structure, as connecting principles. Also, if **P** contains only identities, then the only laws used to explain the laws of $\mathbf{T}_2$ are the laws of $\mathbf{T}_1$, because identities are not laws. Furthermore, by means of the identities in **P**, the elements in the domain of $\mathbf{T}_2$ and the attributes of these elements are seen to be identical with elements, and attributes of elements, in the domain of $\mathbf{T}_1$. We therefore obtain a reduction of the ontology of $\mathbf{T}_2$ to that of $\mathbf{T}_1$. Finally, by means of C_1, we see that the law-sentences of $\mathbf{T}_2$ are *n*-equivalent to law-sentences of $\mathbf{T}_1$. Put very simply, we obtain a complete identification of the laws and ontology of $\mathbf{T}_2$ with at least some of the laws and ontology of $\mathbf{T}_1$. This may appear to be an unnecessarily strong condition for reduction, but it follows naturally from two simple observations: (i) a reduction to $\mathbf{T}_1$ should use no laws other than those in $\mathbf{T}_1$, and (ii) a nomological co-extensionality is a law and needs an explanation, whereas an identity is not a law and is noncausal. I have argued elsewhere that reduction by identities accords with actual scientific reductions. For example, in the reduction of the gas laws, pressure is identified with the average change of momentum of gas molecules per unit area and time, and temperature is identified with the mean translational kinetic energy of a molecule.[8]

The reduction condition which I have just described has another application. $\mathbf{T}_1$ and $\mathbf{T}_2$ can be said to be *equivalent theories* if and only if there is a set **P** of identities such that $\mathbf{T}_1$ and $\mathbf{T}_2$ are reducible to one another using **P** as the set of connecting principles. I also believe that the various concepts developed in this article might be useful in explicating the notion of a unified theory.

CONCLUDING REMARKS

It might be suggested that we could continue to use a set-theoretical language with the addition of a modal operator which would distinguish causal from noncausal sentences. This is a possibility which could be investigated, and its development might produce some interesting technical results. Nevertheless, as chemistry illustrates very explicitly, science often uses the semantics of kinds and attributes, and this semantics is different from that of set theory. Furthermore, since identities are noncausal sentences, it is perfectly natural to interpret a noncausal co-extensionality between two attribute-predicates as an attribute-identity. In addition, attribute-identities seem to be useful in the development of criteria for the identity of events [5].

There are, of course, many questions about attributes. Traditionally, it has been thought that an attribute-identity must be an analytic truth ([9], p. 157), and recently Kripke [6] has argued that all true identities between so-called rigid designators are necessary truths. Unfortunately, these conceptions do not allow synthetic, noncausal, attribute-identities, which at the very least are needed in scientific reductions. I should like to see a formal language which does allow such identities, or at least some substitute for them which would satisfy their scientific functions equally well.

BIBLIOGRAPHY

[1] Achinstein, P., *Law and Explanation*, Oxford University Press, Oxford, 1971.
[2] Bergmann, G., *Philosophy of Science*, University of Wisconsin Press, Madison, 1966.
[3] Causey, R. L., 'Attribute-Identities in Microreductions', *The Journal of Philosophy* **69** (1972), 407–422.
[4] Causey, R. L., 'Uniform Microreductions', *Synthese* **25** (1972), 176–218.
[5] Kim, J., 'Causation, Nomic Subsumption, and the Concept of Event', *The Journal of Philosophy* **70** (1973), 217–236.
[6] Kripke, S., 'Identity and Necessity', *Identity and Individuation*, (ed. by M. K. Munitz), New York University Press, New York, 1971, pp. 135–164.
[7] Nagel, E., *The Structure of Science*, Harcourt, Brace & World, New York, 1961.
[8] Nickles, T., 'Covering Law Explanation', *Philosophy of Science*, **38** (1971), 542–561.
[9] Quine, W. V. O., *From a Logical Point of View*, Harper & Row, New York, 1961.
[10] Sneed, J. D., *The Logical Structure of Mathematical Physics*, D. Reidel, Dordrecht, 1971.

REFERENCES

* This research was partially supported by U.S. National Science Foundation grant GS-39664.

[1] For a discussion of both formal and informal axiomatic systems, as well as the languages used in them, see the first chapter of [10]. In the present article, all terms will be interpreted 'realistically', i.e., I will ignore the so-called observational/theoretical distinction.

[2] Chapter 4 of [7] has a useful discussion of these arguments.

[3] Achinstein writes, "We might say, then, that propositions which express laws are equivalent formulations of the same law if they are either logically equivalent or empirically equivalent. They are logically equivalent if they entail each other. They are empirically equivalent if each can be derived from the other when additional true empirical assumptions are made that are not logically entailed by, and do not entail, either proposition" ([1], p. 16). However, on the same page, he writes further, "There are also formulations of a given law that are neither logically nor empirically equivalent. I shall call these non-equivalent formulations." He considers the concept of law, as used by practicing scientists, to be rather loose (p. 1).

[4] Other examples can be constructed by using other nomological co-extensionalities, e.g., a substance is optically active if and only if it has a dissymmetric molecular structure. Gustav Bergmann calls such nomological co-extensionalities 'cross-sectional laws' and discusses them in [2].

[5] In [8] Nickles discusses the use of what he calls 'identificatory premises' in the explanans of explanatory derivations. But such premises need not be genuine identities; they may be various kinds of biconditionals. This enables Nickles to argue

unjustifiably, I believe, that some explanatory derivations possess what he calls 'strong intensionality'.

[6] For a discussion of this literature, see [4], especially pp. 203–204.

[7] Further discussion of this, and related examples, is in [3], pp. 414–417.

[8] Actual scientific reductions are often microreductions, in which the elements of the domain of $\mathbf{T}_2$ are structured wholes composed of parts which are elements of the domain of $\mathbf{T}_1$. In [4] there is a detailed analysis of microreductions, including discussions of the roles of the part/whole relationships and the thing-identities involved in such reductions.

WOLFRAM HEITSCH

ON LOGICAL ANALYSIS OF METHODS

0.

The present paper is intended to briefly outline an approach to the analysis of methods based on some concepts and procedures which have been discussed in a more elaborate study on the logic of imperatives and norms. For certain details of the formal apparatus (e.g. definitions, theorems, proofs) I refer to my research report, which is available in a German version (Heitsch: 1973).

I. ON THE NOTION OF METHOD FOR THE PURPOSE OF LOGICAL ANALYSIS

In treatises on general methodology and philosophy of science it is occasionally pointed out that up to now there has been no generally accepted definition of the notion 'method' despite its fundamental role in the methodology of sciences. Thus, in the beginning, I would like to express my view on the topic which I will call 'the normative conception of method'.

Scientific research is essentially characterized by three interpenetrating processes—let us call them *finding*, *formulating*, and *solving* problems. This property, together with the fact that problems express an aim—condition—relation in a specific form, provides the basis for methodic procedure in scientific research. Methodic procedure rests on the application of methods. A method, then, may be conceived of as a set of methodic rules which are determined in rather complicated a way by the aim—condition—relation that is expressed by the given problem (or type of problem). Methodic rules represent imperatives referring to actions or operations which are relevant to the process of reaching the aim under given conditions. Imperatives are conceived of as 'performative acts', i.e. they represent acts that are typically executed by verbal utterances. Imperatives thus comprise various acts of referring to a certain action as being permitted, obligatory, prohibited or forbidden, or optional. Norms are established by means of imperatives. A normative sentence asserts the validity of a certain norm, that is, a normative sentence asserts that some action is obligatory, prohibited, permitted or optional. While imperatives, due to their nature as being performative acts, cannot be true or false, normative sentences, on the other hand, can of course be either true or false due to their property of being sentences. Norms which are valid only on certain conditions are called 'conditioned norms'. In cases where the existence or non-existence of some state of affairs (German: Sachverhalt, abbreviated forthwith: *SA*) appears as a condition for some norm, we may speak of '*SA*-conditioned norms'. If, on the other hand, the performance or omission of some specified action figures as a condition, we speak of 'action-conditioned ('A-conditioned') norms'. Actions are thereby described by sentences just like states of affairs are. An *SA*-description is true iff the *SA* described exists, an action-description, however, is true iff the action described is performed.

Since according to our normative conception each method may be represented by a set of imperatives, and since furthermore each set of imperatives can be uniquely assigned to the set of norms established by the given set of imperatives and vice versa, any method can be formalized within the framework of a normative calculus (or calculus of norms or deontic calculus). By means of this approach of formalization of methods we are able to provide an adequate definition of such methodological notions as 'consistency of a method' or 'completeness of a method'. Moreover, we may then proceed to face the problem of justifying the constitution of methods. We shall consider now some aspects of the logic of norms.

II. REMARKS ON THE LOGIC OF NORMS

I want to propose a systematic construction of a logic of norms by extending the propositional calculus.

The inventory of symbols of the logic proposed is to include

— variables for states of affairs = *SA*-variables
— variables for actions = *A*-variables
— propositional functors

— normative operators = *N*-operators
— technical auxiliary symbols.

By inductive definition we can select from the whole set of chains of symbols which can be formed over this inventory the expressions of the logic of norms. Various types of expressions can be characterized. An expression is called an '*SA*-expression' if it does not contain any action-variable or normative operator. An expression is called '*A*-expression' (for: action-expression) if it does not contain any *SA*-variable or normative operator. If in some expression *A* there occurs a normative operator *O*, then we will define that part of *A* which enclosed in brackets follows *O* as the 'scope of the normative operator *O*'. An expression *A* is called a 'normative expression' (*N*-expression) if *A* does not contain any *SA*-variable and if each action-variable occurring in *A* is at all occurrences within the scope of an *N*-operator. An implication, the antecedent of which is formed by an *SA*-expression and the consequent of which is formed by an *N*-expression, is called an '*SA*-conditioned *N*-expression'. If *A* is an *A*-expression, then G(*A*) is an order, that is, tne action described by *A* is obligatory, V(*A*) is a prohibition, that is, the action described by *A* is prohibited, E(*A*) is a permission, and F(*A*) is an expression, qualifying the action described by *A* as optional.

The interpretation of our logic of norms (forthwith: *NL*) is based on the set of truth values $\{t, f\}$. An interpretation *I* of *NL* is an ordered triple $\langle B_{SA}, B_A, M_b \rangle$, where B_{SA} denotes an assignment of truth values to *SA*-variables, B_A denotes an assignment of truth values to *A*-variables, and M_b, finally, is a non-empty set of assignments of truth values to *A*-variables. The notion 'value of the expression *A* at the interpretation *I*' is inductively defined.

This definition of value is justified as follows: An *NL*-interpretation *I*, where $I = \langle B_{SA}, B_A, M_b \rangle$, characteiizes some complex situation. While B_{SA} determines a real state of affairs and B_A determines a state of action, the set of assignments denoted by M_b indirectly expresses an aim to be achieved, insofar, as the set M_b lumps together just those assignments B_A' that determine those states of action which are positively relevant for achieving the aim set. On the basis of this notion of *NL*-interpretation we may derive from the definition of value given above, that in a complex situation an action is obligatory exactly iff it is performed on each state of action which is positively relevant for achieving the aim set.

Semantic notions of our logic of norms can now be correlated with the notion of value. It is thus possible to define the following notions of a logic of norms: 'law in a logic of norms' (*LAW*), 'contradiction in a logic of norms' (*CONTR*), 'equivalence in a logic of norms' (*EQUI*), 'neutrality in a logic of norms' (*NEUTR*), and 'consequence in a logic of norms' (*CONS*). Fundamental theorems of the logic of norms that can be proved with the aid of the definition given above, can be found in my research report. Besides, there is a proposal for a decision procedure to account for laws in a logic of norms.

III. FORMALIZATION OF METHODS IN A CALCULUS OF NORMS (*NL*-CALCULUS)

We are considering here only such methods that are applicable in the course of transforming some system *S* from an initial state into a final state. The proposal is based on the following assumptions:

1. Suppose that there are *n* known *SA*'s relevant for the system *S* and suppose that the descriptions of these *SA*'s are truthfunctionally independent of each other. This leads us to the consequence that the description of each single *SA* relevant for the system *S* can be accounted for by truth-functional connections of descriptions of the *n* known *SA*'s. One should notice that some *SA* is relevant for the system *S* just in case when it *can*, but *need* not, exist in any possible state of the system *S*. The set of the *n* known *SA*'s relevant for the system *S* is denoted by M_S^n.

2. Suppose there are *m* known actions relevant for the system *S* and suppose that their descriptions are truthfunctionally independent of each other. This leads us to the consequence that the description of each single action relevant for the system *S* can be accounted for by truthfunctional connections of the descriptions of the *m* known actions relevant for the system *S*. One should notice that some action *A* is relevant for the system *S* just in case when it *can*, but *need* not, be performed in any possible state of the system *S* and if *A* transforms the system *S* from one state in to another. The set of the *m* known actions relevant for the system *S* is denoted by M_A^m.

If the descriptions of the n known SA's relevant for the system S are represented by the SA-variables $s_1, \ldots, s_n$, and if the descriptions of the m known actions relevant for the system S are represented by the A-variables $a_1, \ldots, a_m$ in the language of the NL-calculus, then a method related to M_S^n and M_A^m is formalized in the language of NL-calculus by means of a set of SA^n-conditioned N^m-expressions. An SA^n-conditioned N^m-expression is an implication, the antecedent of which is formed by an SA^n-expression and the consequent of which is formed by an N^m-expression. An SA^n-expression is an SA-expression which contains as variables at most those n SA-variables $s_1, \ldots, s_n$. An A^m-expression is an A-expression, and an N^m-expression is an N-expression, which as variables contain at most those m A-variables $a_1, \ldots, a_m$. Among SA^n-expressions there is a group of designated expressions formed by 2^n elementary conjunctions in $s_1, \ldots, s_n$, symbolized as $\mathrm{EK}_i(s_1, \ldots, s_n)$, where $1 \leqslant i \leqslant 2^n$, which represent the descriptions of those elementary cases that are relevant for the system S and that are related to M_S^n. Since for each state of the system S there is exactly one elementary case that is relevant for the system S and related to M_S^n, we are able to give a complete and disjoint classification of all states of the system S.

Among the A^m-expressions there is a group of designated expressions formed by 2^m elementary conjunctions in $a_1, \ldots, a_m$, symbolized as $\mathrm{EK}_i\ (a_i, \ldots, a_m)$, where $1 \leqslant i \leqslant 2^m$, which represent the descriptions of those action-states that are relevant for the system S and that are related to M_A^m. Finally, among the N^m-expressions there is a designated group of $2^{(2^m)}$ standard elementary conjunctions in $a_1, \ldots, a_m$, symbolized as $\mathrm{SK}_i(a_1, \ldots, a_m)$, where $1 \leqslant i \leqslant 2^{(2^m)}$, which express a normative evaluation of all action-states that are relevant for the system S and that are related to M_A^m. It should be noticed that a standard conjunction in $a_1, \ldots, a_m$ is an elementary conjunction of elements from the 2^m standard expressions for permissions $\mathrm{E}(\mathrm{EK}_1(a_1, \ldots, a_m)), \ldots, \mathrm{E}(\mathrm{EK}_{2^m}(a_1, \ldots, a_m))$, and that the $2^{(2^m)}$th standard conjunction in $a_1, \ldots, a_m$, where each of the 2^m standard expressions for permissions occurs as negated member of the conjunction, is a contradiction in the NL-calculus.

We may proceed to define the notions of consistency and completeness of some method M related to M_S^n and M_A^m:

(1) A method M related to M_S^n and M_A^m is *consistent* if and only if

for each elementary conjunction $EK_i(s_1, \ldots, s_n)$, where $1 \leqslant i \leqslant 2^n$, there is at most one standard conjunction $\mathrm{SK}_j(a_1, \ldots, a_m)$, where $1 \leqslant j \leqslant 2^{(2^m)}-1$, such that

$$M\ CONS(\mathrm{EK}_i(s_1, \ldots, s_n) \rightarrow \mathrm{SK}_j(a_1, \ldots, a_m))$$

(2) A method M related to M_S^n and M_A^m is *complete* if and only if for each elementary conjunction $\mathrm{EK}_i(s_1, \ldots, s_n)$, where $1 \leqslant i \leqslant 2^n$, there is at least one standard conjunction $\mathrm{SK}_j(a_1, \ldots, a_m)$, where $1 \leqslant j \leqslant 2^{(2^m)}-1$, such that

$$M\ CONS(\mathrm{EK}_i(s_1, \ldots, s_n) \rightarrow \mathrm{SK}_j(a_1, \ldots, a_m))$$

A decision procedure for the properties of consistency and completeness of a method that is related to M_S^n and M_A^m is given in my research report. Besides, the following theorems can be proved:

(1) If some method M that is related to M_S^n and M_A^m is consistent, then in *NL* there is no expression A for which the following holds: *CONTR A* and *M CONS A*

(2) If some method M that is related to M_S^n and M_A^m is not consistent, then in *NL* there is an *SA*-expression A for which the following holds: *NEUTR A* and *M CONS A*.

(3) Some method M that is related to M_S^n and M_A^m is consistent and complete if and only if for each elementary conjunction $\mathrm{EK}_i(s_1, \ldots$ $\ldots, s_n)$, where $1 \leqslant \mathrm{i} \leqslant 2^n$, and for each A^m-expression A^m it holds:

Either $M\ CONS\ (\mathrm{EK}_i(s_1, \ldots, s_n) \rightarrow \mathrm{G}(A^m))$
or $M\ CONS\ (\mathrm{EK}_i(s_1, \ldots, s_n) \rightarrow \mathrm{V}(A^m))$
or $M\ CONS\ (\mathrm{EK}_i(s_1, \ldots, s_n) \rightarrow \mathrm{F}(A^m))$

Due to the last theorem we may say that by means of a consistent and complete method that is related to M_S^n and M_A^m for each elementary case that is relevant for the system S and that is related to M_S^n there is a normative evaluation of all actions that are relevant for the system S and that are related to M_A^m. Since for each state of the system S there is exactly one such elementary case that is relevant for the system S and that is related to M_S^n, we may say that by means of a consistent and complete method that is related to M_S^n and M_A^m for each state of the system S all actions that are relevant for the system S and related to M_A^m are completely regulated.

Now I will take up the problem of justifying the constitution of a consistent and complete method related to M_S^n and M_A^m. We start from the following assumptions:

1. We take for granted that the set of all possible states of the system S is known and that it coincides with the set of all elementary cases that are relevant for the system S and related to M_S^n.

2. A notion of distance is defined over the set of all possible states of the system S. This notion enables us to calculate the distance between any two states of the system S.

A consistent and complete method related to M_S^n and M_A^m, that can be applied in the course of transforming the system S from any arbitrary initial state into some specified final state, can under the assumptions made above be constituted as follows: If some state Z of the system S is given then just those states of action (relevant for S and related to M_A^m) are permitted—provided there is at least one such state of action—which lead to states of the system S for which it holds that their distance to the final state is smaller than the distance of Z to the final state. If there is no such state of action then there is a permission for that state of action which consists in the omission of the actions $a_1, \ldots, a_m$. The remaining states of action relevant for the system S and related to M_A^m are prohibited.

There are, of course, still other possibilities to constitute a consistent and complete method in relation to M_S^n and M_A^m based on the assumptions specified above.

So far we have outlined our proposals concerning the definition of the notions of consistency and completeness with regard to a certain class of methods, and the problem of justified constitution of methods. We may now look for appropriate means to achieve some generalizations of these results.

BIBLIOGRAPHY

Heitsch, W., (1973): *Gedankliche Systeme mit Aufforderungscharakter*, *Forschungsbericht*, Teil 1. Berlin: ITW der AdW der DDR, 1973.

KAREL BERKA

AXIOMATIZATION IN EXPECTED UTILITY THEORY

The paper is concerned with a critical evaluation of formal, semantical, methodological, empirical and philosophical aspect of the axiomatized expected utility theory, developed by J.v. Neumann – O. Morgenstern (6), F. P. Ramsey (7) and their followers with the aim to justify plausibility of Bernoulli's hypothesis according to which a rational individual has to behave in such a manner as to maximize his expected utility. Our analysis will exhibit some problematic results of this mediated approach, implying serious doubts about fruitfulness of this type of axiomatization, characterized rather as a clarification of a concept, the definition of which is difficult to present, than as a unification of a set of statements used so far intuitively and their ordering into a consistent and complete system containing, besides axioms and primitive and defined terms, an extensive subset of theorems as well.

Since our discussion does not only include purely formal aspects of the axiomatized expected utility theory, interpreted either as a theory of normative or descriptive behaviour in risky or uncertain decision situations, or as a theory of economic behaviour of individuals, e.g. consumers, or as a psychological theory of preference ordering and its numerical evaluation in general, we shall not take into account its 'technical' model in the framework of the statistical decision-making theory. The critical part of our discussion will be followed by a proposal of a simplified conception of the theory in question, representing at least its equivalent variant and enabling its immediate verification and practical utilization.

The axiomatized expected utility theory, constructed either in the Neumann-Morgensternian manner or in the Ramseyian one, is, in general, based on the following assumptions.

(1) Under the condition that one appropriately selects a set of axioms expressing some relevant properties of utility, which are intuitively acceptable, more basic than the defining properties of an explicit de-

finition of the concept 'expected utility' and even more easily verifiable than the properties usually ascribed to this concept in its 'pre-axiomatic' stage, it is possible to deduce in a strictly formal way Bernoulli's hypothesis as a theorem (or theorems). The suggested axioms are conceived as necessary and sufficient conditions ensuring the existence of a numerical utility function with the two required properties, namely the 'ordinal' property

(T1) $x \succcurlyeq y$ *implies* $u(x) \geqslant u(y)$,

and the 'cardinal' ('measurable', 'expected') property

(T2) $u(px+(1-p)y) = pu(x)+(1-p)u(y)$,

where x, y represent abstract utilities, $u(x)$, $u(y)$ numerical utilities (numerical values of utility functions), p and $1\text{-}p$ probability values, '$\succcurlyeq$' the binary relation of 'preference/indifference' and '$\geqslant$' the binary relation of 'greater than/equal'. The utility functions is considered to be unique up to any positive linear transformation.

(2) The axioms are at least in principle, empirically significant and verifiable in respect to the actual behaviour of rational individuals.

(3) The axioms, concerned with nonnumerical objects—abstract utilities, desirabilities, prospects, alternatives in decision situations etc.—and expressed by means of quantitatively given objective probabilities (in systems of the Neumann-Morgensternian type) or qualitatively determined subjective probabilities (in systems of the Ramseyian type) enable us to deduce from them statements about numerical utility functions.

(4) The values of the numerical utility functions are quantitatively measurable at least on an interval scale.

The first—*the formal*—assumption implies the logical existence of the numerical utility function: If the axiom set is consistent and complete and the formulas (T1) and (T2) are really theorems of the axiom system in question, than the assertion about the existence of a numerical utility function is logically consistent. The second assumption determining the *empirical* basis of the theory postulates the factual existence of such function: if the axioms are satisfied by the actual behaviour of a given individual, there exists for him a numerical utility function fulfilling the required properties and the individual behaves in risky or uncertain decision situations according to Bernoulli's hypothesis. The third assumption, based theoretically on the correspondence between empirical and numerical relational systems, characterizes its *methodological* background: the axiomatization has as its goal to deduce from nonnumerical axioms theorems concerned with numerical utility functions in such a manner that the left-hand sides of the formulas (T1) and (T2) represent empirical counterparts of the numerical concepts on the right-hand sides. The fourth assumption, connected with the *operational* fulfilment of the expected utility theory implies that we can exhibit in a significant manner the necessary conditions of measurability of intensive magnitudes on interval scales—an arbitrary origin, an arbitrary, yet objectively reproducible unit, and the invariance of the scale form in respect to any linear transformation.

The interpretation of (T1)—namely 'If an individual prefers or is indifferent to the alternative (prospect, utility) before y, than he assigns to the utility function $u(x)$ a numerical value which is greater than (or equal to) the numerical value assigned to the utility function $u(y)$'—does not make any serious troubles in respect to the methodological assumption, determining the basic aim of the axiomatization in this domain. But the interpretation of (T2)—namely 'The expected utility (the expected utility function) of the probability combination of two alternatives (prospects, utilities) x and y—of the mixed alternative $px+(1-p)y$—is identical with (equal to) the expected utility of their mean value'—does not satisfy this postulate. It is not based on a satisfactory empirical explication of the combination operation. Various interpretations, suggested so far as a way out of this situation, cope neither with the assumed 'natural', empirical nature of this operation, nor with its relationship to the operation of numerical addition, expressed by the double occurence of the sign '+' in the formula (T2). Neither the original interpretation, attempted by J.V. Neumann and O. Morgenstern by introducing to the operation of numerical addition as its empirical counterpart an operation of forming the 'center of gravity' of x and y with two alternative weights,[1] nor the purely formal transcriptions introduced by their followers for the left-hand side of (T2), e.g. $a\alpha b$, $(P, Q; p, 1-p)$, $\langle px, (1-p)y\rangle$, $h(x, \alpha, y)$, confining the occurence of the sign '+' only to the right-hand side, the 'numerical' one, can be considered as really satisfactory solutions of this fundamental problem.

The fact, as it seems to us, that the formula (T2) does not satisfy the assumed correspondence between empirical and numerical concepts, being the necessary foothold for the empirical nature of the axiomatized expected utility theory, is supported by other syntactical features of this conception and their semantical interpretations as well. The occurence of the sign '=' as its main functor and the occurence of the expression *u*( ) on its both sides, are further arguments in favor of our conception. Since the sign '=' is in the original version of the axiomatized expected utility theory interpreted as 'true identity' ([6], 617), both sides must have the same denotatum: either in respect to meaning or in regard to numerical value. If we interpret this sign as expressing 'equality' of both sides only, they must still have at least the same meaning, if not the same sense. How can we then explain an identity (or equality) of two heterogeneous expressions—a nonnumerical one and a numerical one? Our doubts about the empirical, nonnumerical nature of the left-hand side is further strengthened by the occurence of the expression *u*( ), which is in all interpretations always understood as a number, as a numerical utility function.

It seems to us that the just mentioned arguments sufficiently prove that both sides of the formula (T2)—in contradistinction to the formula (T1)—are numerical expressions. The formula (T2) is either an equation determining the equality of two numerical expressions or a definition of the expected utility concept, defining the numerical value of a mixed prospect (a probabilistic combination of risky or uncertain alternatives, etc.) as being equal to the weighted mean value of its components. The value in question can be computed in exactly the same way as the mathematical expectation of a random magnitude. This result is obviously in agreement with the original characterization of the expected utility concept: 'We have practically defined numerical utility as being the thing for which the calculus of mathematical expectation is legitimate' ([6], 28).

Another distinction between (T1) and (T2) appears in respect to probability: the ordinal property of the expected utility concept is formulated in an unprobabilistic context, whereas the intrinsic feature of the second property consists just in its probabilistic nature. Even this methodologically and philosophically important discrepancy is 'solved' in a purely formal manner. A simple (certain) alternative is expressed as a mixed combination of two alternatives with the probability value p equal 1, e.g. in the system of Herstein-Milnor ([5], 291) by the condition

$$1a+(1-1)b = a^2$$

of their axiomatic definition of the mixture set concept. In fact, this purely syntactical convention does not overbridge an objectively given difference between both basic properties of the expected utility concept.

The weakness of the discussed theory seems therefore to lie in an insufficient clarification of its fundamental ideas from a semantical and empirical point of view, connected with an unjustified overemphasis put on its purely syntactical aspects. This fact explains to a certain degree the reasons for the predominant importance given by the followers of the axiomatized expected utility theory to the formal justification of Bernoulli's hypothesis. Are the fundamental theorems of the Neumann-Morgenstern system, namely the existence theorem (A:V) combining both the basic properties (T1) and (T2) and the uniqueness theorem (*A*: *W*) according to which "utility is a number up to a linear transformation" ([6], 628), as well as other intermediate theorems really rigorously derived from the axioms? Such a claim would be true if and only if the axiomatic system in question were deductively closed. This requirement is, however, incompatible with the methodological assumption—with the main goal of the assumed mediated justification of Bernoulli's hypothesis. If the axiomatic system were really deductively closed, the set of its theorems would be—in agreement with the *metalogical closure postulate* ([10], 363n)—contained in the set of its axioms. In this case, we cannot, on the other hand, assert the deducibility of theorems dealing with numerical utility functions from axioms formulated for abstract, nonnumerical utilities. The correctness and appropriateness of the axiomatization in the expected utility theory is therefore for metatheoretical reasons burdened with the following dilemma:

Either: the fundamental theorems are not rigorously deduced from the axioms which implies that their proof is formally incorrect,

or: the numerical utilities are already tacitly assumed in the axioms against the explicitly formulated assumption that they are concerned with nonnumerical utilities, which implies that the whole derivation is burdened with the logical error of begging the question and is therefore metodologically incorrect.

Following the singular steps in deducing the fundamental theorem in any axiomatic system of the expected utility theory both horns of this dilemma become obvious: The derivation is not exclusively based on the axioms, but rather on further numerical conventions implicitly contained in the axioms and on definitions introducing new concepts not implied by the primitive terms.

This principal objection against the axiomatized expected utility theory will be exemplified by discussing the system of Chernoff and Moses ([3], 81n, 350nn).[3] Chernoff and Moses postulate the derivation of both properties of the expected utility function

(P1)′ $u(P_1) > u(P_2)$ if and only if the individual prefers P_1 to P_2,

and

(P2)′ If P is the prospect where, with probability p, the individual faces P_1, and with probability $1-p$ he faces P_2, then $u(P) = pu(P_1)+ +(1-p)u(P_2)$[4]

from the following four axioms:

(Ax.1) of *comparability*:

$$\textit{either } P_1 \succ P_2 \textit{ or } P_1 \sim P_2 \textit{ or } P_2 \succ P_1$$

(Ax.2) of *transitivity*:

$$\textit{If } P_1 \succcurlyeq P_2 \textit{ and } P_2 \succcurlyeq P_3, \textit{ then } P_1 \succcurlyeq P_3$$

(Ax.3) of *continuity*:

$$\textit{If } P_1 \succ P_2 \succ P_3, \textit{ then there exist a } p \textit{ such that } (P_1, P_3; p, 1-p) \succ P_2 \textit{ and a } q \textit{ such that } P_2 \succ (P_1, P_3; q, 1-q)$$

(Ax.4) of *independence*:

$$\textit{If } P_1 \succ P_2, \textit{ then } (P_1, P_3; p, 1-p) \succ (P_2, P_3; p, 1-p).$$

The alleged existence of the expected utility function, which has to be proved by derivation on the basis of these axioms, is actually dependent on the following preaxiomatic conventions and further definitions:

(C1) The basic 'interval' of prospects $[P_1, P_0]$, where $P_1 \succ P_0$, is selected in such a way that it is isomorphic with the numerical interval of probability values p $[1, 0]$; the indices of the prospects corresponding to the probability values denote values of the utility function $u(P)$ as well.

(C2) Any mixed prospect P such that $P_1 \succ P \succ P_0$, can be expressed as a p-probability combination of both extreme prospects

$$P_p = (P_1, P_0; p, 1-p).$$

(D1) For the preference ordering $P_1 \succ P \succ P_0$, where P lies *in* the basic interval $[P_1, P_0]$, $u(P) = p$ holds.

(D2) For the preference ordering $P \succ P_1 \succ P_0$, where P lies *before* the basis interval $[P_1, P_0]$, $u(P) = \frac{1}{p}$ holds.

(D3) For the preference ordering $P_1 \succ P_0 \succ P$, where P lies *behind* the basic interval $[P_1, P_0] u(P) = -\frac{p}{1-p}$ holds.

From these definitions, representing for every mixed prospect P three possible preference orderings, it follows that (D1) is the fundamental one; the other two can be conceived as its transformations.

Using the convention (C1) one can further easily prove that the definition (D1), characterizing the operationalization and calculation of the numerical utility function, is a consequence of (T2)′ and not one of the assumptions by means of which this basic property of the expected utility function has to be proved. Setting $u(P_1) = 1$ and $u(P_2) = 0$ we obtain from the formula (P2)′ the numerical expression

$$\text{(T2)} \qquad u(P) = p \cdot 1 + (1-p) \cdot 0,$$

from which it immediately follows that $u(P) = p$. Similarily, we can show in agreement with the preference ordering $P \succ P_1 \succ P_0$ according to (D2) by transforming the formula

$$u(P_1) = pu(P_1) + (1-p)u(P_0)$$

that in this case $u(P) = \frac{1}{p}$. An analogous result will be obtained by transforming the formula

$$u(P_0) = pu(P_1) + (1-p)u(P)$$

according to (D3) in agreement with the preference ordering $P_1 \succ P_0 \succ P$ that in this case $u(P) = -\frac{p}{1-p}$.

Our remarks concerning the axiomatic approach of Chernoff and Moses, which could be easily confirmed by analysing any other axiomatic system of the expected utility theory, yield the conclusion that the logical existence of a numerical utility function of the discussed type is not sufficiently justified. This result throws doubts on its factual existence as well.

What arguments can be raised against the assumption of its factual existence? First of all, we can point to the limited applicability of the expected utility theory, determined by the requirement of consistency of any individual preference ordering and the need to make clear what is meant by *rational* behaviour of people in risky or uncertain decision-situations. The concept of rational behaviour which has to be defined independently of the requirements of rationality, formulated by the axioms, remains till now an unclear notion. Even if one adopts some intuitive conception of rational behaviour, we can in practice find various cases of behaviour which seem to be rational, nevertheless are inconsistent with the theory in question or are falsifying some of its assumptions. It is well known that prudent people when gambling on very small stakes prefer to gain with certainty smaller amounts of money to the prospect of obtaining with a very high probability only a slightly greater amount. But in other decision situation, e.g. when buying lottery-tickets the very same individuals are likely to prefer a very unprobable prospect of winning a great amount of money. There exist people who take the advantages of insurance since they feel the need of security. Yet at the same time they accept bets, thus prefering risk to certainty. This kind of behaviour contradicts the preceding one where they prefer certainty and do not mind to suffer a minor financial loss by paying the premium. Many people prefer a certain prospect although its expected utility is smaller than the expected utility of an uncertain probability mixture of prospects.

These facts together with the negative results following from the experimental verification of the axiomatized expected utility theory stipulated a modification of the initial conception. The axioms of the expected utility theory are now considered rather as normative principles which are only approximately fulfilled by the actual behaviour of men in decision situations than descriptive laws of human behaviour. The set of axioms is then split into two subsets. The subset of behavioural postulates (axioms of rationality) contains the axiom of ordering, i.e. the axioms of comparability and transitivity in the system of Chernoff and Moses, and the axiom of independence. The subset of purely structural axioms includes then the axiom of continuity. From this point of view it is only required that the axioms of the first subset should have an empirical significant interpretation and it is expected that they will be, at least, approximately satisfied in decision-making.

Both basic functions usually ascribed to the axiom of independence in making consistent preference judgments, namely to discover preferences between more complex alternatives on the basis of preferences between simpler alternatives and to uncover inconsistencies of preference ordering, exemplified only by artificially constructed examples ([4], 109f) which do not characterize the actual behaviour of a men in real decision situations, cannot be considered as satisfactory for elucidation of its empirical significance.

Against the empirical nature of this axiom, formulated in various axiomatic systems in different versions, e.g. as the 'strong independence axiom' by P. A. Samuelson ([8], 672) or as the 'sure-thing principle' by L. J. Savage ([9], 21ff), serious objections were raised in the famous critique by M. Allais, who pointed out that the axiomatic justification of Bernoulli's hypothesis is based on false evidence ([2], 505). His justified objections against this axiom which "is regarded by many as the core of expected, utility theory, for without it the expectation part of expected utility vanishes" ([4], 108), were however based only on fitting counter-examples. For these reasons, we are considering the following proof that this axiom is, in fact, a tautology more convincing.

Let us adopt a modified formulation of axiom 4, written in the initial, more transparent Neumann-Morgenstern's symbolization,

$$\text{(Ax.4)}' \quad \text{if } x \succ y, \text{ then } px+(1-p)z \succ py+(1-p)z.$$

If we now interpret the variables x, y, z and p as propositional variables, say p, q, r, and s, the relation '$\succ$' as the functor of implication, '+' as the functor of inclusive disjunction, '(1−)' as the functor of negation and the concatenation as the functor of conjunction, we can express the logical form of this axiom as follows

$$\text{(Ax.4)*} \quad (p \rightarrow q) \rightarrow [(sp \vee \neg sr) \rightarrow (sq \vee \neg sr)].$$

This formula yields after known transformations

$$(\text{Ax.4})^{**}\quad (p \to q) \to (p \vee r \to q \vee r),$$

which is obviously a tautology. If we choose as the starting-point of our proof another variant, namely

$$(\text{Ax.4})''\quad \text{if } x \sim y, \text{ then } px+(1-p)z \sim py+(1-p)z,$$

it suffices to substitute (according to the antecedent) x for y (or vice versa), thus obtaining in the consequent

$$px+(1-p)z \sim px+(1-p)z.$$

The independence axiom is therefore also a structural axiom without any empirical content. For its tautological nature it cannot be considered as a normative principle of rational behaviour either. There remains as a behavioral postulate only the axiom of ordering. Solely this axiom, in spite of grounded doubts concerning the empirical realization of the transitivity of the preference and indifference relations or the completeness of the preference ordering, reflects actual behaviour of a man in a decision situation. This axiom is not however sufficient for the axiomatic justification of Bernoulli's hypothesis, since only the first property of expected utility is its necessary consequence. For this reason we can actually admit factual existence only for a utility function satisfying the ordinal property.

What empirical justification can be ascribed to its expected property? In discussing this problem, closely connected with the assumption that utility is a measurable magnitude, it is necessary to analyse the way of calculating the numerical values of utility functions determined by its special operationalization, various forms of selecting numerical or rather quasinumerical values assigned to abstract utilities, and the role of probability in this theory.

Under the assumption of some preference ordering, say $P_1 \succ P \succ P_0$, in agreement with definition D, operationalization is based on the indifference between the certain alternative P and the probability mixture $pP_1+(1-p)P_0$. If the alternatives are concerned with money or are in mediated way expressible in money or by other quantitative data, it is very easy to calculate by means of this procedure the probability value p, for which a 'rational' individual will be indifferent. For example, if some individual is indifferent between the gain of 600 \$ (P) with certainty and the prospect to win 1000 \$ (P_1) with probability p or 500 \$ (P_0) with the complementary probability $1-p$, he will assign p the value 0.2. And vice versa, for any objectively given or subjectively estimated probability value p of some mixed prospect $pP_1+(1-p)P_0$ it is possible to calculate the numerical value of the alternative P having for a 'rational' individual the same expected utility as the combined prospect. In our case, e.g. for $p = 0.4$, the individual would have to be indifferent between the above mentioned combination and the prospect to win with certainty 700 \$ (P).

This procedure cannot be applied to purely qualitative alternatives which are mediatedly expressible neither by means of money nor by other quantitative data. Taking into account Bernoulli's distinction between the 'mathematical' and 'moral' expectation of money which is for the lack of an objective comparability of two different preference orderings practically inefficacious, it seems necessary to adopt even in the first type concerned with objectively quantitative alternatives a modified version. The preference ordering $P_1 \succ P \succ P_0$ is then represented in a quasinumerical interval [1, 0] with two extreme values: 1 assigned to the most desired alternative and 0 assigned to the least desired one, whereby both values are interpreted as numerical utilities. In this version we have not to do with the expected amounts of money, but with their expected utility. For the corresponding numerical preference ordering, namely $1 > u(600) > 0$, we obtain, in agreement with the usual interpretation, that $u(600) = p$. Just for this reason the probability value of the mixed alternative is the basic property of the expected utility theory and its application. If we took into account only the equality $u(P) = p$, based on an isomorphism between the numerical interval of probability values [1, 0] and the interval of abstract utilities $[P_1, P_0]$, respectively the quasi-numerical interval of utility functions $[u(P_1), u(P_0)]$, disregarding their semantical differences, we could conclude that the concept 'numerical utility' has exactly the same meaning as the concept 'probability value'.

Such an interpretation ignores the difference between elements of the expected utility function interval, representing desirabilities (abstract utilities) of heterogeneous alternatives, and the elements of the probability interval, representing their probable occurrence. This problem is usually 'solved' by assuming that the concept of expected utility is measurable on an interval scale, what has at the same time to guarantee—in analogy with the measurement of temperature by means of the Celsius or Fahrenheit scale—that the quasi-numerical values of utility functions are at least 'numbers up to linear transformations'. This analogy, as can be easily shown, fails, since for utility, two basic properties characterizing all true instances of intensive magnitudes measurable on an interval scale are not satisfied. So, in contradistinction to temperature, we have not till now found a unit of utility, neither some acceptable pseudo-unit.[5] For this reason we have in the case of 'utility scales' no significant interpretation of the known transformation formula

$$y = ax+b \quad a > 0, \quad b \geqslant 0,$$

according to which it is possible to assign to any numerical scale a value, e.g. on the Fahrenheit scale, a univocal value on the Celsius scale, and vice versa. These temperature scales are mutually linearily transformable, because both constants have a significant interpretation: a represents the unit and b the origin. Such interpretation, in absence of a unit, does not hold for two utility scales.

Our discussion which could be extended by further remarks induces us to suggest a simplified version, based on a direct approach to the expected utility theory, on an immediate empirical interpretation and verification of the required consequences. Instead of attempting to test its axioms we may equally well try to verify its basic theorems. Thus it can be immediately shown whether at all, or in what degree Bernoulli's hypothesis of maximizing expected utility really characterizes the behaviour of rational men in risky or uncertain decision situations. This approach is from the methodological standpoint, as used in deductive sciences, without any doubt correct and logically equivalent with the reverse one, of course, under the assumption that the axiomatic system in question is deductively closed. This proviso does not matter, since all advocates of the axiomatic justification of Bernoulli's hypothesis are convinced about the fulfilment of this condition and must therefore, if being consistent, fully admit—at least from the formal point of view—that both ways must yield the same result. There remains, of course, an objection that from the psychological point of view both procedures are not equally plausible. This objection does not stand the proof because the axioms are definitely not more plausible than both basic theorems. For most people rather the opposite may be true. Both basic properties of the expected utility concept seem to be more obvious than the axioms of continuity or independence.

The straightforward procedure, accompanied by a simple operationalization and calculation, does not lead to illusions as to the value of the axiomatic method in an empirical theory, especially when limited to a very narrow purpose or applied without sufficient semantical foundations, neither to a philosophically doubtful hope that an axiomatization of intensive magnitudes, i.e. utilities, may yield numerical consequences. Nor does it force us to pay our attention mainly to purely formal aspects of the theory in question, but orientates further research rather to various informal problems which are in the case of any empirical theory equally, if not more important.

BIBLIOGRAPHY

[1] Adams, E. W., 'Survey of Bernoullian Utility Theory', in H. Solomon (Ed.), *Mathematical Thinking in the Measurement of Behavior*, Glencoe, Ill. 151–268 (1960).

[2] Allais, M., 'Le comportement de l'homme rationnel devant le risque: critique des postulats et axiomes de l'ecole americaine', *Econometrica*, **21** (1953), 503–546.

[3] Chernoff, H., and Moses, L. E., *Elementary Decision Theory*, New York-London -Sydney (1959).

[4] Fishburn, P. C., *Utility Theory for Decision Making*, New York (1970).

[5] Herstein, I. N., and Milnor, J., 'An axiomatic approach to measurable utility', *Econometrica*, **21**, (1953), 291–297.

[6] Neumann, J. von., and Morgenstern, O., *Theory of Games and Economic Behavior*, Princeton, N. J., 1944, 1947, 1953.

[7] Ramsey, F. P., 'Truth and Probability', in F. P. Ramsey, *The Foundations of Mathematics and Other Logical Essays*, London-New York (1931), 156–168.

[8] Samuelson, P. A., 'Probability, utility and the independence axiom', *Econometrica*, **20** (1952), 670–678.

[9] Savage, L. J., *The Foundations of Statistics*, New York (1954).

[10] Tarski, A., 'Fundamentale Begriffe der Methodologie der deduktiven Wissenschaften, I', *Monath. f. Math. u. Phys.* **37** (1930), 361–404.

REFERENCES

[1] This holds similarly for other purely verbal empirical analogues, e.g. 'the combination of abstract utilities with alternative probabilities', 'a combined option interpreted as a lotery (a gamble) on which one option is obtained with one probability and the other with its complementary probability', 'a interval average', 'a mixture (a mixture set, a mixed prospect) of pure and mixed prospects', 'a direct linear combination of simple probability measures' etc.

[2] In this formula a and b are abstract utilities.

[3] Their system will be slightly modified in accordance with other formally more complicated axiomatizations.

[4] The number $u(P)$ is the expected utility function of the given prospect and denotes the same as $u(P_1, P_2; p, 1\text{-}p)$.

[5] Some authors try to overcome this difficulty by introducing the concept 'utile', but without defining or even explaining its content. Certain others adopt a more neutral standpoint: any number may serve as a *unit* for measuring utilities. In respect to the above mentioned isomorphism they usually select the greater extreme element of the interval [0, 1] ([1], 165), or inconsistently consider as unit ([1], 183) the difference between 1 (assigned to the most desired alternative) and 0 (assigned to the least desired one).

P. ROBINSON

A LOGICAL MODEL FOR GAME-LIKE SITUATIONS AND THE TRANSFORMATION OF GAME-LIKE SITUATIONS

Game theory models which have been in vogue for the last 30 years are essentially static even though they may model situations without equillibria, and may be studied from the points of view of changing strategies, coalitions and of perfect information or the lack of it, and even though any number of players, strategies and outcomes may be modeled by using larger matrices. Game theory models are inadequate for modeling even simple alternative logical relationships among rule applications which are present in simple actual games and they are inadequate for representing any changes in game structure. The addition of logical relationships among rules is an initial step in remedying the first inadequacy and the addition of a temporal order in the application of the logical relationships allows changes in game structure to be represented. The adding of *social rules* or *causal factors* external to the game to account for the changes completes the initial steps for a model of *game-like situations* which are no longer static.

For an example of a game theory matrix of a two-person, non-zero sum game with two strategies:

		Player II	
		A	*B*
Player I	*a*	0,0	10,5
	b	5,10	0,0

Game matrices may also be formulated as a set of re-write rules in which the arrow may be understood as symbolizing a necessary condition. Whether or not it is a necessary condition depends on what factors are taken to be modeled by the rules. I am considering it from the position that the outcomes cannot be gained except by playing by the rules. For example a king's pawn cannot be moved to king's pawn four except by applying a rule of chess. But the rules alone are not sufficient for the outcomes. The player has to act and to make that move. Together the rule and the act according to the rule are sufficient.

$$Aa \rightarrow 0, 0$$
$$Ba \rightarrow 10, 5$$
$$Bb \rightarrow 0, 0$$
$$Ab \rightarrow 5, 10$$

Other logical relations might be considered as holding between the application of strategies and outcomes. For example indetermined outcomes might be modeled as follows:

$$Ba \rightarrow (10, 5) \vee (5, 10)$$

In actual games or game-like situations the rules of the game are logically interrelated in more complex ways than by the simple exclusive or ($\vee$) of the game theoretical matrix as in: $(Aa \rightarrow 0, 0) \vee \vee (Ba \rightarrow 10, 5) \vee (Bb \rightarrow 0, 0) \vee (Ab \rightarrow 5, 10)$. Rules may be required to be applied together (the logical $\wedge$) or be related by the inclusive or ($\vee$). Some rules may negate the application of other rules. Rules may also be required to be applied sequentially. Actual games and game-like situations in ethical, social and ecological relationships have this structure of interrelationships among their rules. The rules themselves are taken to be only some of the necessary conditions for the procedure of the game. Logical interrelationships among the rules are another set of conditions. The modeling of the systematic structural relations among rules which make up an actual game could be done in ordinary logic or it could be done more graphically using components similar to those invented by Sydney Lamb to model linguistic networks. The important difference in my use of Lamb's symbols is that I am modeling a one-way process so that I do not employ the feature of Lamb's components that processes can move through them in two directions.

The following is a table of logical components for modeling either some logical relationships between the initiation of rules or strategies and their outcomes or the logical interrelationships among rules in a game-like system.

①, and ③ are considered as rules to be applied when the components are used to logically order the application of rules in an actual

Table of Components

downward unordered ***AND***

(1) ⟶ (2) ∧ (3)

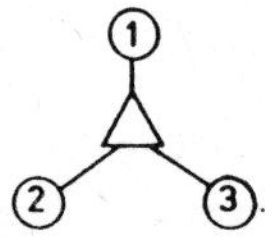

upward ordered ***AND***

((1) at time 1) ∧ ((2) at time 1) ⟶ (3) at time 2

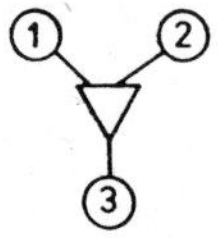

upward unordered ***AND***

(1) ∧ (2) ⟶ (3)

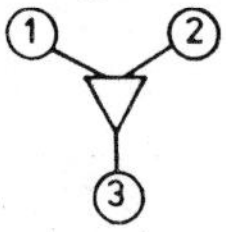

downward ordered ***AND***

(1) at time 1 ⟶ ((2) at time 1) ∧ ((3) at time 2)

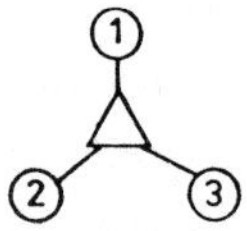

downward unordered ***OR***

(1) ⟶ (2) ∨ (3)

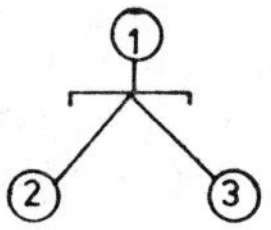

upward ordered ***OR***

((1) at time 1) ∨ ((2) at time 2) ⟶ (3) at time 2

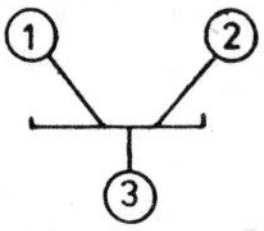

upward unordered ***OR***

(1) ∨ (2) ⟶ (3)

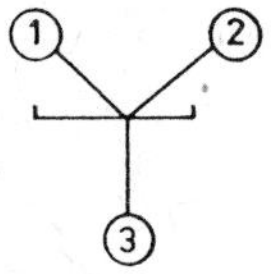

downward ordered ***OR***

((1) at time 1) ⟶ ((2) at time 1) ∨ ((3) at time 2)

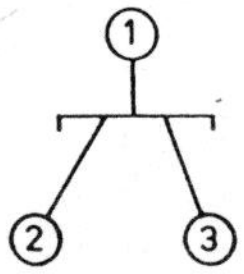

inhibitory ***AND***

(1) ∧ ¬ (2) ⟶ (3) (1) ∧ (2) ⟶ 0

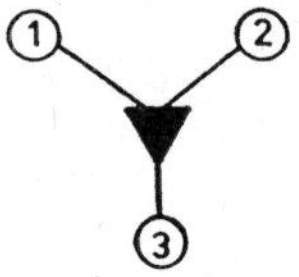

game-like system. (The components could also be used to model the structure of a single rule when the relationship between the strategy and the outcome requires a more complex model. In this case ① or ① ∧ ② or ① ∨ ② would be strategies applied and ③ or ②∧③ or ② ∨ ③ would be outcomes rather than rules).

0 is either that no further rule need be applied or that no outcome is necessitated so that:

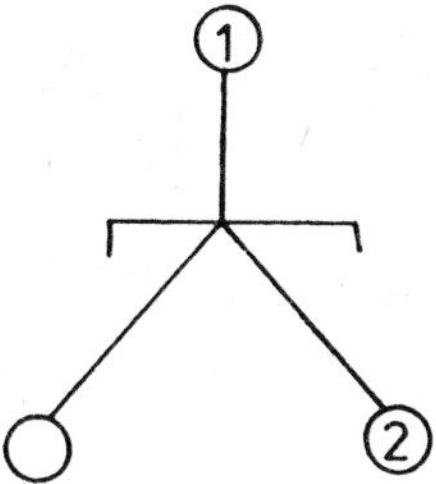

means that either no new rule need be applied or rule ② is applied or else that no new outcome is attained or outcome ② is attained.

The interpretation of the arrow in these definitions is still that of symbolizing strategies as necessary conditions for outcome or applications of rules as necessary conditions for applications of further rules although other stronger logical relations might be considered (sufficient condition or equivalency).

The modeling of a game theoretical matrix using the components is overly simple. For example see following diagram, page 224.
An example of a simple game such as Parcheesi more fully utilizes the logical relationships among game rules.

The matrix of a game theoretical model may model the structure of a game-like social situation. That structure may be imposed upon the behavior of individuals, as in the case of the Prisoner's Dilemma when it is imposed on actual prisoners, or it may be immanent in their behavior. In either case the structure is a causal factor in the action of individuals. The other important causal factor is the acting will of the individual. Both the formal structure of the game and the action of individuals in the game are necessary causes of the specific social situation. Since the matrix describes a necessary condition for a specific

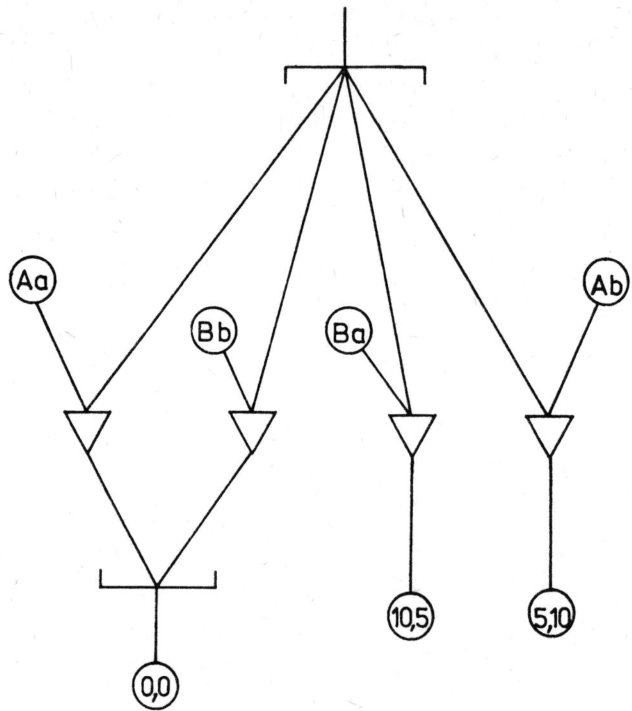

social game, each combination of strategies which results in an outcome is a necessary condition for that outcome in that social situation.

Empirical laws of social situations may be written in terms of descriptions of actual social interactions which fit game theoretical models and which are lawfully related to resultant social interactions when the activity of individuals acting in accordance with the rules modeled is included in the description and a *ceteris paribus* clause is included. However the events described by these laws may only constitute a n ecessary but not sufficient condition for changes in the structure o social interactions.

If a game as a complete system including the actions of playing the game were functional, dysfunctional or developmental, the game as played would be a sufficient condition for its own stability, breakdown or development into a new game. Such game systems may be modeled. However actual game systems are not functional, dysfunctional or developmental in themselves. Other factors such as a player having only a finite amount of money to lose, make a game in which that player must continue to lose eventually to breakdown. Therefore the game is under those external circumstances dysfunctional and leading to its own collapse. Another feature of a game which may make it dysfunctional is that it may be simple enough that players may understand it and therefore refuse to play it. But the understanding is an external factor to the game and this makes it dysfunctional. Game structure alone is neither necessary nor sufficient to produce a new game structure nor is the game as played unless it is part of a larger social or ecological system in which other causal factors necessitate the creation of a new game. Other factors besides merely social ones may result in a new game structure. Game x played under certain conditions of resources, information and cooperation may be sufficient to result in game y. Information about the structure of a game is a necessary but not a sufficient condition for a meta-game. But the conscious activity of individuals can be both necessary and sufficient to devise and play a new game.

Changes in the formal structure of game theoretical models can be the result of changes in the number of players, number of strategies, or in the payoffs. Changes in the number of players are represented by additions or deletions of columna in a game matrix. Changes in the number of strategies are represented by changes in the number of rows in a matrix. A special case of changes in the number of strategies by conditionalizing a player's strategies according to the strategies of another player is a meta-game. In perfect information games changes in number of players would be represented by changes in the number of nodes in a tree and changes in number of strategies by the number of branches in a tree. Both changes in the numbers of strategies and of players would result in changes in the number of rules in a rule representation. Changes in the payoffs are represented by changes in the designation in the payoff column, or in end states of rules.

Transformation of one game into another would include changes in the strategies, number of players and the outcomes and would involve additions, deletions and changes in the rules. Transformations of games would also involve changes in the logical interrelationships among the rules. Additional rules or causal factors must also be included in an adequate model of game changes in order to account for the process of transformation. To model a transformation of one game into another an ordered *AND* is added to the relationships among the rules at points of change and the changes in the rules and their relationships which taken together make up the new game are connected to the model so that after the original sequence of rules is applied the new rules in their interrelationships are applied, thereby representing the shift from one game to another. The ordered *AND* could be specified with a time delay in order to make a significant temporal distinction between the games. This amounts to no more than describing the first game, saying, 'and then' followed by a description of the second game. Causal factors or rules necessitating the change are also connected by *AND* components to the connections between the games so that the original game and the causes of change result in the new game and explain the change. Further mathematical description of the relationships between the games would be necessary for a formal scientific description of the change. Representations of transformations of standard game matrices into new games or representation of them as self-maintaining or functional is overly simple using logical components since the transformation is not gradual through a sequence of changes in parts of the game and there are no complex interrelationships among the rules.

The following are models of a functional game, a developmental game, and a dysfunctional game.

A functional game with 4 strategies:

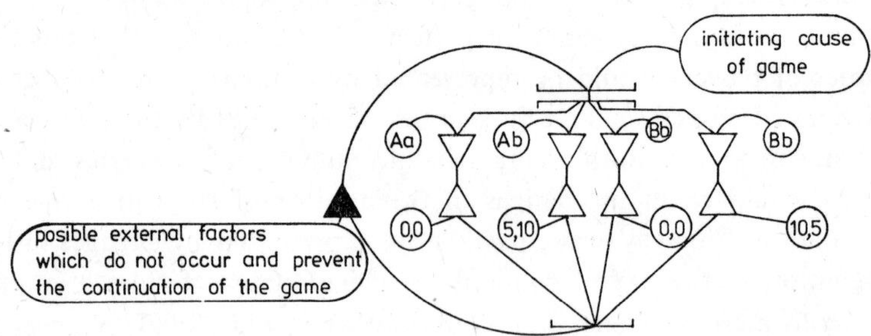

A developmental game with 4 strategies which after *n* plays is transformed into a new game with 8 strategies (either a new player with 2 strategies enters a 2 × 2 game or 4 new conditionalized strategies are employed by one of the two players after n times, transforming it into a metagame:

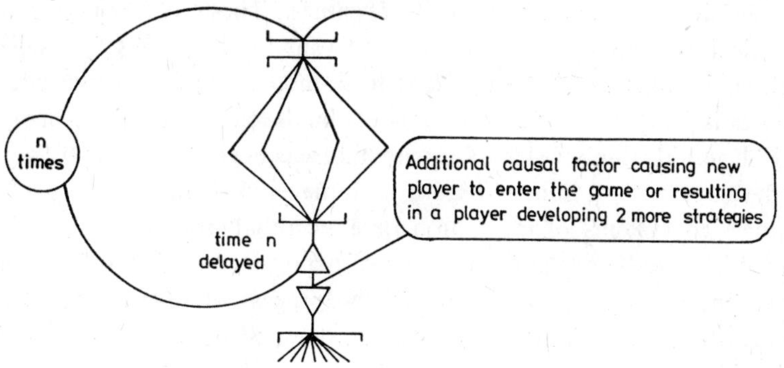

A dysfunctional game:

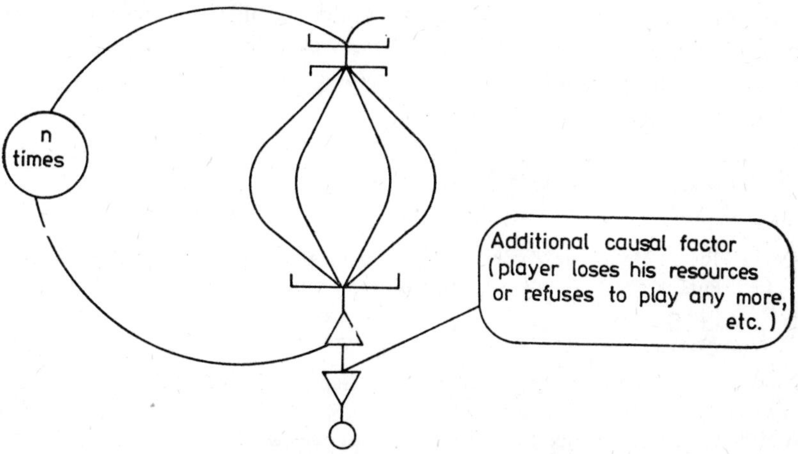

Transformations of games are necessary for adequate models of man's ethical, economic, and ecological interactions. Games in themselves are static and incapable of modeling biological, social or psychological change. Even when more logically complex and sequential relationships are used to model game-like structures, games are essentially ephemeral phenomena because they are limited by external causal, cybernetic or dialectical systems and by an internal structure which when put into action destroys them or transforms then into a new game. The extra-

-game conditions of biological, environmental, technological, social, and ideological change and additional information about the game itself are the important determinants of games. These determinants are historical and involve time delays irrelevant in the games themselves.

Even the placing of game theoretical models within larger causal systems is inadequate if the analysis of the development of the mental, ethical, and ideological side of human nature is neglected or the biological development of human nature and the environment is neglected. Changes in systems of evaluation or conceptual organization in man may ultimately be explained in terms of neurophysiological information processing systems, but a theory of the process of re-evaluation of a game and its pay-offs goes beyond any present scientific model of man. Value judgments are taken as ultimate in game theoretic and systems models. If values change the new system cannot be compared with the old. The only justification for changing human valuations could be made only if all outcomes of all games under the old values were negative because from the point of view of the old values in any new game system with new values all of the outcomes will be indeterminate.

BIBLIOGRAPHY

Game theory:

Davis, Morton D.: *Game Theory: A Nontechnical Introduction*, N.Y.: Basic Books, Inc., Pub. 1970.

Meta-games:

Howard, Nigel: 'The Mathematics of Meta-Games', and 'The Theory of Meta-Games', in *General Systems*, Vol. 11, part. **5**, 1966, pp. 187–200 and 167–185.

Rapaport, Anatol: 'Escape from Paradox', *Scientific American*, July 1967, Vol. **217**, No. 1, pp. 50–56.

Linguistic Networks:

Lockwood, David G.: *Introduction to Stratificational Linguistics*, N.Y., Narcourt Brace Jovanovich. Inc. 1972.

Makkai, Adam and Lockwood, David G.: *Readings in Stratificational Linguistics*, University of Alabama Press, 1973.

P. M. WILLIAMS

INDETERMINATE PROBABILITIES

I. INTRODUCTION

It has been objected against the subjectivistic interpretation of probability that it assumes that a subject's degree of belief $P(E)$ in any event or proposition E is an exact numerical magnitude which might be evaluated to any desired number of decimal places. This is believed to follow from the fact that if a system of degrees of belief expressed by the function P can interpret the exact quantitative probability calculus, it must be the case for every real number p and every event E under consideration that either $P(E) = p$ or $P(E) \neq p$. The same argument, however, would appear to show that no empirical magnitude can satisfy laws expressed in the classical logico-mathematical framework, so long as it is granted that indeterminacy, to a greater or lesser extent, is present in all empirical concepts. It would apply, for instance, to the concept of length. More to the point, it would apply to any empirical interpretation of probability whether subjectivistic or not.

The general problem of indeterminacy of empirical concepts has already received attention in the methodology of empirical sciences.[1] One solution is as follows.[2] Suppose the non-logical constants of a language $\mathscr{L}$ consist of just the predicate expressions $P_1, \ldots, P_n$. A (two--valued) *structure for* $\mathscr{L}$ can be represented as a relational system of the form

$$\mathfrak{M} = \langle U, R_1^{\mathfrak{M}}, \ldots, R_n^{\mathfrak{M}} \rangle$$

where U is a non-empty set and $R_i^{\mathfrak{M}}$ is a k-ary relation on U whenever the corresponding predicate is k-ary. It is supposed to be understood what it means for $\mathfrak{M}$ to be a *model* of a sentence σ of $\mathscr{L}$ (or, informally, for σ to be true in $\mathfrak{M}$.) If $\mathfrak{M}$ is a structure for $\mathscr{L}$, the pair $L = \langle \mathscr{L}, \mathfrak{M} \rangle$ is regarded as an interpreted language and a sentence σ is said to be *true in* L just in case $\mathfrak{M}$ is a model of σ. It follows then that for every sentence σ, either σ is true in L or its negation not-σ is true in L.

The structure $\mathfrak{M}$ assigns a unique denotation to every predicate expression of $\mathscr{L}$. On the other hand, if some of the expressions of $\mathscr{L}$ are thought to be more or less indeterminate under their intended interpretations, it is not unreasonable to regard the interpretation of $\mathscr{L}$ as provided formally by a family $\mathscr{M}$ of structures for $\mathscr{L}$ each of the above form. Such a language is represented by the pair $L = \langle \mathscr{L}, \mathscr{M} \rangle$. Each structure $\mathfrak{M} \in \mathscr{M}$ represents an exact interpretation of the expressions of $\mathscr{L}$ compatible with its semantical rules. Two relations on U can then be associated with each predicate P_i of $\mathscr{L}$, namely

$$R_i^+ = \bigcup \{R_i^{\mathfrak{M}} \colon \mathfrak{M} \in \mathscr{M}\}$$

$$R_i^- = \bigcap \{R_i^{\mathfrak{M}} \colon \mathscr{M} \in \mathscr{M}\}$$

Suppose, for simplicity, that P_i is a one-place predicate. Then it is natural to regard the statement that an individual a of U has the property P_i as true just in case a belongs to R_i^-, and as false just in case a does not belong to R_i^+. If a belongs to R_i^+ but not to R_i^-, the statement in question is to be regarded as lacking either truth-value since, loosely speaking, it is true on one interpretation and false on another. More generally, a sentence σ of $L = \langle \mathscr{L}, \mathscr{M} \rangle$ is said to be *true in L* if and only if every structure $\mathfrak{M}$ of $\mathscr{M}$ is a model of σ. Unless all structures in $\mathscr{M}$ happen to be elementarily equivalent, there exists a sentence σ such that neither it nor its negation is true in L. Notice, however, that every thesis of classical logic is true in L, for instance 'either σ or not-σ' even though, possibly, neither σ nor not-σ is true in L.

The above remarks principally concern the indeterminacy of predicate expressions. It is clear, however, that they may be modified to apply also to function symbols. Rather than treat this question in any generality, we shall turn directly to the special case of the probability function.

It would be a complex task to analyse in any detail the full language of the probability calculus. This is unnecessary, however, as long as attention is restricted to the indeterminacy of just the probability function. The following discussion, therefore, concerns a family of structures assigning a unique denotation to all expressions of the language in question other than possibly the probability function symbol. If it is assumed that the axioms of probability are true statements of this language, the object of study is a family $\mathscr{P}$ of probability distributions over a definite collection of events $\mathscr{E}$.[3] To each event $E \in \mathscr{E}$ there corresponds the set of real numbers

$$I_E = \{P(E)\colon P \in \mathscr{P}\}.$$

It seems reasonable to suppose these to form a sub-interval of the unit interval whose limits $P_*(E) = \inf I_E$ and $P^*(E) = \sup I_E$ indicate the bounds of indeterminacy in the individual's degree of belief in the occurrence of E. It might also be expected that any probability distribution assigning to each event E in $\mathscr{E}$ a value that lies between values assigned to E by some two members of $\mathscr{P}$ also belongs to $\mathscr{P}$ or, more exactly, that if $\mathscr{P}$ is a probability distribution on $\mathscr{E}$ and, for each $E \in \mathscr{E}$, there exist $P_1, P_2 \in \mathscr{P}$ such that $P_1(E) \leqslant P(E) \leqslant P_2(E)$, then $P \in \mathscr{P}$. This implies the first condition. If the bounds $P_*(E)$ and $P^*(E)$ are attained for every event E, the second condition is equivalent to the requirement that every probability distribution on $\mathscr{E}$ assigning to each event E a value lying between these bounds belongs to $\mathscr{P}$.[4]

If the possible values of $P(E)$ form a closed interval, the statement 'the probability of E is p' is true, according to the proposals made above, if and only if

$$P_*(E) = p = P^*(E).$$

It is false if and only if $p > P^*(E)$ or $p < P_*(E)$ and otherwise it is indeterminate. Nonentheless, the statement 'either the probability of E is p or it is not p' is always true.[5] More generally, the statement 'the probability of E lies between p_1 and p_2, will be true, false or indeterminate according to the realtion of overlap between the intervals (p_1, p_2) and $(P_*(E), P^*(E))$. The general rule that determines whether or not a probability statement is true, however, is whether or not it holds for each of the distributions in $\mathscr{P}$. This leaves the question of the structure of $\mathscr{P}$ open.[6]

II. UPPER AND LOWER PROBABILITIES

The treatment throughout will be elementary to the extent that it deals only with the properties of finitely additive probability distributions. If Ω is a non-empty set and $\mathscr{E}$ is a non-empty field of subsets of Ω, a (finitely additive) probability distribution on $(\Omega, \mathscr{E})$ is a real-valued set function P satisfying

(K1) $P(\Omega) = 1$;

(K2) $P(E_1 \cup E_2) = P(E_1) + P(E_2)$ *for any disjoint* $E_1, E_2 \in \mathscr{E}$;

(K3) $P(E)$ *is non-negative for any* $E \in \mathscr{E}$.

We shall not restrict attention to fields of events, however, since such a restriction is superfluous for present purposes. Suppose, therefore, that $\mathscr{E}$ is any collection of subsets of Ω subject only to the requirement that $\Omega \in \mathscr{E}$. What conditions should be regarded as characterising a probability distribution on $(\Omega, \mathscr{E})$? (K1-3) are insufficient. For if $\mathscr{E}$ contains no union of disjoint sets, or *a fortiori* if it contains no disjoint sets, (K2) is vacuously satisfied. Hence (K1-3) are insufficient to ensure, for example, the monotonicity of P. But if P is such a non-monotone function, it cannot be extended to a probability distribution on any field containing $\mathscr{E}$. In addition, if the values of $P(E)$ are interpreted as betting rates in the usual way, such a system of rates will be incoherent in the sense that it will be possible to bet at these rates in such a way as to ensure a gain. These remarks suggest two ways in which (K1-3) might be strengthened. It has been shown by de Finetti, however, that they are equivalent; that is to say, the requirement that it should be possible to extend P to a finitely additive function on a field containing $\mathscr{E}$ is equivalent to the usual requirement of coherence. In order to explain the meaning of the conditions to which these are in turn equivalent certain further preliminaries are needed.

Let Ω be a non-empty set and $\mathscr{E}$ a family of subsets of Ω containing Ω. Such a pair $(\Omega, \mathscr{E})$ is referred to as a *covered set*. For each $E_i \in \mathscr{E}$ the corresponding lower case letter e_i will denote the *indicator function* of E_i, namely the function defined on Ω which assumes the value 1 in E_i and 0 outside. 1 will denote the indicator function of Ω, viz. the constant function 1, and 0 that of the null set $\emptyset$. ($\emptyset$, however, need not belong to $\mathscr{E}$.) Let $L(\mathscr{E})$ be the real linear space consisting of all finite linear combinations of the form

$$x = \sum_{i=1}^{n} \alpha_i e_i$$

where $\alpha_1, \ldots, \alpha_n$ are any real numbers and $e_1, \ldots, e_n$ are the indicator functions of some $E_1, \ldots, E_n \in \mathscr{E}$. Some of the random quantities in $L(\mathscr{E})$, besides $e_1, e_2, \ldots,$ may be indicator functions of sets. Thus if

E_1 and E_2 are disjoint, e_1+e_2 is the indicator function of their union (which may or may not belong to $\mathscr{E}$.) If E_1 and E_2 are not disjoint, e_1+e_2 is not the indicator function of a set. $L(\mathscr{E})$ is also endowed with an order structure expressed by writing $x < y$ to mean that $x(\omega)$ is strictly less than $y(\omega)$ for each $\omega \in \Omega$. $x \leqslant y$ has a similar meaning. $x > y$ and $x \geqslant y$ have the same meanings as $y < x$ and $y \leqslant x$ respectively.[7] We also define $X^{(+)} = \{x \in L(\mathscr{E}) : x > 0\}$.

If $(\Omega, \mathscr{E})$ is a covered set, then we say, following de Finetti, that a real-valued set function P defined on $\mathscr{E}$ is a *probability distribution* on $(\Omega, \mathscr{E})$ if and only if

(F1) $P(\Omega) = 1$;

(F2) *For any* $E_1, \ldots, E_n \in \mathscr{E}$ *and for any real numbers* $\alpha_1, \ldots, \alpha_n$

$$\sum_{i=1}^{n} \alpha_i(e - P(E_i)1) \notin X^{(+)}.$$

The second condition can be understood in terms of betting. It states that it should not be possible to select events $E_1, \ldots, E_n$ and stakes $\alpha_1 \ldots \alpha_n$ (positive or negative depending on whether the bets are on or against the events in question) so that a combination of bets relative to these events, at the rate $P(E_i)$ for event E_i, will ensure a positive gain. (Of course it would be the same thing to require that no such quantity should be uniformly negative.) If $\mathscr{E}$ is a field, (F1-2) are equivalent to (K1-3). When $\mathscr{E}$ is not a field, (F1-2) are necessary and sufficient for it to be possible to extend P onto any field containing $\mathscr{E}$ so as to satisfy (K1-3).[8]

Given P satisfying (F1-2), a linear functional p can be defined on $L(\mathscr{E})$ by setting

$$p(x) = \sum_{i=1}^{n} \alpha P(E_i)$$

whenever $x = \sum_{i=1}^{n} \alpha_i e_i$. (F2) ensures that the value of the sum is independent of the representation of x and hence that p is well-defined. $p(x)$ is of course just the expectation of x. It satisfies

(F1′) $p(1) = 1$;

(F2′) $p(x) < 0$ *whenever* $x < 0$.

Conversely, given any linear functional p satisfying (F1′-2′), the set function defined for each $E \in \mathscr{E}$ by

$$P(E) = p(e)$$

satisfies (F1-2).

Suppose now that a family $\mathscr{P}$ of probability distributions on a covered set $(\Omega, \mathscr{E})$ is given. We are interested, for each $E \in \mathscr{E}$, in the bounds of the values $P(E)$ as P ranges over $\mathscr{P}$:

$$P^*(E) = \sup\{P(E) : P \in \mathscr{P}\}$$
$$P_*(E) = \inf\{P(E) : P \in \mathscr{P}\}.$$

It is convenient to employ in addition two associated functions defined on $R \times \mathscr{E}$. For each real number α and for each $E \in \mathscr{E}$, define

$$\begin{aligned} P^*(\alpha, E) &= \alpha P^*(E) \text{ if } \alpha \geqslant 0, \\ &= \alpha P_*(E) \text{ if } \alpha < 0; \\ P_*(\alpha, \mathrm{E}) &= \alpha \mathrm{P}_*(E) \text{ if } \alpha \geqslant 0, \\ &= \alpha P^*(E) \text{ if } \alpha < 0. \end{aligned}$$

Then we have the following:

Theorem 1. If P is a non-empty family of probability distributions on a convered set $(\Omega, \mathscr{E})$, then:

(A1) $\quad P^*(\Omega) = P_*(\Omega) = 1;$

(A2) *For any* $E_0, \ldots, E_n \in \mathscr{E}$ *and for any real numbers* $\alpha_0, \ldots, \alpha_n$

$$(\alpha_0, e_0 - P_*(\alpha_0, E_0)1) + \sum_{i=1}^{n} (\alpha_i e_i - P^*(\alpha_i, E_i)1) \notin X^{(+)}.$$

Proof. (A1) is trivial. Suppose (A2) fails to hold. Then there exists a uniformly positive random quantity x of the form indicated. Since x assumes only finitely many values, there exists $\alpha > 0$ such that $x - \alpha 1 > 0$ and, correspondingly, there exists $P \in \mathscr{P}$ such that

$$\alpha_0 P(E_0) < P_*(\alpha_0, E_0) + \alpha.$$

Since, for each $i = 1, \ldots, n$.

$$\alpha_i P(E_i) \leqslant P^*(\alpha_i, E_i),$$

it follows, contrary to the hypothesis of the theorem, that

$$\sum_{i=0}^{n} \alpha_i(e_i - P(E_i)1) > \underline{x} - \alpha 1 > 0.$$

By suitable choice of the α_i's and E_i it can easily be seen that (A2) implies such relations as the following:

$$0 \leqslant P^*(E) \leqslant P^*(E) \leqslant 1 \text{ for any } E \in \mathscr{E};$$

$$P^*(E_1 \cup E_2) \leqslant P^*(E_1) + P^*(E_2) \textit{ if } E_1, E_2 \textit{ are disjoint and } E_1 \cup E_2 \in \mathscr{E};$$

$$P^*(E_1 \cup E_2) \geqslant P^*(E_1) + P_*(E_2) \textit{ if } E_1, E_2 \textit{ are disjoint and } E_1 \cup E_2 \in \mathscr{E};$$

together with the corresponding inequalities for P_*.[9] We remark also that just as (F1-2) have the simpler corresponding forms (F1′-2′) so, as we shall see, (A1-2) have simpler forms concerning functionals on $L(\mathscr{E})$.

The question arises whether, given functions P^* and P_* defined on $(\Omega, \mathscr{E})$, there exists a family $\mathscr{P}$ of probability distributions on $(\Omega, \mathscr{E})$ relative to which the upper and lower probabilities of each $E \in \mathscr{E}$ are just $P^*(E)$ and $P_*(E)$. This is answered affirmatively by

Theorem 2. Let P_*, P^* be real-valued set functions satisfying (A1-2) defined on a covered set $(\Omega, \mathscr{E})$. Then there exists a family $\mathscr{P}$ of probability distributions on $(\Omega, \mathscr{E})$ such that for any $\mathrm{E} \in \mathscr{E}$:

$$\sup\{P(E)\colon \mathrm{P} \in \mathscr{P}\} = P^*(E)$$

$$\inf\{P(E)\colon \mathrm{P} \in \mathscr{P}\} = P_*(E)$$

Proof. Let $\mathscr{E}_1$ be any collection of subsets of Ω such that $\mathscr{E} \subseteq \mathscr{E}_1$. Define the functional p^* on $L(\mathscr{E})$, for any $x \in L(\mathscr{E})$, by

$$p^*(x) = \inf\{\Sigma_i P^*(\alpha_i, E_i)\colon \Sigma_i \alpha_i e_i = x;\ E_1, \ldots, E_n \in \mathscr{E}\}.$$

(From now on we omit reference to the range of values of the index i which is always finite). Define p_1^* on $L(\mathscr{E}_1)$ by setting

$$p_1^*(x) = p^*(x) \text{ if } x \in L(\mathscr{E});$$

$$p_1^*(x) = \inf\{\Sigma_i P^*(\alpha_i, E_i)\colon \Sigma_i \alpha_i e_i \geqslant x;\ E_1, \ldots, E_n \mathscr{E}\} \text{ if } x \notin L(\mathscr{E}).$$

It follows that

(o) $-\infty < p_1^*(x) < +\infty$ for any $x \in L(\mathscr{E}_1)$;

(i) $p_1^*(x+y) \leqslant p_1^*(x) + p_1^*(y)$ for any $x, y \in L(\mathscr{E}_1)$;

(ii) $p_1^*(\alpha x) = \alpha p_1^*(x)$ for any $x \in L(\mathscr{E}_1)$ and any real number $\alpha \geqslant 0$;

(iii) $p_1^*(\alpha e) = P^*(\alpha, E)$ for any $E \in \mathscr{E}$ and any real number α;

(iv) $p_1^*(\alpha e) \leqslant 0$ for any $E \in \mathscr{E}_1$ and any real number $\alpha \leqslant 0$.

(o) follows from (A2). (i) and (ii) follow from the definitions except in the case $\alpha = 0$. It must be shown therefore that $p_1^*(0) = 0$. Since $\Omega \in \mathscr{E}$ and $0 = 1 - 1$, we have $p_1^*(0) \leqslant P^*(\Omega) - P_*(\Omega) = 0$. On the other hand, if $0 = \Sigma_i \alpha_i e_i$ for some $E_1, \ldots, E_n \in \mathscr{E}$, it follows from (A2), in the case $\alpha_0 = 1$ and $e_0 = 1$, that $\Sigma_i P^*(\alpha_i, E_i) \geqslant 0$. Hence $p_1^*(0) \geqslant 0$. Concerning (iii), $e = e$ implies $p_1^*(e) \leqslant P^*(E)$. On the other hand, if $e = \Sigma_i \alpha_i e_i$, we have $P^*(E) \leqslant \Sigma_i P^*(\alpha_i, E_i)$ from (A2) in the case $\alpha_0 = -1$ and $e_0 = e$. Hence $p_1^*(e) \geqslant P^*(E)$. It follows from (ii) that $p_1^*(\alpha e) = P^*(\alpha, E)$ for $\alpha \geqslant 0$. The case $\alpha < 0$ can be dealt with similarly. (iv) follows from the fact that $-e \leqslant 1 - 1$ for every $E \in \mathscr{E}_1$.

Now let $\mathfrak{P}_1$ be the family of all linear functionals p_1 on $L(\mathscr{E}_1)$ dominated by p_1^*, i.e. satisfying

$$p_1(x) \leqslant p_1^*(x) \text{ for any } x \in L(\mathscr{E}_1).$$

We show that any such linear functional is a probability distribution on $L(\mathscr{E}_1)$; that is to say, it satisfies (F1′-2′). Notice first that (i) and (ii) imply that $0 \leqslant p_1^*(x) + p_1^*(-x)$ and hence for any $p_1 \in \mathfrak{P}_1$

$$-p_1^*(-x) \leqslant p_1(x) \leqslant p_1^*(x).$$

In particular

$$1 = P_*(\Omega) = -p_1^*(-1) \leqslant p_1(1) \leqslant p_1^*(1) = P^*(\Omega) = 1.$$

Concerning (F2′), suppose that $\mathscr{E}_1$ is a field. Then if $x \in L(\mathscr{E}_1)$ and $x < 0$, x has an expression as $x = \Sigma_i \alpha_i e_i$ where each $\alpha_i < 0$ (e.g. the corresponding sets form a partition of Ω.) It follows from (iv) that for any $p_1 \in \mathfrak{P}_1$, $p_1(e_i) \geqslant -p_1^*(-e_i) \geqslant 0$ for each $i = 1, \ldots, n$; hence $p_1(x) < 0$.

It remains to show that p_1^* is the upper envelope of $\mathfrak{P}_1$. For any $x \in L(\mathscr{E}_1)$, define the linear functional p_0 on the subspace $L(x)$ of $L(\mathscr{E}_1)$ by

$$p_0(\alpha x) = \alpha p_1^*(x) \text{ for any real number } \alpha.$$

Since $x = 0$ implies $p_1^*(x) = 0$, p_0 is well defined. Then $p_0(\alpha x) \leqslant p_1^*(\alpha x)$ for any real number α. For if $\alpha \geqslant 0$, $p_0(\alpha x) = p_1^*(\alpha x)$ and if $\alpha < 0$, $p_0(\alpha x) = \alpha p_1^*(x) = -p_1^*(\alpha x) \leqslant p_1^*(\alpha x)$. It follows by the Hahn-Banach Theorem[10] that p_0 can be extended to a linear functional p_1 on $L(\mathscr{E}_1)$ dominated by p_1^* throughout $L(\mathscr{E}_1)$, i.e. to a linear functional $p_1 \in \mathfrak{P}_1$. Hence p_1^* is the upper envelope of $\mathfrak{P}_1$. Let $\mathfrak{P}$ be the family of all restrictions of members of $\mathfrak{P}_1$ to $L(\mathscr{E})$. Then, clearly, p^* is the upper envelope of $\mathfrak{P}$. Let $\mathscr{P}$ be the corresponding family of set functions on $(\Omega, \mathscr{E})$, i.e. all those functions P such that for some $p \in \mathfrak{P}$

$$P(E) = p(e) \text{ for any } E \in \mathscr{E}.$$

Since each $p \in \mathfrak{P}$ satisfies (F1′-2′), the corresponding $P \in \mathscr{P}$ is a probability distribution on $(\Omega, \mathscr{E})$. Furthermore, for any $E \in \mathscr{E}$,

$$\sup\{P(E){:}P \in \mathscr{P}\} = \sup\{p(e){:}p \in \mathfrak{P}\} = p^*(e) = P^*(E),$$

$$\inf\{P(E){:}P \in \mathscr{P}\} = \inf\{p(e){:}p \in \mathfrak{P}\} = -\sup\{p(-e){:}p \in \mathfrak{P}\} = = -p^*(-e) = P_*(E).$$

Corollary 2.1. (A1-2) imply the following for any $x \in L(\mathscr{E})$:

(i) $x < 0$ *implies* $\inf\{\Sigma_i P^*(\alpha_i, E_i){:}\ \Sigma_i \alpha_i e_i = x\} < 0$;

(ii) $x > 0$ *implies* $\sup\{\Sigma_i P_*(\alpha_i, E_1){:}\ \Sigma_i \alpha_i e_i = x\} > 0$.

Proof. Since $x \in L(\mathscr{E})$, the quantity on the right in (i) is the value of $p^*(x)$ defined in the proof the the theorem. But, as we have seen, (A1-2) imply, for any $x \in L(\mathscr{E})$, the existence of a probability distribution p on $L(\mathscr{E})$ such that $p(x) = p^*(x)$. Hence if $x < 0$, $p^*(x) < 0$. (ii) follows by applying (i) to $-x$.

In view of Theorems 1 and 2 and of the subsequent consideration of betting, (A1-2) will be regarded as the axioms for upper and lower probabilities corresponding to (F1-2) as the axioms of probability *simpliciter*. Just as (F1-2) have simpler forms in terms of linear functionals on the space $L(\mathscr{E})$, (A1-2) likewise have simpler corresponding forms. The situation is as follows.

Given the covered set $(\Omega, \mathscr{E})$ and the real linear space $L(\mathscr{E})$ of finite linear combinations of indicator functions of events in $\mathscr{E}$, a *probability distribution on* $L(\mathscr{E})$ is a functional p such that

(F0.1′) $p(x+y) = p(x)+p(y)$ *for any* $x, y \in L(\mathscr{E})$;

(F0.2′) $p(\alpha x) = \alpha p(x)$ *for any* $x \in L(\mathscr{E})$ *and any real number* α;

(F1′) $p(1) = 1$;

(F2′) $p(x) < 0$ *whenever* $x < 0$.

An *upper probability distribution on* $L(\mathscr{E})$ is a functional p^* such that

(A0.1′) $p^*(x+y) \leqslant p^*(x)+p^*(y)$ *for any* $x, y \in L(\mathscr{E})$;

(A0.2′) $p^*(\alpha x) = \alpha p^*(x)$ *for any* $x \in L(\mathscr{E})$ *and any real number* $\alpha \geqslant 0$;

(A1′) $p^*(1) = -p^*(-1) = 1$;

(A2′) $p^*(x) < 0$ *whenever* $x < 0$.

We can say therefore that a functional f on $L(\mathscr{E})$ satisfying

$$f(\alpha 1) = \alpha \textit{ for any real number } \alpha$$
$$f(x) < 0 \textit{ whenever } x < 0$$

is a probability distribution, or an upper probability distribution, on $L(\mathscr{E})$ according as it is linear or sublinear.

There is a unique correspondence between real-valued set functions P on $(\Omega, \mathscr{E})$ satisfying (F1-2) and functionals p on $L(\mathscr{E})$ satisfying (F0′-2′). The correspondence is given by

$$p(e) = P(E) \textit{ for any } E \in \mathscr{E}.$$

Analogously, given any real-valued functions P_*, P^* on $(\Omega, \mathscr{E})$ satisfying (A1-2), there exists an upper probability distribution p^* on $L(\mathscr{E})$ extending P_* and P^* in the sense that

(*) $p^*(\alpha e) = P^*(\alpha, E)$ *for any* $E \in \mathscr{E}$ *and any real number* α.

Conversely, given any upper probability distribution p^* on $L(\mathscr{E})$, the real-valued set functions P_*, P^* on $(\Omega, \mathscr{E})$ defined by (*) satisfy (A1-2).

In general, however, the correspondence expressed by (*) is not unique. On the other hand, given P_*, P^*, any subadditive functional p^* on $L(\mathscr{E})$ satisfying (*) will satisfy, for any $x \in L(\mathscr{E})$,

$$p^*(x) \leqslant \inf\{\Sigma_i P^*(\alpha_i, E_i) \colon \Sigma_i \alpha_i e_i = x\}.$$

The proof of Theorem 2 and its Corollary shows that, quite generally, there exists a (necessarily unique) upper probability distribution on $L(\mathscr{E})$ which everywhere attains this bound.

Given a family of probability distributions on $(\Omega, \mathscr{E})$, the associated upper and lower probabilities satisfy (A1-2). Conversely, given upper and lower probabilities on $(\Omega, \mathscr{E})$ satisfying (A1-2), there exists a family of probability distributions on $(\Omega, \mathscr{E})$ of which these are the upper and lower bounds. Correspondingly, given a family of probability distributions on $L(\mathscr{E})$, its upper envelope is an upper probability distribution on $L(\mathscr{E})$. Given an upper probability distribution p^* on $L(\mathscr{E})$, any linear functional p on $L(\mathscr{E})$ dominated by p^* is a probability distribution on $L(\mathscr{E})$. Moreover, p^* is the upper envelope of all the linear functionals (probability distributions) which it dominates.

III. EXTENSIONS OF THE DOMAIN OF DEFINITION

The failiure to restrict attention to a field of events, apart from this restriction being superfluous for many purposes, has a particular merit from the point of view of the subjectivistic interpretation of probability. It means that the domain of events concerning which it must be assumed that the subject has some more or less well-determined degree of belief, i.e. concerning which he has some more or less definite opinion, is unrestricted. In particular it is not necessary to assume it to be so extended as to satisfy such closure conditions as those of a field. However, once probabilities are admitted, in any case, to be more or less indeterminate quantities, arbitrarily extended domains of definition can be admitted without giving rise to difficulties in the intended interpretation.

Given the collection of events $\mathscr{E}$ concerning whose occurrence the subject has some more or less definite degrees of belief, and any further collection of events $\mathscr{E}'$ concerning which he has no particular opinion, it is not unnatural to regard the degrees of belief concerning the events in $\mathscr{E}'$ as wholly undetermined or, rather, determined to no further extent than that to which the degrees of belief concerning the events in $\mathscr{E}$ already require. Let $\mathscr{E}_1 = \mathscr{E} \cup \mathscr{E}'$. The proof of Theorem 2 shows how, given the upper and lower probabilities P^*, P_* concerning events in $\mathscr{E}$, the functional p_1^* can be defined throughout $L(\mathscr{E}_1)$. (It makes no difference whether or not $\mathscr{E}_1$ is a field.) The corresponding set functions P_1^* and P_{*_1} might then be regarded as expressing the upper and lower limits of the subject's degrees of belief in the events in $\mathscr{E}_1$. Concerning the events in $\mathscr{E}$, they agree with the old assignments. Concerning any event $E' \in \mathscr{E}'$, it is not difficult to see that no higher value could be assigned to the upper probability of E' than $P_1^*(E')$, without coming into conflict with (A1–2), nor any lower than $P_{*_1}(E')$. Thus P_1^* and P_{*_1} express the most extreme cases of indeterminacy which the existing evaluations permit. Another approach is possible. One could begin with a family $\mathscr{P}$ of probability distributions on $(\Omega, \mathscr{E})$, representing the evaluations concerning the events in $\mathscr{E}$, and regard the family $\mathscr{P}_1$ of all extensions of members of $\mathscr{P}$ to probability distributions over $(\Omega, \mathscr{E}_1)$ as representing the evaluations over the extended set. The upper and lower probabilities would then be defined in terms of $\mathscr{P}_1$.[11] These approaches are equivalent, however, if $\mathscr{P}$ is *complete* in the sense that, for any probability distribution P on $(\Omega, \mathscr{E})$, $P \in \mathscr{P}$ whenever the condition

$$\inf\{P(E)\colon P \in \mathscr{P}\} \leqslant P(E) \leqslant \sup\{P(E)\colon P \in \mathscr{P}\}$$

holds for every $E \in \mathscr{E}$.

IV. BETTING

It will be shown how, under certain assumptions, the proposed axioms for upper and lower probabilities can be derived from considerations of betting. A *bet* concerning an event E is an arrangement whereby a sum $\alpha\beta$ is exchanged for a sum α if E occurs or 0 if it does not. The bet is said to be *on* or *against* E according as $\alpha > 0$ or $\alpha < 0$. (If β lies in the unit interval, a bet on E yields a net gain if E occurs and a net loss if it does not; a bet against E yields a net gain if E does not occur and a net loss if it does). β is called the *betting rate* ad α the *stake*. If $\alpha = 0$ we have just the null bet.

Let e be the indicator of the event E and 1 the indicator of the sure event Ω as before. Then a bet concerning E is a random quantity of the form $\alpha(e - \beta 1)$. Given the covered set $(\Omega, \mathscr{E})$ and $E_1, E_2 \in \mathscr{E}$, the bets $\alpha_1(e_1 - \beta_1 1)$ and $\alpha_2(e_2 - \beta_2 1)$ concerning events E_1 and E_2 can be combined to form the compound bet

$$\alpha_1(e_1 - \beta_1 1) + \alpha_2(e_2 - \beta_2 1)$$

whose outcome depends on both E_1 and E_2. Generally any random quantity of the form

$$x = \Sigma_i \alpha_i(e_i - \beta_i 1)$$

will be regarded as a bet. Since every random quantity $x \in L(\mathscr{E})$ can be expressed in this form, $L(\mathscr{E})$ becomes the space of all bets concerning a finite number of events in $\mathscr{E}$.

A bet is said to be *acceptable* to the subject in question if he is willing for his fortune to change in accordance with its outcome. Let $A \subseteq L(\mathscr{E})$

denote the class of acceptable bets.[12] It will be assumed that whenever x and y are acceptable bets, the combined bet $x+y$ is acceptable; also, that the acceptability of a bet does not depend on the absolute values of the stakes concerned (loosely speaking, it depends only on the betting rates). On the other hand, it will not be assumed that it makes no difference whether the bet is on or against the events in question. Precisely, these two assumptions are made concerning A:

(B1) $x, y \in A$ implies $x+y \in A$ for any $x, y \in L(\mathscr{E})$;

(B2) $x \in A$ implies $\alpha x \in A$ for any $x \in L(\mathscr{E})$ and any real number $\alpha > 0$.[13]

Given A and any event $E \in \mathscr{E}$, it is of interest to consider the smallest rate at which the subject is prepared to bet against E at unit stake and the largest rate at which he is prepared to bet on E, again at unit stake, viz.

$$P^*(E) = \inf\{\beta\colon \beta 1 - e \in A\};$$
$$P_*(E) = \sup\{\beta\colon e - \beta 1 \in A\}.$$

If, for some event E, the subject declines to accept any bet of the form $\beta 1 - e$ or if he declines to accept any bet of the form $e - \beta 1$, $P^*(E)$ and $P_*(E)$ are understood to be determined in accordance with the rule that $-\infty < \beta < +\infty$ for every real number β. Hence $P^*(E) = +\infty$ and $P_*(E) = -\infty$ in the two cases. Arithmetic operations involving $+\infty$ and $-\infty$ are understood to be determined in accordance with similar rules. The expressions $(+\infty - \infty)$ and $(-\infty + \infty)$, however, are regarded as undefined.

Given A, several situations can arise. First, either A contains a uniformly and strictly negative quantity, i.e. a bet from which the individual is bound to lose (strictly) whatever happens, or it does not. In the latter case A will be said to satisfy the *negative condition of coherence*:

(C⁻) $A \cap X^{(-)} = \emptyset$

where

$$X^{(-)} = \{x \in L(\mathscr{E})\colon x < 0\}.^{14}$$

Secondly, either A contains every strictly and uniformly positive quantity, i.e. a bet from which the subject is bound to gain (strictly) whatever happens, or it does not. In the former case A will be said to satisfy the *positive condition of coherence*:

$$(\mathrm{C}^+) \qquad X^{(+)} \subseteq A$$

where

$$X^{(+)} = \{x \in L(\mathscr{E}) \colon x > 0\}.$$

Writing (C) as the conjunction of (C^-) and (C^+), viz.

$$(\mathrm{C}) \qquad A \cap X^{(-)} = \emptyset \text{ and } X^{(+)} \subseteq A,$$

we have

Theorem 3. Given (B1–2), (C) implies (A1–2).

Proof. Define the functional p^* on $L(\mathscr{E})$ for any $x \in L(\mathscr{E})$ by

$$p^*(x) = \inf\{\beta \colon \beta 1 - x \in A\}.$$

Given (B1–2) and (C), $p^*(x)$ is everywhere finite and satisfies (A0′–2′), as may easily be verified. It follows that there exists a family of probability distributions on $(\Omega, \mathscr{E})$ with respect to which for any event $E \in \mathscr{E}$, $P^*(E)$ and $P_*(E)$ are the upper and lower bounds. The result follows from Theorem 1.

The question arises whether (A1–2) are sufficient to establish (C). (A1–2) are in fact sufficient for (C^-), as will be shown, but there is an exception to (C^+). The question hinges on whether or not the subject is willing to bet on the sure event; in other words, whether or not $1 \in A$. If $1 \notin A$, it can be shown that (A1–2) imply $P_*(E) = P^*(E) = P(E)$, say, for every event $E \in \mathscr{E}$, where P is a probability distribution on $(\Omega, \mathscr{E})$. It can be shown to follow that $A = \{\Sigma_i \alpha_i e_i \colon \Sigma_i \alpha_i P(E_i) = 0\}$; whence $A \cap X^{(-)} = \emptyset$ but also $A \cap X^{(+)} = \emptyset$. This is the situation of a subject whose upper and lower betting rates coincide and who accepts a bet just in case it has zero expectation.[15]

The situation can be expressed as follows:

Theorem 4. Given (B1–2), if $1 \notin A$, (A1–2) imply that $A = \{x \in L(\mathscr{E})\colon p(x) = 0\}$ for some probability distribution p on $L(\mathscr{E})$; hence (C^-) holds.

The proof is straightforward. On the other hand:

Theorem 5. Given (B1–2), if $1 \in$ A, (A1–2) imply (C).

Proof. (A2) implies that $-\infty < P_*(E), P^*(E) < +\infty$. Hence for any $x = \Sigma_i \alpha_i e_i$, and for any $\varepsilon > 0$, there exist $\beta_1, \ldots, \beta_n$ such that

$$\alpha_i \beta_i > P_*(\alpha_i, E_i) - \varepsilon/n$$

$$\alpha_i(e_i - \beta_i 1) \in A \qquad (i = 1, \ldots, n).$$

It follows from (B1) that $x-\beta 1 \in A$ where $\beta = \Sigma_i \alpha_i \beta_i > \Sigma_i P_*(\alpha_i, E_i) - -\varepsilon$. Hence

$$\sup\{\beta\colon x-\beta 1 \in A\} \geqslant \sup\{\Sigma_i P_*(\alpha_i, E_i)\colon \Sigma_i \alpha_i e_i = x\}.$$

Therefore by Corollary 2.1 (ii), $x > 0$ implies $\sup\{\beta\colon x-\beta 1 \in A\} > 0$. Thus there exists $\alpha > 0$ such that $x-\alpha 1 \in A$. Since $1 \in A$ implies that $\alpha 1 \in A$ for all $\alpha > 0$, it follows that $x \in A$. On the other hand, if $x < 0$ then $\alpha 1 - x > 0$ for some $\alpha < 0$ and hence $\alpha 1 - x \in A$. But $\alpha < 0$ implies $\alpha 1 \notin$ A by (A1). Therefore $x \notin A$.

V. CONCLUSION

Let us suppose that, for whatever reason, the class A of bets acceptable to some subject satisfies (C). Then the limits P^*, P_*, defined in terms of A, satisfy (A1–2). We know then that there exists a family $\mathscr{P}$ of probability distributions over the events in question such that $p^*(x) = \inf\{\beta\colon \beta 1 - x \in A\}$ is the upper envelope of the expectations of the bets in $L(\mathscr{E})$ relative to this family. Equivalently, $p_*(x) = \sup\{\beta\colon x-\beta 1 \in A\}$ is the lower envelope. But, given (C), $\alpha 1 \in$ A if $\alpha > 0$ and $\alpha < 0$ implies $\alpha 1 \notin A$. Therefore $p_*(x) > 0$ implies $x \in A$ and $p_*(x) < 0$ implies $x \notin A$.

This situation can be interpreted as follows. If (C) holds, it is possible to explain why some of the bets that are neither strictly positive nor strictly negative are acceptable to the subject concerned, and why others are not, by attributing to the subject a certain quantity called his "degree of belief" in the events in $\mathscr{E}$. This quantity, however, need not be regarded as exactly determined. That is to say, when we speak in this way, it should not be forgotten that—as in all empirical discourse—we employ a language that may not be exact, one in which not every statement need have a determinate truth value. This can be taken into account by treating the symbol designating the subject's degrees of belief to be interpreted by the family $\mathscr{P}$, as explained in the introduction.[16] The subject's behaviour is then at least partially explained by noticing that he accepts any bet of which it is true that it has positive expectation and he declines any bet of which it is not true that it has non-negative expectation (certain borderline cases being undecidable on this basis). Some such statements concerning the expectation values of bets may be neither true nor false. But the axioms of probability *simpliciter* are true in any case. However indeterminate may be the probabilities to which they refer, when conjoined to true premises, they can only yield true conclusions.

BIBLIOGRAPHY

Dempster, A. P., [1968]: 'A Generalization of Bayesian Inference', *Journal of the Royal Statistical Society* Series **B, 30**, 205–232

Dunford N., J. T. Schwartz [1958]: *Linear Operators. Part I: General Theory.* Interscience Publishers, New York de Finetti, B. [1972]: *Probability, Induction and Statistics*, John Wiley and Sons, London

Good, I. J., [1962]: 'Subjective Probability as the Measure of a Non-Measurable Set' in: E. Nagel, P. Suppes, A. Tarski (eds.), *Logic, Methodology and Philosophy of Science*, Stanford University Press, Stanford, pp. 319–329.

Koopman, B. O., [1940]: 'The Bases of Probability', *Bulletin of the American Mathematical Society*, **46**, 763–774; reprinted in: H. E., Kyburg, H. E. & Smokler (eds.), *Studies in Subjective Probability,* John Wiley & Sons, New York, pp. 161–172.

Przełęcki, M., [1969]: *The Logic of Empirical Theories*, Routledge Kegan Paul, London

Smith, C. A. B. [1961]: 'Consistency in Statistical Inference and Decision', *The Journal of the Royal Statistical Society* Series **B, 23**, 1–37.

Williams, P. M. [1974]: 'Certain Classes of Models for Empirical Systems', *Studia Logica*, 33, 73–90.

Wójcicki, R. [1973]: 'Basic Concepts of Formal Methodology of Empirical Science', *Ajatus*, **35**, 168–196.

REFERENCES

[1] See, for example, Przełęcki [1969], Wójcicki [1973], Williams [1974].

[2] Q. v. Przełęcki [1969, pp. 18–23].

[3] It would be of interest to extend this treatment to the case of indeterminate events, e.g. the event that a person has a height of 180 ± 10^{-10}cms—an event of which it might be neither true nor false to say that it has occurred.

[4] These conditions are mentioned at this point only as an aid to informal explanation. They are not employed as general presuppositions of the formal treatment below.

[5] It is interesting to find the difference between 'σ' and 'σ is true' already clearly expressed in Koopman [1940] with his distinction between "contemplated propositions" and "asserted propositions". Koopman also states emphatically, though in a somewhat different context (p. 166), that "The distinction between an asserted disjunction and a disjoined assertion is fundamental: $(u \vee v) = 1$ must never be confused with $(u = 1) \vee (v = 1)$."

[6] It may be appropriate here to mention an apparent difficulty which is sometimes raised. It might be said that whilst an indeterminacy is admitted in the values of $P(E)$, it is assumed that the values of $P_*(E)$ and $P^*(E)$ are exactly determined, or—what comes to the same thing—that $\mathscr{P}$ is uniquely determined. But this is not the case. The use of a function symbol or a class term in accordance with classical logico-mathematical rules does not presuppose that the denotation of that symbol is exact, at least if the above proposals concerning semantical indeterminacy are accepted, since these rules hold as well for determinate as for indeterminate concepts. It is not possible to tell from any assertion made within a language itself whether a symbol is being regarded by the language user as determinate or indeterminate. In the same way, in the theory of sets, it is not possible to decide on the basis of any assertion expressible within the theory itself whether the intended model is of the classical two-valued type or of the more general Boolean-valued type. (In fact the present treatment of indeterminacy can be regarded as dealing with a Boolean-valued model by suitably evaluating the statements of the probability calculus in the power set of $\mathscr{P}$). If one so wished, the same treatment that is now being applied to 'P' could be applied, at a higher level, to '$\mathscr{P}$'. There is of course no bound to this process.

[7] The elements e_i spanning $L(\mathscr{E})$ are taken to be indicator functions of *sets*, corresponding to the events under consideration, in order to conform to the more familiar approach. Otherwise it would be appropriate to treat $L(\mathscr{E})$ in its own right as the linear space spanned by indicators of the events themselves, viz. the random quantities which assume the values 1 or 0 according as the events concerned do or do not occur. (In that case it would be unnecessary, from the mathematical point of view, to distinguish between the events and their indicator functions: cf. de Finetti [1972, pp xviii–xxiv]).

[8] Q.v. de Finetti [1972, pp. 77–79].

[9] Cf. Good [1962] from which the present notation is borrowed. The important works of Smith [1961] and Dempster [1968], in which similar ideas are advanced and a very wide range of related topics discussed, unfortunately came to my attention too late to be taken into account.

[10] Q. v. Dunford & Schwartz [1958, p. 62].

[11] The case where $\mathscr{P}$ has a single member way already discussed by de Finetti [1972, p. 106].

[12] It is not necessary to suppose that A is, in some sense, an exact class: cf. footnote 6, p. 231.

[13] This differs from the more usual treatment precisely in not postulating closure under negative scalar multiplication. Of course, there one can restrict attention to arbitrarily small non-zero stakes, but closure under negative scalar multiplication is required within these limits. Similarly it would be sufficient to assume here that A is the intersection of a convex cone with a set of the form $\{x \in L(\mathscr{E}) : ||x|| \leqslant \alpha\}$ for some $\alpha > 0$, where $||x|| = \max\{/x(\omega)/ : \omega \in \Omega\}$. Details are left to the interested reader.

[14] Geometrically, the problem raised by (C^-) alone is that of separating two convex cones by a hyperplane. Since $X^{(-)}$ is open and contains$(-)1$, if the hyperplane exists

it can be characterised by a probability distribution on $L(\mathscr{E})$ with respect to which every acceptable bet has non-negative expectation.

[15] It should be emphasized, however, that the coincidence of the upper and lower rates does not require the subject to decline any of the strictly positive bets though, given (A1-2), it does require him to accept all or none.

[16] According to this proposal the values of $P(E)$ as P ranges over $\mathscr{P}$ form a closed interval. It is left up to the subject, however, whether or not he is willing to bet at the extreme rates.

FREDERICK SUPPE

THEORETICAL LAWS[1]

Just as the computer who wants his calculations to deal with sugar, silk, and wool must discount the boxes, bales, and other packings, so the mathematical scientist, whan he wants to recognize in the concrete the effects which he has proved in the abstract, must deduct the material hindrances, and if he is able to do so, I assure you that things are in no less agreement than arithmetical computations. The errors, then, lie not in the abstractness or concreteness, not in geometry or physics, but in a calculator who does not know how to make a true accounting.

-Galileo Galilei, *Dialogue Concerning Two Chief World Systems*[2]

I. INTRODUCTION

Michael Scriven has argued plausibly that the various standard attempts to characterize laws of nature are inadequate—in part because they fail to recognize that the key property of physical laws is their inaccuracy.[3] He then suggests that this insight can be accommodated by an analysis along the following lines: "typical physical laws express a relationship between quantities or a property of systems which is the *simplest useful approximation* to the true physical behavior and which appears to be theoretically tractable."[4] More generally, recognition of the pervasive inaccuracy of physical laws has suggested to a number of philosophers that an adequate analysis of theories might involve recourse to some notion of *approximate truth.*[5] While the approximate truth approach has plausibility, my suspicion is that it is not the most promising way to deal with the inaccuracy of laws; for I suspect that the inaccuracy of physical laws is a manifestation of relatively deep structural and epistemological properties of laws and theories which will be obscured or missed if one attempts to analyze laws as approximately true generalizations (or other sorts of propositions). Accordingly, in this paper I explore the nature of physical laws in a manner that avoids recourse to the notion of approximate truth.

Exploiting the fact that laws often, if not always, are components of scientific theories, I will investigate the properties of laws occurring in a scientific theory; as such my concern here is restricted to *theoretical laws*. Whether there are non-theoretical laws and if so how they differ from theoretical laws are issues which will not be considered here. Section 2 of the paper briefly sketches the *Semantic Conception of Theories* which will be assumed in the analysis of theoretical laws. Then, in Section 3, I will look at various types of theoretical laws, and determine a number of their more significant properties. Section 4 considers the nature of teleological and functional laws. Section 5 summarizes the findings of this paper and indicates in what sense it is true that a key feature of (theoretical) physical laws is their inaccuracy.

II. LAWS ON THE SEMANTIC CONCEPTION OF THEORIES

Although the term 'law' can be used in science to refer to the most elementary isolated empirical generalizations (e.g., "Ducks have webbed feet"), clearly its most central use is with reference to entities occurring within theories—i.e., to *theoretical laws*. As such it is reasonable to expect that considerable light can be shed on the nature of laws by investigating the nature and function of theoretical laws within theories. To do so requires presupposing some analysis of the nature of scientific theories. I will presuppose here the *Semantic Conception of Theories* developed recently by Beth, van Fraassen, and myself. Since the Semantic Conception has been developed in considerable detail elsewhere,[6] I will confine my discussion here to a brief heuristic sketch.

On the Semantic Conception, theories are concerned with specifying the behaviors of systems of entities, where the behavior is construed as changes in a specified finite set of parameters characteristic of these entities. At a given time the simultaneous values of these parameters determine the *state* of the system. The *intended scope* of a theory is some class of physically possible systems (e.g., the class of all causally possible mechanical systems comprosed of a finite number of interacting bodies). It is the job of a theory to characterize that class of causally possible systems by indicating all and only those time-directed sequences of states which correspond to behaviors of possible systems within its intended scope.

What is the structure of a theory, and how does that structure enable it to indicate which time-directed sequences of states correspond to behaviors of possible systems within its intended scope? In essence a theory is a general model[7] of the behavior of the systems within its scope. The model is a *relational system* whose domain is the set of all logically possible state occurrences, and whose relations determine time-directed sequences of state occurrences which correspond to the behavior of possible systems within its intended scope and indicate which changes of state are physically possible. These sequencing relations are the *laws* of the theory.[8] A variety of different sorts of laws are possible which differ in the ways they determine possible sequences of state occurrences. For example, if the theory has just *deterministic laws of succession* (as Newton's mechanics does), then the laws (sequencing relations) will determine unique sequences of successive states systems may assume over time. If the theory has just *statistical theories of succession*, the relations will be akin to those characteristic of branching trees governed by a Markov process, each path indicating a different sequence of states a system may subsequently assume over time, and assigning conditional probabilities to each state change. If the theory has just *deterministic laws of coexistence*, (as do microeconomic supply-demand equilibrium theories) the laws partition the class of state occurrences, and the possible time-directed sequences of state occurrences are those which can be formed from a single partition class. A number of other types of laws are found in science, many of which will be investigated in subsequent sections. The predictive capabilities of theories vary depending upon what sorts of laws they possess: For example, deterministic laws of succession enable one to predict unique subsequent states of a system; statistical laws of succession only enable one to predict that at subsequent time *t* the system will be in one of a specified set of states, with a probability of being in that state assigned to each state; deterministic laws of coexistence only enable one to predict that at subsequent time *t* the system will be in some state "equivalent" to what it was at earlier times, and no probabilities are assigned to these states.

A theory thus models the behaviours of the possible systems in its intended scope by determining sequences of state occurrences which correspond to the behaviors of all possible such systems. However, as is the case generally with iconic models, this correspondence need not be one of identity. For in specifying possible changes in state, the theory tacitly assumes that the only factors influencing the behavior of a system are those which show up as state parameters in the theory, whereas in fact the values of these parameters often are influenced by outside factors which do not show up as parameters of the theory. As such the nature of the correspondence is as follows: *The sequences of states determined by the theory indicate what the behaviors of the possible systems within the theory's scope would be were it the case that only the parameters of the theory exerted a nonnegligible influence on those behaviors.*[9] That is, the theory characterizes what the possible behaviors of systems are under idealized circumstances wherein the values of the parameters do not depend on any outside influences, and thus relates counterfactually to many actual systems within its intended scope. An example will help here. Consider a microeconomic theory concerning the interplay of supply and demand schedules for goods. According to this theory, goods will be priced at an equilibrium price wherein supply equals demand; supply, demand, and price may change over time, but according to the theory these changes will always be such that there is an equilibrium of supply and demand at a given price. As such the theory is one whose variables are price, quantity of supply, and quantity of demand, with deterministic laws of coexistence such that for a given price supply quantity equals demand quantity. The theory in effect says that the behavior of supply-demand-price market system will be such that the system always will be in a state where supply equals demand at the given price. But this is a highly *idealized* picture of the market which assumes that there is perfect competition, that cobweb situations do not occur, that there is no time-lag in the reaction of prices to changes in supply or demand situations, etc.; that is, it is an idealized situation wherein the behavior of the system depends *only* on supply, demand, and price. This idealized situation often fails to obtain in the market; but that does not show the theory to be wrong. For the theory here *only* purports to describe what the supply-demand-price behavior *would be* for that market *were it the case that these idealized conditions were met*. When the idealized conditions are met the theory directly can predict what will happen; when the idealized conditions are not met, it still maybe possible to use the theory to predict what will happen via recourse to *auxiliary theories*: Suppose we have a supply-demand-price system which does not satisfy the idealized conditions, but we also have another auxiliary theory which tells us how the other factors operant in the system (e.g. cartels) influence the supply quantity, demand quantity, and price. Then using our supply-demand-price theory we can predict what *would be* the subsequent state of the system *were* the idealized conditions met, then use the auxiliary theory to determine how the actual situation will deviate from the idealized situation, the two theories together yielding an accurate prediction of what the actual behavior will be.[10]

According to Semantic Conception of Theories, then, scientific theories are relational systems functioning as iconic models which characterize all the possible changes of state the systems within its scope could undergo under idealized circumstances. And the theory will be *empirically true* if and only if the class of possible sequences of state occurrences determined by the theory is identical with the possible behaviors of systems within its intended scope under idealized conditions. Whenever a system within the theory's intended scope meets the idealized conditions, the theory can predict the subsequent behavior of the system (the preciseness of the prediction depending on what sorts of laws the theory has, as indicated above); and when the idealized conditions are not met, the theory can predict the behavior of the system if used in conjunction with a suitable auxiliary theory.[11]

III. LAWS OF SUCCESSION, COEXISTENCE, AND INTERACTION

On the Semantic Conception of Theories, laws are relations which determine possible sequences of state occurrences over time a system within its intended scope may assume. We now investigate a number of types of such laws which commonly are encountered in scientific theories and establish various theorems about them. We begin by considering the types of laws most commonly discussed in the theories literature—i.e., deterministic and statistical laws of succession, coexistence, and interaction.[12]

Depending on the theory, time is construed as being either *discrete* or *continuous*. In discrete-time theories, time usually is construed as having the order properties of the natural numbers (i.e., as an ω sequence) or the integers (i.e., as an $\omega^* + \omega$ sequence).[13] Continuous-time theories construe time as having the order properties of the real numbers (i.e., as an λ sequence). Let α be a variable whose only values are ω, $\omega^* + \omega$, or λ; any simple-ordering of times having the order properties of an ω, $\omega^* + \omega$, or λ order type will be known as an *α-time sequence.* The laws of a theory are relations which determine possible simple-orderings of states having the order properties of an α-time sequence.

The defining parameters $p_1, \ldots, p_n$ $(n \geqslant 1)$ of a theory are variables ranging over attributes and/or probability distribution functions over attributes; time, t, may or may not be a defining parameter of the theory. Let $p_i(t)$ be the value of parameter p_i at time t. Then the *state* s of a system at time t is $\langle p_i(t), \ldots, p_n(t) \rangle$. (Where needed, $q_i(t)$ and $r_i(t)$ — with or without primes or subscripts — will be construed analogously.) Since the same state s may occur at more than one time (i.e., in distinct state occurrences), we will use '$s(t)$' to designate the occurrence of state s at time t. D will be the set of all logically possible state occurrences for a theory.[14]

At each time t a physical system within the intended scope of a theory assumes some particular state, and the *behavior* of that system consists in its changes in states over time. That behavior is represented by a simply-ordered set of state-occurrences wherein $s(t) < s'(t')$ if and only if $t < t'$ (t and t' ranging over some α-time sequence). It is the job of the laws of the theory to characterize all (and only) those possible behaviors of systems within the theory's intended scope by determining those simple-orderings of state occurrences which represent the idealized behaviors of systems within that intended scope; the class of simple-orderings so determined is known as the class of *theory-induced simple-orderings.* The *laws* of a theory T are relations $R(s(t), s'(t'))$ [or $R(s(t), s'(t'), p)$] holding between state occurrences [or between state occurrences and real numbers p]. Suppose T is a theory with only one law, R, and let O be a simple-ordering of state-occurrences for T of order-type α, and let t, t' range over the α-time sequence associated with T; then O is a theory-induced simple-ordering of state-occurrences if and only if for every s(t): s'(t'), (1) $s(t) < s'(t')$ under O if and only if $t < t'$ under the α-time sequence, and (2) either $\langle s(t), s'(t') \rangle \in R$ or $\langle s(t), s'(t'), p \rangle \in R$ for some real number p such that $0 \leqslant p \leqslant 1$. If T has more than one law, condition (2) must be satisfied for each law R.

The foregoing preliminaries our of the way, we now define a variety of different kinds of laws.

Definition 1. Let t, t' be variables ranging over an α-time sequence. Then a *classical deterministic law of succession* is a relation $R(s(t), s'(t'))$ meeting the following condition for every t and t' such that $t < t'$:

$$\textit{if } \langle s(t), s'(t')\rangle \in R \textit{ and } \langle s(t), s''(t')\rangle \in R, \textit{ then } s' = s''.$$[15]

An example of such a classical deterministic law is that specified by the equations of motion for classical particle mechanics.

Definition 2. Let t, t' be variables ranging over an α-time sequence. Then a *classical statistical law of succession* is a relation $R(s(t), s'(t'), p)$ meeting the following conditions for each t and t' such that $t < t'$:

$$\langle s(t), s'(t'), p\rangle \in R \quad \text{iff} \quad P(s'(t'), s(t)) = p,$$

where P satisfies the axioms for the conditional probability operator, and

$$\textit{for each } s(t), \sum_{s'(t')\in D} P(s'(t'), s(t)) = 1.$$

An example of such a classical statistical law of coexistence is that determined by a finite Markov process. Intuitively, a classical deterministic law of succession is one where the current state determines unique subsequent states; and a classical statistical law of succession is one where for a given current state the law allows that the system may assume a number of different subsequent states and assigns conditional probabilities that these various subsequent states will be assumed. Other laws of succession are possible. In particular, there could be *non-classical* deterministic laws of succession where the present state does not determine unique subsequent states, but the states assumed over some time interval do determine unique subsequent states. History sometimes is alleged (e.g., by Thucydides) to be deterministic in this non-classical way.[16]

Next we define two types of laws of coexistence.

Definition 3. Let t, t' and t'' be variables ranging over an α-time sequence. Then a *deterministic law of coexistence* is a relation $R(s(t), s'(t'))$ meeting the following conditions for any t, t', t'' such that $t < t'$

and $t' < t''$:

$$\begin{aligned}&\textit{for every } s, \langle s(t), s'(t')\rangle \in R;\\ &\textit{for every } s, s', \textit{ if } \langle s(t), s'(t')\rangle \in R, \textit{ then } \langle s'(t), s(t')\rangle \in R;\\ &\textit{for every } s, s', \textit{ and } s'', \textit{ if } \langle s(t), s'(t')\rangle \in R \textit{ and}\\ &\quad \langle s'(t'), s''(t'')\rangle \in R, \textit{ then } \langle s(t), s''(t'')\rangle \in R.\end{aligned}$$

The Boyle-Charles gas laws are examples of such laws.

Definition 4. Let t, t' be variables ranging over an α-time sequence. Then a *statistical law of coexistence* is a relation $R(s(t), s'(t'), p)$ meeting the following condition for every t and t' such that $t < t'$:

$$\langle s(t), s'(t'), p\rangle \in R \quad \textit{iff} \quad P(s(t)) = P(s'(t')) = p,$$

where P satisfies the axioms for the unconditional probability operator; and for any t, $\sum_{s(t)\in D} P(s(t)) = 1$.

A special case example of such a law (viz. where for any t, $P(s(t)) = P(s'(t))$ for all $s'(t) \in D$) is the Boltzmann hypotheses that all states of a gas are equi-probable; more generally, such laws intuitively say that a system is such that it subsequently may assume only equi-probable states.

Although such statistical laws of coexistence typically are cited in the literature as a distinct type of law,[17] in fact they are a special variety of statistical laws of succession.

Theorem 1. Every statistical law of coexistence is a classical statistical law of succession.

Proof. Let $R(s(t), s'(t'), p)$ be a statistical law of coexistence. (i) Then for every t, t' such that $t < t'$, $\langle s(t), s'(t'), p\rangle \in R$ iff $P(s(t)) = P(s'(t')) = p$. But, then, for any $s(t)$, $s'(t')$, and p such that $\langle s(t), s'(t'), p\rangle \in R$, $P(s'(t'), s(t)) = P(s(t)) = p$. So, $\langle s(t), s'(t'), p\rangle \in R$ iff there is some real number p such that $P(s'(t'), s(t)) = p$. (ii) Since R is a statistical law of coexistence, for any t, $\sum_{s(t)\in D} P(s(t)) = 1$. But via (i), $P(s(t)) = p = P(s'(t'), s(t))$ for any $s'(t')$ such that $\langle s(t), s'(t'), p\rangle \in R$, So $\sum_{s'(t')\in D} P(s'(t'), s(t)) = 1$. Q.E.D.

It should be noted that the non-statistical analogues to statistical laws of succession are just deterministic laws of coexistence. Hence the laws defined by definitions 1, 2 and 3 (together with their various expansions described above) are exhaustive of the actual laws of succession and coexistence.

Given the usual taxonomy of laws, it would seem reasonable next to consider laws of interaction. However, it will prove expedient if, instead, we consider two types of laws which apparently have not been identified as such in the literature. These are deterministic and statistical *laws of quasi-succession*. As will emerge, the so-called teleological and functional laws are special cases of such laws.

Definition 5. Let t and t' be variables ranging over an α-time sequence, let $p_1, \ldots, p_n$ be the basic parameters of the theory, and let $i_1, \ldots, i_k$ ($k \leqslant n$) be a non-redundant listing of parameters among $p_1, \ldots, p_n$; for simplicity, suppose that $s(t) = \langle i_1(t), \ldots, i_k(t), p_{k+1}(t), \ldots, p_n(t)\rangle$. Then $R(s(t), s'(t'))$ is a *classical deterministic law of quasi-succession* iff the following condition is met for any t, t' such that $t < t'$:

$$\text{if } \langle s(t), s'(t')\rangle \in R \text{ and } \langle s(t), s''(t')\rangle \in R, \text{ then}$$
$$i_1'(t') = i_1''(t'), \ldots, i_k'(t') = i_k''(t').^{18}$$

Similar expansions to non-classical deterministic laws of quasi-succession can be made as were made for deterministic law of succession. Such expansions being made, it is obvious that a deterministic law of succession is a deterministic law of quasi-succession where k = n. Hence,

Theorem 2. Every deterministic law of succession is a deterministic law of quasi-succession.

A statistical analogue to deterministic laws of quasi-succession is possible.

Definition 6. Let t and t' be variables ranging over an α-time sequence, let $p_1, \ldots, p_n$ be the basic parameters of the theory, and let $i_1, \ldots, i_k$ ($k \leqslant n$) be a non-redundant listing of the parameters among $p_1, \ldots, p_n$; for simplicity, suppose that $s(t) = \langle i_1(t), \ldots, i_k(t), p_{k+1}(t), \ldots, p_n(t)\rangle$. Then a *classical statistical law of quasi-succession* is a relation $R(s(t), s'(t'), p)$ meeting the following condition for each t and t' such that $t < t'$, $\langle s(t), s'(t'), p\rangle \in R$ iff $\in \sum_{s^*(t')} P(s^*(t'), s(t)) = p$, where $s^*(t')$ is any state occurence just like $s'(t')$ except possibly for the values of $p_{k+1}(t'), \ldots, p_n(t')$, for each $s^*(t')$ there is some real number q

such that $P(s^*(t'), s(t)) = q$, where P satisfies the axioms for the conditional probability operator; and for each $s(t)$,

$$\sum_{s'(t')} \sum_{s^*(t')} P(s^*(t'), s(t)) = 1.$$

Clearly, a statistical law of quasi-succession is a statistical law of succession where k = n. Hence,

Theorem 3. Every statistical law of succession is a statistical law of quasi-succession.

Laws of interaction concern the behavior resulting from the interaction of two physical systems; one of the methodologically most significant sorts of interaction is perhaps that in which one of the interacting systems is a measurement device allowed to interact with a system, where the outcome of the measurement system is indicative of the value of one or more parameters characteristic of the measured system. But interactions are of wider interest since the auxiliary theories which enable the application of theories to non-isolated systems within their intended scopes often are governed by laws of interaction.

We begin by considering deterministic laws of interaction.

Definition 7. Let t and t' be variables ranging over an α-time sequence, let $s_1(t) = \langle p_1(t), \ldots, p_m(t)\rangle$, $s_2(t) = \langle q_1(t), \ldots, q_n(t)\rangle$, let $s_{12}(t) = \langle p_1(t), \ldots, p_m(t), q_1(t), \ldots, q_n(t)\rangle$, and let the $i_1, \ldots, i_k$ ($k \leqslant m+n$) be distinct elements of the list $p_1, \ldots, p_m, q_1, \ldots, q_n$. Then a *classical deterministic law of interaction* is a relation $R(s_{12}(t), s'_{12}(t'))$ satisfying the following condition for every t, t' such that $t < t'$:

> *if* $\langle s_{12}(t), s'_{12}(t')\rangle \in R$ *and* $\langle s_{12}(t), s''_{12}(t'), \rangle \in R$, *then* $s'_{12}(t')$ *and* $s''_{12}(t')$ *have the same value for the* $i_j(t')$ ($1 \leqslant j \leqslant k$).

An example of such a law of interaction is that governing a Wheatstone bridge when it is used to test the unknown resistance of a resistor.

From Definitions 5 and 7 we immediately obtain

Theorem 4: Every classical deterministic law of interaction is a classical deterministic law of quasi-succession.

It is obvious that Definition 7 can be extended to non-classical cases in the manner discussed above. and when so extended the non-classical extension of Theorem 4 will be obtained.

A restricted class of deterministic laws of interaction underlie classical measurement processes involving apparatus.[19] For in such cases we typically have a system s in state $s_1(t)$ and we bring it into 'contact' with a measurement apparatus in state $s_2(t)$ in order to ascertain the value of some parameter $i_1(t)$. As a result of that interaction, the system assumes some state $s_{12}(t')$ where $t' = t+\Delta t$, such that the combined systems will be in a state $s'_{12}(t')$ having $i'_2(t'), \ldots, i'_k(t')$ as parameter values only if $i_1(t)$ was a specific unique value. Thus, via the law of interaction governing the combined systems, the determination of $i'_2(t'), \ldots, i'_k(t')$ enables one to determine $i_1(t)$. If S is governed by a classical deterministic law of succession R, the interacting combined systems are governed by a classical deterministic law of interaction R', and R and R' are such that for a given $s_1(t)$ and $s_{12}(t)$ R and R' determine $s'_1(t')$ and $s'_{12}(t')$ such that i_1 was the same value in both states for all $t' > t$, then the measurement process is *non-disturbing with respect to* i_1; if this condition is met for all the parameters characteristic of states in S, then the measurement process is *non-disturbing*. When these conditions are not met for i_1 or all the parameters characteristic of states in S, then the measurement process is, respectively, *disturbing with respect to* i_1 or *disturbing*.[20] Definition 7 is sufficiently general as to encompass all these types of interacting and non-interacting measurement procedures, as well as other sorts of interactions. Finally, we note the possibility that the measured system could be one governed by statistical laws of quasi-succession and the measurement apparatus such that their interaction is governed by a classical deterministic law of interaction.

In many cases, the interaction of a system governed by a statistical law of succession and a measurement system (governed either by statistical or deterministic laws) will result in a combined system governed by a statistical law of interaction.

Definition 8. Let t and t' be variables ranging over an α-time sequence and let $s_1(t) = \langle p_1(t), \ldots, p_m(t)\rangle$, $s_2(t) = \langle q_1(t), \ldots, q_n(t)\rangle$, and $s_{12}(t) = \langle p_1(t), \ldots, p_m(t), q_1(t), \ldots, q_n(t)\rangle$. Then a *classical statistical law of interaction* is a relation $R(s_{12}(t), s'_{12}(t'), p)$ satisfying the following conditions for every t, t' such that $t < t'$:

> $\langle s_{12}(t), s'_{12}(t'), p\rangle \in R$ *iff for each* $s_{*12}(t')$ *having the same values for* $i_1(t'), \ldots, i_k(t')$ $(k \leqslant m+n)$ *as* $s'_{12}(t')$, *where the*

$i_1(t'), \ldots, i_k(t')$ are distinct elements on the list $p_1(t'), \ldots$ $\ldots, p_m(t'), q_1(t'), \ldots, q_n(t')$, there is a real number q such that $P(s^*_{12}(t'), s_{12}(t)) = q$ and $\sum_{s^*_{12}(t')} P(s^*_{12}(t'), s_{12}(t)) = p$, where P satisfies the axioms for the conditional probability operator, and for each $s_{12}(t)$, $\sum_{s_{12}(t)} \sum_{s^*_{12}(t')} P(s^*_{12}(t'), s_{12}(t)) = 1$.

From Definitions 6 and 8 it immediately follows that

Theorem 5. Every classical statistical law of coexistence is a classical statistical law of quasi-succession.

Obviously Definition 8 and Theorem 5 admit of the sorts of non--classical extensions mentioned above.

In manners analogous to those discussed for classical deterministic laws of interaction, a system S and an interacting measurement apparatus M such that the interacting system SM is governed by a classical statistical law of interaction can be used to obtain measurements of a parameter i characteristic of S-provided that the statistical law of interaction is such that from selected parameters in $s_2(t)$ and $s'_{12}(t')$ one is able to determine a conditional probability p that $i(t)$. Repeated measurement interactions with type S systems and recourse to stochastic techniques may enable the obtaining of further data about the behavior of S-type systems.

Our survey of the standard classification of laws (including even their non-classical extensions) has determined that there are only three basic types of laws-deterministic laws of quasi-succession, statistical laws of quasi-succession, and deterministic laws of coexistence. Other sorts of laws may be possible; in particular, we need to consider the perennial question whether the teleological and functional laws found in the social sciences and in biology constitute new and additional sorts of laws. This will be considered in Section 4.

Before we address ourselves to those specific questions, we need to consider another possible source of additional varities of laws: Nothing in the Semantic Conception of Theories restricts a theory to the possession of just one law. However, if a theory possesses more than one law, the theory will sanction only those sequences satisfying *all* the laws. Since laws are state-occurrence simple-ordering relations the conjunctive combination of several laws in effect is the intersection of several relations; and the result of such an intersection is itself a relation. Accordingly, theories generated by multiple laws will be theories governed by a single law. This raises the possibility that combinations of the basic three types of laws we have discovered thus far may yield laws of new sorts. Whether they do is an open-question to which we have no answer-though our conjecture is that they do not. Regardless of the closure properties on the combinations of laws, it is the case that the combination of laws to characterize local regularities is a commonplace in science which plays a central and apparently essential role in the scientific enterprise.[21]

IV. TELEOLOGICAL AND FUNCTIONAL LAWS

Teleological and functional laws are often claimed to constitute a distinct variety of laws differing from those found in physics (this is the so-called *separatist analysis*). We now investigate whether this is so.

By a *teleological system* we mean a system governed by a *teleological law*. What is characteristic of teleological systems is that they *tend towards some goal state or set of states*.[22] Crucial to analyzing teleological systems, hence teleological laws, is understanding 'tends'. We begin our investigation of 'tends' by considering an example of an archetypal teleological system-a servo-mechanism. Our servo-mechanism will be a simplified TV antenna rotor, which can move the antenna to one of four positions, labeled 1, 2, 3, and 4, viz.:

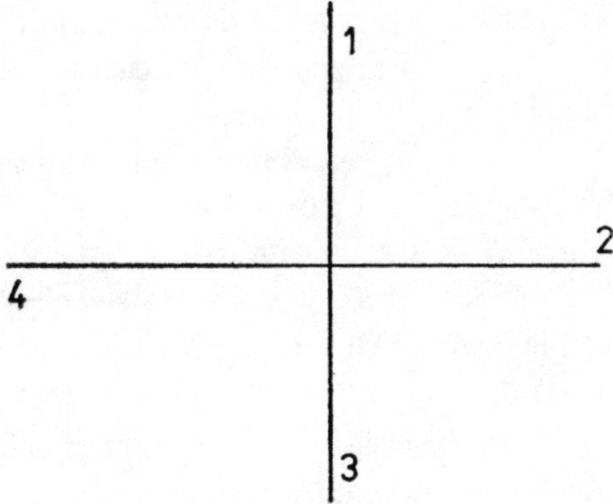

The servo-mechanism has an input device consisting of a dial which can be turned to one of four numbers 1, 2, 3, 4. It the antenna is at the location indicated on the dial, then the antenna remains in its current position. If the antenna is in a location other than that indicated on the dial, the antenna moves towards the position indicated on the dial, passing through intermediate positions in ascending or descending numerical order (thus it cannot pass directly from position 1 to 4 or 4 to 1), moving to an adjacent location each time step until such time as the dialed location is reached. The law governing this system, $R(s(t), s'(t'))$, uses an ω-time sequence, and is such that $s(t) = \langle d(t), l(t)\rangle$, where $d(t)$ is the number indicated on the dial at time t and $l(t)$ is the location of the antenna at time t. Sixteen different $s(t)$ are possible, and R is such that

$$\langle s(t), s'(t+1)\rangle \in R \quad \textit{if and only if}$$

$$l(t+1) = \begin{cases} l(t) \text{ if } d(t)-l(t) = 0 \\ 1+l(t) \text{ if } l(t) < d(t) \\ l(t)-1 \text{ if } l(t) > d(t) \end{cases}$$

regardless what d(t+1) *is.*

Several observations are in order. First, R is a deterministic law of quasi-succession. Second the set of goal states are those such that $d(t) = l(t)$.

We can now see the way in which this system is such that it 'tends' towards a set of goal states. So long as the goal state d remains unchanged over a time interval $[t, t']$, its behavior will be such that $|d(t)-l(t)|$ will be progressively reduced over time until it is minimized--which will be the case only if a goal state is reached. However, if $d(t) \neq d(t+l)$, the behavior may be such that $|d(t)-l(t)| < |d(t+1)-l(t+ +1)|$. The crucial general point here is that teleological systems typically are systems whose behavior is in part a function of its interaction with an environment (here d) outside its control; and the tendency towards a goal state is an essential characteristic of its behavior only when reacting with a stable environment. If the environment is unstable, the goal state may never be reached or even closely approached.

The foregoing example is fairly typical of a deterministic teleological system,[22a] and can be generalized. For simplicity, we will restrict our generalization to the case where the state parameters of the system all are measurable.

Definition 9. Let $R(s(t), s'(t'))$ be a classical deterministic law of quasi-succession where $s(t) = \langle i(t), \ldots, i_k(t), p_{k+1}(t), \ldots, p_k(t)\rangle$, where the $p_i(t)$ and $i_j(t)$ are as in Definition 5 and are all measurable. Let $f(i_1(t), \ldots, i_k(t), p_{k+1}(t), \ldots, p_n(t))$ be a function whose values are real numbers. Let $G = \{s(t) | f(s(t)) = \max(f(s(t)))\}$. Then R is a *classical deterministic teleological law with respect to goal* G just in case the following condition is satisfied for every $s(t)$ whenever each $p_j(t')$ $(k+1 \leqslant j \leqslant n)$ is constant for all $t' > t$:

$$\operatorname*{limit}_{t' \to \infty} f(s'(t')) = \max(f(s(t))).^{23}$

The statistical analogue to the notion of deterministic 'tending towards a goal state' intuitively is that in a stable environment the probability of the system being in a goal state will approach unity in the limit.

Definition 10: Let $R(s(t), s'(t'), p)$ be a classical statistical law of quasi-succession where $s(t) = \langle i_1(t), \ldots, i_k(t), p_{k+1}(t), \ldots, p_n(t)\rangle$, where the $p_i(t)$ and $i_j(t)$ are as in Definition 6 and are all measurable. Let $f(i_1(t), \ldots, i_k(t), p_{k+1}(t), \ldots, p_n(t))$ be a function whose values are real numbers. Let $G = \{s(t) | f(s(t)) = \max(f(s(t)))\}$. Then R is a *classical statistical teleological law with respect to goal* G just in case the following condition is satisfied for every $s(t)$ whenever each $p_j(t')$ $(k+1 \leqslant j \leqslant n)$ is constant for all $t' > t$:

$$\operatorname*{limit}_{t' \to \infty} \left(\sum_{s'(t')} P(s'(t'), s(t)) f(s'(t')) \right) = \max(f(s(t))).$$

An R which determines an absorbing Markov chain with set G of absorbing states is an example of such a classical statistical teleological law (where $k = n$) with respect to G. The notion of a boundary condition[24] provides an interesting perspective on teleological laws: For a given goal determined by a function f, a teleological law relative to that goal is nothing other than a law of quasi-succession such that the imposition of a boundary condition to the effect that the non-i parameters remain constant yields a new law satisfying the tendency conditions given in Definition 9 or Definition 10. *Inter alia*, this indicates that teleological laws do not constitute some new form of law not reducible to the traditional sorts of laws encountered in physics. Since it is well known that functional laws are just a species of teleological law,[25] it follows that functional laws likewise are not some new form of law non-reducible to the traditional sorts of laws encountered in physics. The separatist analysis is false.

V. CONCLUSIONS

In this paper we have investigated the nature of theoretical laws on the Semantic Conception of Theories. On that view, laws are just relations which determine temporal simple-orderings of state occurrences of systems. We characterized the standard sorts of laws found in science, viz. deterministic and statistical laws of succession, coexistence, and interaction and teleological laws. We also introduced two new sorts of laws, deterministic and statistical laws of quasi-succession. We showed that deterministic and statistical laws of quasi-succession and deterministic laws of coexistence are more basic than the others in the sense that all the other standard sorts of laws turn out to be reducible to one of these three sorts of law. We also showed that teleological and functional laws are special sorts of laws of quasi-succession; as such the separatist thesis is false.

Before concluding our discussion, a possible objection to our analysis of theoretical laws needs to be anticipated and answered, however. It is philosophical commo nplace that laws must support counterfactual inferences; and from this it usually has been concluded that in order to analyze or characterize laws, essential resort to the counterfactual conditional must be taken; and that in the absence of an adequate analysis of the counterfactural conditional, little in the way of analyzing laws is possible. Our analysis of various types of theoretical laws here has been set-theoretical and at no place has the counterfactual conditional been resorted to. To many this is sufficient to render the analysis offered here suspect. Such suspicions deserve serious but brief consideration.

The reason we have been able to proceed set-theoretically and avoid recourse to the counterfactual conditional in our investigation is that we have confined our attention to theoretical laws—i.e., to laws which occur within theories. On the Semantic Conception of Theories, the laws of a theory describe what the behavior of the systems within its intended scope *would be were* the systems isolated and/or certain idealized conditions met. It does so by *non-counterfactually* describing the behavior of systems under these idealized and isolated conditions, and asserting that these idealized systems stand in a counterfactual relation to most actual systems within its scope. As such a theory does support counterfactual inferences, and by extension so do its laws. But the counterfactual element in a theory is localized not in the actual statement of the laws, but rather in the *physical interpretation* of the theory—i.e., in the relation between the idealized systems described by the laws and the actual typically non-idealized circumstances obtaining in the world.[26] Differently put: To be law-like is to support counterfactual inferences. And on the Semantic Conception of Theories, this amounts to being a state-sequencing relation occurring in a theory that simple-orders state occurrences, which theory must relate counterfactually to the physical systems in its intended scope. But such simple-ordering relations themselves (as opposed to their counterfactual interpretation) can be characterized set-theoretically without recourse to the counterfactual conditional. As such the line of objection sketched fails.

This discussion of how the counterfactual component enters into theories and their laws also indicates in what sense it is true that a fundamental characteristic of (theoretical) laws is their inaccuracy: Theoretical laws precisely characterize the behavior of systems under isolated and/or idealized conditions which usually do not obtain in actual systems within the theory's intended scope. As such, the behavior of real-world systems within a theory's intended scope often will deviate from that specified by the theory. But that does not make the theory or its laws inaccurate. For if the theory is empirically true, its laws are a totally accurate description of what the behaviors of the real-world systems within its scope *would be were* the systems isolated and/or various idealized conditions met. And given that this is the descriptive function of a theory and its laws, the characterizations provided by an empirically true theory are totally accurate. Thus a fundamental feature of theoretical laws is that they idealize their phenomena; but doing so does not make them *ipso facto* inaccurate. Indeed, on the Semantic Conception of Theories, the only inaccurate laws are those characteristic of empirically false theories. This finding vindicates our feeling, mentioned at the outset, that resort to a notion of approximate truth is not required for an adequate analysis of theoretical laws.[27]

REFERENCES

[1] Support for the writing and presentation of this paper was provided by NSF Grant GS-39677 and travel grants from the ACLS, the IUHPS, and the University of Maryland. I would like to thank professors Raymond Martin and Lars Svenonious for helpful comments on the draft of this paper.

[2] Translated by Stillman Drake, second edition, (Berkeley: University of California Press, 1970), pp. 207–208.

[3] 'The Key Property of Physical Laws-Inaccuracy', pp. 91–101 in H. Feigl and G. Maxwell (eds.), *Current Issues in the Philosophy of Science* (New York: Holt, Rinehart, and Winston, 1961); a commentary on this paper by Henryk Mehlberg and a rejoinder by Scriven follow on pp. 102–104.

[4] *Ibid.*, p. 100; defects in this proposal are presented in Mehlberg's commentary following Scriven's paper.

[5] Several papers at this conference deal with the issue of approximate truth.

[6] The Semantic Conception of Theories was first suggested by Evert Beth in his 'Towards an Up-to-Date Philosophy of the Natural Sciences' *Methodos*, **1** (1949): 178–185; in my 'The Meaning and Used Models in Mathematics and the Exact Sciences' (*Ph. D. Dissertation, The University of Michigan*, 1967), I independently advanced the Semantic Conception and developed it for a large class of theories. Further development of the Semantic Conception has been done by Bas van Fraassen in his 'On the Extension of Beth's Semantics of Physical Theories', *Philosophy of Science*, **37** (1970): 325–339, and his 'A Formal Approach to the Philosophy of Science', pp. 303–366 in R. Colodny (ed.), *Paradigms and Paradoxes* (Pittsburgh: University of Pittsburgh Press, 1972); and by me in Part II of 'What's Wrong With the Received View on the Structure of Scientific Theories?', *Philosophy of Science*, **39** (1972): 1–19, 'Some Philosophical Problems in Biological Taxonomy and Speciation', pp. 190–243 in J. Wojciechowski (ed.), *Conceptual Basis of the Classification of Knowledge* (Munich: Verlag Dokumentation, 1974), pp. 221–230 of 'The Search for Philosophic Understanding of Scientific Theories' (pp. 3–241 in F. Suppe (ed.), *The Structure of Scientific Theories*, Urbana: University of Illinois Press, 1974; second edition, 1977), 'Theories, Their Formulations, and the Operational Imperative', *Synthese*: **25** (1972): 129–164, and 'Theories and Phenomena', 45–92 in W. Leinfellner and W. Köhler, *Developments in the Methodology of Social Science* (Dordrecht: Reidel, 1974). van Fraassen's and my developments of the Semantic Conception differ in some respects, since he views theories instrumentally whereas I am a realist. In what follows my realistic version of the Semantic Conception will be followed. Although differing in a number of respects, the Semantic Conception is closely related to some of Patrick Suppes' work; cf. especially his 'What is a Scientific Theory?', pp. 55–67 in S. Morgenbesser (ed.), *Philosophy of Science Today* (New York: Basic Books, 1967) and 'Models of Data', pp. 252–261 in E. Nagel *et al*, *Logic, Methodology, and the Philosophy of Science* (Stanford: Stanford University Press, 1962). For a discussion how Suppes' views differ from the Semantic Conception, cf. van Fraassen, 'A Formal Approach to the Philosophy of Science', *op. cit.*

[7] The word 'model' must be used with extreme care since it can mean a number of different things in science. Here we are using 'model' to mean *iconic model*—an entity which is structurally similar to the entities in some class (as, e.g., a model airplane is a model of the real planes in the class of F-4H fighters). For good discussions of the various meanings 'model' can take on in science, cf. Patrick Suppes, 'A Comparison of the Meaning and Uses of Models in Mathematics and the Empirical Sciences' (pp. 163–177 in H. Freudenthal, *The Concept and the Role of the Model in Mathematics and Natural and Social Sciences* (Dordrecht: D. Reidel, 1961)) and Chapter 7 of P. Achinstein, *Concepts of Science* (Valtimore: Johns Hopkins Press, 1968); our discussion follows Suppe, *The Meaning and Used Models in Mathematics and the Exact Sciences* (Ph. D. Dissertation: The University of Michigan, 1967), pp. 80–107.

[8] Laws typically are taken to be linguistic entities. The legitimacy of construing laws as extra-linguistic entities is argued for in Peter Achinstein, *Law and Explanation* (Oxford: Oxford University Press, 1971), Chapter 1. M. Bunge also construes laws extra-linguistically in his *Causality: The Place of the Causal Principle in Modern Science* (New York: Meridian, 1963). For the notion of a relational system, cf. A. Tarski and R. Vaught, 'Arithmetic Extensions of Relational Systems', *Composito Mathematicae*, **13** (1957), 81–102. Unpublished results recently obtained demonstrate that if suitably broad classes of automata are assumed, the relational system of a theory always will be an automaton; the class of automata in question includes non-deterministic and non-finite automata. See B. Zeigler, *Theory of Modelling and Simulation* (New York: Wiley-Interscience, 1976) for characterizations of such automata.

This heuristic account greatly oversimplifies how it is that laws determine sequences of state occurrences; a more detailed exposition of the account is given in the next section. If the laws are statistical, the domain of the relational system will contain real numbers as well.

[9] This correspondence is the heart of the physical interpretation of theories on the Semantic Conception. For a full and detailed discussion of the physical interpretation of theories, cf. my 'Theories, Their Formulations, and the Operational Imperative', *op. cit.* A much more detailed discussion of the physical interpretation of theories is presented in my forthcoming book, *Facts, Theories, and Scientific Observation*, Chapters 7 and 10.

[10] The auxiliary theory must be such that in the particular circumstances its idealized conditions are met. Auxiliary theories can be 'local theories' applying only to a very limited variety of circumstances; see Suppe, 'Theories and Phenomena', *op. cit.* for a detailed discussion of the foregoing claims.

In order to predict a unique state in the case just summarized, the auxiliary theory would also have to impose some rather severe boundary conditions as well.

[11] Cf. Suppe, 'Theories and Phenomena', *op. cit.*, for a discussion of suitable auxiliary theories; for additional discussion of the empirical truth conditions for theories, cf. Suppe, 'Theories, their Formulations, and the Operational Imperative', *op. cit.*

In case the theory contains statistical laws, the various probabilities with which state changes occur in the sequences determined by the theory also must be empirical probabilities characteristic of such changes for corresponding phenomena under idealized conditions.

[12] E. g., such a classification is used by van Fraassen, 'An Extension of Beth's Semantics...', *op. cit.*, myself in most of the works cited above in note 6, and partially by Hempel on p. 352 of his 'Aspects of Scientific Explanation' (pp. 331–496 in Hempel, *Aspects of Scientific Explanation* (New York: The Free Press, 1965)).

[13] Other discrete orderings of time are possible-e. g., as having the order properties of the rational numbers—but will be ignored here as science does not seem to have resorted to them.

[14] The notation, $s(t)$, for the occurrence of s at t is somewhat misleading in that it suggests that s is a function of t, which it is not; rather state occurrences are ordered pairs, $\langle s, t\rangle$. The use of ordered pairs $\langle s, t\rangle$ to indicate state occurrences in the developments which follow proves to be notationally quite messy, and so I have opted for the (somewhat non-standard) $s(t)$ notation.

[15] Note that this definition employs a cut procedure when $\alpha = \lambda$; since laws are extra-linguistic entities, this definition is not liable to the sorts of defects in defining 'deterministic laws' that Richard Montague pointed out in his 'Deterministic Theories', pp. 325–370 in Washburne, *Decisions, Values, and Groups* (New York: Pergamon Press, 1957).

[16] The deterministic laws of succession (including both classical and non-classical ones such as those just discussed) intuitively seem to include the so-called causal laws. Whether they do in fact depends upon whether Hume, Reichenbach, and others are correct that cause must preceed effect, or whether Kant was correct that cause and effect can be simultaneous. If Hume is correct, then Definition 1 and the expansion to non-classical cases would encompass all causal laws; if Kant is correct, they do not. For a recent defense of Kant's position, see A. Melnick, *Kant's Analogies of Experience* (Chicago: University of Chicago Press, 1974).

If Kant and Melnick are correct, Definition 1 is adequate for those cases where cause preceeds effect; the extension to cases where cause and effects are simultaneous involve major difficulties which we will not go into here. Whether they are correct or not is to my mind an open question, the resolution of which will have a crucial bearing on the problem of explanation and deserves serious investigation.

[17] Cf. note 12 above for examples.

[18] For appropriate caveats cf. notes 15 and 16 above.

[19] For the formulation of a realistic (vs. conventionalistic) theory of measurement, and a discussion of the role of laws of interaction in their design, cf. Suppe, 'Theories and Phenomena', *op. cit.*, esp. Section V.

[20] For simplicity I have confined the characterization of interferring and non-interferring measurements to systems S governed by classical deterministic laws of succession; analogues to these notions for S governed by deterministic or statistical laws of quasi-succession and/or statistical measurement apparatus are possible; cf. below for related discussion. For further discussion, cf. also my 'Theories and Phenomena', *op. cit.*, Section VII.

[21] Cf. Suppe, *ibid.*, for further discussion of such procedures.

[22] Cf. Richard Rudner, *Philosophy of Social Science* (Englewood Cliffs: Prentice Hall, 1966), Chapter 5, especially pp. 94–95.

[22a] There are respects in which it is atypical-viz. that the goal state d is part of the environment (in many cases this will not be so) and no hysteresis occurs in the tending towards relation under a stable environment. Our formal definitions allow for these possibilities.

[23] Notice that this definition allows considerable hysteresis in the tending towards relation, as does the next definition.

[24] The phrase 'boundary condition' tends to be used with two different meanings in science: (1) conditions concerning the extent to which the system in question meets the idealization conditions of the theory, and (2) local regularities characteristic of a system within the scope of a theory which further restrict the possible behaviors allowed by the laws of the theory-as, e.g., when a parameter in a theory whose only law is a deterministic law of coexistence is allowed variable values, but outside constraints on system S are such that the parameter maintains a constant value. I am using 'boundary condition' in the second sense here.

[25] Cf. Rudner, *op. cit.*, Section 19.

[26] For further discussion of this localization of the counterfactual element, cf. my 'The Search for Philosophic Understanding of Scientific Theories', *op. cit.*, pp. 42–45. A more profound defense of the claims just made is provided in my *Facts, Theories, and Scientific Observation, op. cit.*

[27] Approximate truth may play a role, however, in assessing whether measured data (error-laden as all such data must be due to the limitations in accurracy of measurement apparatus) are in agreement with the predictions afforded by the theory. Differently put, although approximate truth plays no role in determining the empirical truth or falsity of a theory (hence its laws), it may figure centrally in deciding which theories to accept as true. Also, work on the growth of scientific knowledge indicates that in assessing theories one often is less concerned with the empirical truth of a theory than with its *promise* for subsequent development into a true theory; here approximate truth sometimes plays a role. For discussion of this latter point, cf. Section III–B and especially Section III–C of my "Afterword--1976" in the second edition of my *The Structure of Scientific Theories, op. cit.*

RAIMO TUOMELA

CAUSALITY, ONTOLOGY, AND SUBSUMPTIVE EXPLANATION

I

David Hume distinguished four main elements in the relation of causation: constant conjunction, contiguity in space and time, temporal priority, and necessary connection. In this paper I will be mainly concerned with the first of these properties; and I will be mostly interested in some ontological aspects of event-causality. As Jaegwon Kim and others have shown, the ontological aspects of causality are very problematic and at least the subsumptivistic view of event-causation seems to be in deep ontological trouble (see e.g. Kim [1969] and [1973]). My main purpose in the paper is to try to present some arguments to defend the subsumptivistic or deductive-nomological approach.

Speaking in general terms, in my view causation is semantically a "mongrel" notion. It is logically or conceptually related both to the external world and to our conceptual framework. Accordingly, it is singular objective events that primarily are causes and effects but they are so only in virtue of being conceptualized in a certain way, viz. in virtue of being instantiations of some suitable generic events which are nomically connected.

Singular claims concerning event-causation have to be justified by means of 'backing' laws which subsume the cause and effect candidates in question. I take it that of the above Humean features of causality spatiotemporal contiguity and temporal priority seem to be properties primarily of singular events. On the other hand, constant conjunction and nomic necessity make sense only in connection with generic events. As I think the backing causal laws have to 'speak about' generic events, causal laws are to be elucidated in terms of their accounting for constant conjunction and, seemingly contrary to Hume, of nomic connection, too.

To illustrate what I have in mind consider the following familiar example. If I strike a dry match under normal circumstances (presence of oxygen, etc.) the match will light. Then we say that this striking of the match caused its lighting. This is an example of a causal relation holding between two singular events. But this causal claim has to be justified by making reference to a backing law connecting suitable generic events. In this case the law would connect, I think, not the generic observable events of striking and lighting but rather two theoretical physicochemical (complex) properties.

As recent discussion has shown, the notion of causality is seriously troubled by some ontological issues. First, there is no general agreement as to what kind of ontological entities qualify as causes and effects. Are causes and effects events, states, conditions, processes, properties, facts, propositions, or what? A final answer to this problem is not central to my concerns in this paper. I shall tentatively assume that primarily the category of *events*—broadly understood—is the type of entities which qualify as causes and effects. It should be clear, however, that some differences in views concerning causality are affected by what is meant by events (cf. e.g. Kim [1973] and our discussion below). It makes a relevant difference to our discussion below whether one views singular events as 'bare' and unstructured particulars or as having some structure. For instance, Davidson thinks of singular events as unstructured whereas I give them structure (cf. Davidson [1967] and [1969]). Thus in my view an event consists of a locus-object, a point of time, and a generic property exemplified by the locus-object. The property in question can be called the *aspect* property (or the *constitutive* property) of the event, and it is to be construed intensionally. To take a simple monadic example, the event of Socrates drinking hemlock at time t is represented by $\langle\langle \text{Socrates}, t\rangle, \text{drinks hemlock}\rangle$ or, more formally, by $\langle\langle x, t\rangle, P\rangle$. The property P is the aspect or constitutive property of this event. On the other hand, this event is said to *merely* exemplify the property of consuming poison, say Q, which is not constitutive of the event of Socrates drinking hemlock at t. The identity condition for monadic events now may be construed as follows: $\langle\langle x, t\rangle, P\rangle = \langle\langle y, t'\rangle, Q\rangle$ if and only if $x = y$, $t = t'$, and $P = Q$. As in our above example $P \neq Q$, the events in question are deemed different.[1]

It should be noticed that I intend my treatment to be applicable also to the causation of human actions and that this can become possible in part because I think actions can be regarded as a special case of events, viz. events brought about by human agents.

The second hard problem in discussing the ontology of causality is this: What kind of statements qualify as *event-descriptions*? If we want to make sense of locutions like a 'redescription of an event' this problem has to be solved. Davidson, one of the main theoreticians in this field, has argued (partly on the basis of a paradoxical result discussed in the next section) that the attempt to characterize event-describing statements is futile and that (singular) events should rather be referred to by nominalizing verb phrases (singular terms) (see e.g. Davidson [1969]). Many ordinary language statements which may seem to describe particular events are on a closer look existential Thus, for example, the correct logical form of the action sentence 'Shem Kicked Shaun' is not according to Davidson anything like 'Kicked (Shem, Shaun)', which sentence is based on a two-place predicate, but rather (Ex) Kicked (Shem, Shaun, x), which is based on a three place predicate incorporating the variable x for singular events. According to Davidson's suggestion, if the kicking of Shem by Shaun occurred at midnight this is taken care of simply by (Ex) (Kicked (Shem, Shaun, x) $\wedge$ At midnight (x)). But what if this kicking was intense? If we add the conjunct 'Intense (x)' we are saying that the event was intense whereas what we wanted to do was to modify the verb 'kicked'. However, it seems that problems like this can be handled, and attempts to develop a theory of verb-modification have been made. Therefore the Davidsonian framework seems to us acceptable provided the structured nature of events can properly be accounted for.

Singular events are referred to and picked out by singular terms of the scientific language employed. Therefore we may employ singular causal claims such as: e caused e'. (Here 'e' names the singular event e and e' names e'.) But singular causal claims like this do not display the structure of events and hence not the generic properties in virtue of which the singular events can be shown to be causes or effects. The justification of causal claims is to be made by reference to a backing theory which in our view is best formulated linguistically. It follows that we need *event-describing statements*. Of their form and nature we can here say at least the following.

If event $e = \langle\langle x, t\rangle, P\rangle$ in most cases of interest the property P can be designated by a predicate 'P' in our scientific language. Then the occurrence of e can be represented linguistically by '$P(e) \wedge (Ex)\ (x = e)$'. (It is not necessary here to consider linguistically the locus and the time of the occurrence of e.) The true statement $P(e)$ is a *constitutive* statement for e, and it may be considered (in a sense) analytically true of e (see Tuomela [1974a]). Notice, however, that the backing theory need not contain P at all as the subsumption can be effected by reference to other predicates true of e.

If what I just said can be accepted we may move from a singular causal statement like "e caused e'" to a corresponding linguistic formulation of the type "That e has P caused that e' has Q". Thus we seem to be able to go back and forth between ontological or material talk and linguistic fact-talk. But there are troubles here.

First, we may not know under which predicates e and e' are causally related even if we may have reasons for claiming that e caused e'.

Secondly, the linguistic formulation must be taken to entail the truth of both the antecedent and the consequent, I think. But then it is immediately seen to be non-truthfunctional (switching antecedent and consequent does not preserve the truth of the causal claim). If 'caused' then represents a non-truth-functional conditional, what could its logical form be? I shall not investigate this problem here, but I think we are about to confuse two separate issues. When we speak of e.g. match-striking causing match-lighting we think, or should think, of a *general* statement connecting these generic properties (or, rather, some underlying physico-chemical counterparts). Formally, we then deal with a law of e.g. the form $(x)\ (y)\ (A(x) \wedge P(x, y) \rightarrow B(y))$ (cf. Section IV), which some singular events e and e' may instantiate and which hence satisfy the singular conditional '$A(e) \wedge P(e, e') \rightarrow B(e')$'. This situation is to be distinguished from the case where a singular event e is, or is claimed to be, a cause of another singular event e'. We are here mainly concerned with this latter case. The former case is concerned with the nature of the backing law, which we cannot here discuss in any detail.

II

Let us now go on to consider a certain analysis of singular causal claims. It is the most sophisticated version of those discussed and criticized in Kim [1969]. For the purposes of this section we consider a purely extensional set-theoretic construal of events (thus aspect properties become sets and relations). The present causal analysis can be formulated as follows:

(C) *Event e is a cause of event e′ if and only if there are singular statements D and D′ such that D and D′ describe e and e′ respectively, and D′ is a logical consequence of D together with some causal law S but not of D alone.*

It may seem plausible to assume that event-describing statements are referentially transparent and that the class of event-describing statements is closed with respect to logical deduction. This amounts to accepting the following two plausible looking principles (cf. Kim [1969], pp. 206–207):

(I) *If a statement D′ is obtained from a statement D by replacing some referring expression in D by a co-referential expression, then D and D′ describe the same event.*

(L) *If a statement D entails another statement D′ the event described by D is said to include the event described by D′.*

As an immediate corollary of (L) we then have

(L′) *Logically equivalent statements describe the same event.*

But the acceptance of these principles leads to the result that singular causal statements are truth-functional, after all. Hence we get an 'ontological' paradox. This well known paradox has many faces, but we shall be mainly interested in the following version of it:

Theorem 1. Any cause can be described by means of any arbitrary true event-describing statement, and any effect can be described by means of any arbitrary true event-describing statement.

We shall here prove this theorem only for the simple case where the causal law S can be written as $(x)\,(A(x) \rightarrow B(x))$, where '$\rightarrow$' may denote a suitable nomological implication or merely the ordinary material implication. (The proof goes through for other kinds of causal laws such as laws of the form $(x)\,(y)\,(A(x) \wedge P(x, y) \rightarrow B(y))$ to be discussed in Section IV). Let us now assume that an object named a has both A and B. Therefore the events described by 'a has A' and 'a has B' become subsumed under our law S according to the principle (C). But now consider an arbitrary true event-describing statement 'b has K'. Let then 'I' stand for the definite description operator and consider the following statements:

(1) 'a has A'

(2) 'a has B'

(3) 'b has K'

(4) '$(Ix)\,(x = b \wedge a$ has $A)$ has K'

(5) '$(Ix)\,(x = a \wedge b$ has $K)$ has A'

Now, by principle (I), (1) and (5) describe the same event. Similarly, (3) and (4) describe the same event. As (4) entails (1), the event described by (4) includes the event described by (1). Analogously, the event described by (5) is seen to include the event described by (3). But then (L) tells us that 'a has A' and 'b has K' describe the same event. Hence, according to (C), S subsumes the event-descriptions 'b has K' and 'a has B' so that the event described by the former is a cause of the event described by the latter.

Quite analogously we can prove that the effect of a given cause can be described by means of any arbitrary true event-describing statement.

Our Theorem 1 can in fact be 'ontologically' strengthened if we, in accordance with our earlier considerations, require that singular causal statements entail the existence of the cause-event and the effect-event. For then we get the following result: If an event e causes an event e' then e causes any arbitrary event e'' and e' is caused by any arbitrary event e'''.

Our above observation is of course disastrous also for deductive (causal) explanation, if explanation is identified with plain deduction. For it means that given an explanans (a law S and an event-description D) we can explain any singular explanandum statement D'. Likewise any true D' can be explained by means of any true initial condition statement D and a suitable law statement.

Still another aspect of our paradox can be proved by means of (I) and (L′) and stated as follows:
All true (event-describing) statements describe one and the same event and all false statements describe the same thing, whatever that thing is. (This well known version of our paradox has been proved in e.g. Quine (1966), pp. 160–161).

Obviously something is deeply wrong with our referential apparatus, but what? Recall that we were discussing some singular statements which were supposed to be event-describing. But it is by no means clear that all singular statements, not even those we have been considering, are event-describing. Thus, even if we may plausibly assume that we start in (C) with some event-describing statements there is no quarantee that the application of conditions (I) and (L) always yields event-describing statements. For instance, Kim [1969], pp. 209–213, suspects that the class of event-describing statements is not closed under logical deduction, and hence that (L) is not acceptable for this reason. He also has some doubts concerning (I) related to the use of *predicative expressions* (e.g. 'the tallest man in this room') as compared with *genuine referring* expressions (e.g. 'John') in this condition. I agree with Kim in his doubts and critical remarks concerning (I) and (L). But below my main attack will be different and somewhat indirect, as I shall concentrate on different aspects of the situation, i.e. on (C).

But perhaps it would not matter so much if (I) and (L) occasionally give us odd looking event-descriptions, a critic may voice. However, as our answer we may try to refer to Theorem 1, the proof of which does not seem to involve any very odd event-descriptions. Indeed, Davidson [1967] for one considers (I) and (L) (or at least (L′), if that matters) and hence the resulting paradoxical theorem, acceptable. He then concludes that causal connections between events cannot be described by means of singular statements, but that rather it has to be done by means of a two-place causal predicate. For the reason mentioned earlier, and for some additional reasons related to partial causation (see e.g. Davidson [1967], pp. 698–699), I accept this conclusion even if I do not accept Davidson's main reason (i.e. singular causal claims treated by means of *statements* are 'shown' to be both truth-functional and non-truthfunctional, which 'result' is absurd).

We may notice here that our rejection of (I) and (L) means that at least the linguistic notion of eventhood is construed as an *intensional* (i.e. non-transparent and ultra-intensional) notion.

III

As I indicated, my approach to solving the paradox expressed in Theorem 1 consists of something neither Kim nor Davidson have tried. That is, I reject condition (C). But, taken in conjunction with some plausible semantical assumptions, then also (L) will have to be rejected.[2]

I will try below to analyze causality in terms of nomological causal explanation, and this requires some modifications in (C). Let us now write $C(e, e')$ for "e caused e'". Then, instead of (C), I now propose the following principle:

(C*) '$C(e, e')$' *is true only if there are singular statements D and D' such that D and D' describe e and e' respectively, and there is a causal law (or theory) S which jointly with D deductively explains D' (but D alone does not explain D').*

Condition (C*) gives a necessary condition for the truth of the causal predicate $C(e, e')$. To give a sufficient condition, as well we have to add some *general* pragmatic conditions which pertain to the use of theory S, and some singular conditions concerning primarily the spatio-temporal relations between e and e' (see the discussion in Tuomela [1974b]). Here I shall not discuss these additional conditions nor what is required of the causal theory S as to its form and content (to e.g. avoid the apparent circularity in (C*)).

To clarify my new proposal and to show why it does not lead to ontological paradoxes of the above kind I shall start by sketching the account of the logical (i.e. deductive-inferential) aspects of deductive explanation I have in mind for (C*).[3]

One central requirement for deductive explanations is that they should not be more circular than necessary. (Of course there must be some common content between the explanans and the explanandum; otherwise there could not be a deductive relationship between them.) As recent discussion on explanation has shown, the following kind of general condition has to be accepted: In an explanation the components of the explanans and the explanandum should be noncomparable. We say that two components or statements P and Q are *noncomparable* exactly when not $\vdash P \rightarrow Q$ and not $\vdash Q \rightarrow P$. (See e.g. Ackermann [1965], Omer [1970], and Tuomela [1972] for a discussion leading to the acceptance of this general condition.) Actually our analysis of noncomparability needs some refinement, mainly because of the vagueness of the notion of a component. To accomplish this we use the notions of a sequence of truth functional components of an explanans and of a set of ultimate sentential conjuncts of an explanans (cf. Ackermann and Stenner [1966]).

A sequence of statemental well formed formulas $\langle W_1, W_2, \ldots, W_n \rangle$ of a scientific language $\mathscr{L}$ is a *sequence of truth functional components* of an explanans ("theory") T if and only if T may be built up from the sequence by the formation rules of $\mathscr{L}$, such that each member of the sequence is used exactly once in the application of the rules in question. The W_i's are thus to be construed as tokens. The formation rules of $\mathscr{L}$ naturally have to be specified in order to see the exact meaning of the notion of a sequence of truth functional components of a theory finitely axiomatized by a sentence T. A *set of ultimate sentential conjuncts Tc* of a sentence T is any set whose members are the well formed formulas of the longest sequence $W_1, W_2, \ldots, W_n$ of truth functional components of T such that T and $W_1 \wedge W_2 \wedge \ldots \wedge W_n$ are logically equivalent. If T is a set of sentences then the set Tc of ultimate conjuncts of T is the union of the sets of ultimate sentential conjuncts of each member of T. We may notice here that although by definition the Tc-sets of two logically equivalent theories are logically equivalent they need not be the same. (Also notice that there are no restrictions which would exclude e.g. the use of a causal or nomic implication in $\mathscr{L}$. Still we do not necessarily need such an implication below.)

Now we are ready to state a better version of the noncomparability requirement for a Tc of a theory T constituting an explanans (cf. Tuomela [1976]): For any Tc_i in the largest set of truth-functional components of T, Tc_i is noncomparable with the explanandum.

In addition to this condition we require that the explanans and the explanandum of an explanation are consistent, that the explanans logically implies the explanandum, and that the explanans contains some universal laws.

Finally, there is a nontrivial logical condition for our explanation relation, call it E, which condition guarantees that an explanans provides a proper amount of relevant information. This is condition (5) below. (The reader is referred to Tuomela [1972] and [1976] for a discussion of its acceptability.) Now we can state our model of explanation (termed the *weak DEL-model* in Tuomela [1972]). Let T be a statement, Tc a set of ultimate sentential components of T (or actually a conjunction of components in the context $E(L, Tc)$), and L a singular statement to be explained. Then we say that the relation $E(L, Tc)$ satisfies the logical conditions of adequacy for the deductive explanation of (singular or general) scientific statements if and only if

(1) $\{L, Tc\}$ is consistent;

(2) $Tc \vdash L$;

(3) Tc contains at least some universal statements;

(4) for any Tc_i in the largest set of truth functional components of T, Tc_i is noncomparable with L;

(5) it is not possible, without contradicting any of the previous conditions for explanation, to find sentences $S_i, \ldots, S_r$ ($r \geqslant 1$) at least some of which are essentially universal such that for some $Tc_j, \ldots, Tc_n$ ($n \geqslant 1$):

$$Tc_j \ldots Tc_n \underset{p}{\vdash} S_i \wedge \ldots \wedge S_r$$

$$\text{not } S_i \wedge \ldots \wedge S_r \vdash Tc_j \wedge \ldots \wedge Tc_n$$

$$Tc_s \vdash L$$

where Tc_s is the result of the replacement of $Tc_j, \ldots, Tc_n$ by $S_i, \ldots, S_r$ in Tc, and '$\underset{p}{\vdash}$' means 'deducible by means of predicate logic but not by means of universal or existential instantiation only'.

Condition (5) is not quite unambigously formulated as it stands. The reader is referred to Tuomela [1976] for its clarification (note especially conditions (V) and (Q) discussed in that paper) and also for an alternative interpretation of '$\underset{p}{\vdash}$'. See Tuomela [1972] for a detailed discussion of the formal properties of the notion of explanation that this model generates. Here it must suffice to make the following general remarks only.

In the above model of explanation an explanandum may have several explanantia differing in their quantificational strength (depth). On each quantificational level, however, only the weakest explanans-candidate qualifies. Our model thus generates an explanation-tree for each explanandum such that the explanantia in different branches may be incompatible whereas the explanantia within the same branches are compatible and increasingly stronger.[4]

More exactly, the weak *DEL*-model has the following central logical properties (see Tuomela (1972)):

Theorem 2. The explanatory relation of the *DEL*-model has the following logical properties:

(a) $E(L, Tc)$ is not reflexive.

(b) $E(L, Tc)$ is not symmetric.

(c) $E(L, Tc)$ is not transitive.

(d) If $E(L, Tc)$ and if, for some Tc', $\vdash Tc' \rightarrow Tc$ (assuming not $\underset{p}{\vdash} Tc' \rightarrow Tc$), then $E(L, Tc')$, provided every $Tc'_i \in Tc'$ is noncomparable with L.

(e) $E(L, Tc)$ is not invariant with respect to the substitution of either materially or logically equivalent explanantia nor explananda.

(f) If $E(L, Tc)$ and if, for some T' such that $\vdash T \equiv T'$, T and T' possess identical sets of ultimate sentential components (i.e. $Tc = Tc'$), then $E(L, Tc')$.

(g) If $E(L, Tc)$ and for some L', $\vdash L \equiv L'$, then $E(L', Tc)$, provided that, for all Tc_i in Tc, Tc_i and L' are noncomparable.

(h) If $E(L, Tc)$ and $E(L, Tc')$, then it is possible that Tc and Tc' (and hence the corresponding theories T and T') are mutually (logically) incompatible.

What interests us especially much in this context is the property (e). The lack of linguistic invariance exhibited by it shows or expresses the fact that explanation is a *pragmatic* notion: How you *state* your deductive argument may make a great difference concerning the explanatory value of the argument.

Let us see now how far we have come. We have a (hopefully) workable model for the central logically explicable aspects of explanation. (We shall not here discuss other aspects of explanation.) I have elsewhere (Tuomela [1972] and [1973]) emphasized some other interesting features concerning contexts where scientific laws or generalizations are explained by means of theories. But here we are more interested in the explanation of some aspects of singular events by means of causal theories according to our principle (C*). Therefore we assume that the singular explanandum L is an event-describing statement. Furthermore the explanans T (and hence Tc) will contain an event-describing statement ("initial-condition") which describes the cause (in the case of *causal* explanation). One thing that may be mentioned here is that our explanations concern only some aspects of (singular) events, viz. those aspects the explanatory theory deals with. It does not make much sense to ask for an explanation of a singular event *in toto*, even if the causal predicate $C(e, e')$ of course deals with singular events in all their concreteness. For, first, explanations are primarily concerned with linguistic items and all the aspects of a singular event can never be described, in generic terms. Secondly, the aim of science is to study generic aspects of the world in the first place. The individual and idiosyncratic features of singular events are therefore of secondary importance for theoretical science. In any case it should be noticed that our principle (C*) allows for a transition between "ontological" causal statements like $C(e, e')$ and linguistic causal statements like "D is a cause of D'" with e.g. $D = P(e)$ and $D' = Q(e')$.

Whether an argument is an explanation should *prima facie* be decided on its own right, so to speak, and not on the the basis of its having been somehow derived from another acceptable explanatory argument. This is the case in our model of explanation. If an argument satisfies our logical arguments it becomes a potential explanation. To qualify as a materially valid explanation still some other requirements have to be imposed. For instance, the generalization (or theory) used in the explanans should be lawlike, which we take to entail that it contains only predicates expressing natural kinds. Furthermore, the explanans should be true and accepted as true (or something analogous), etc. Some of these further demands will have to refer to pragmatic features of the explainer and his scientific community.

Related to general aspects of the situation the validity of principle (L) of Section II can now be discussed. To show conclusively that within our account of singular causality (L) is invalid would require the development of a full blown semantical theory. This task cannot be undertaken here (see e.g. the discussion in Tuomela [1974a]). Still something can be said about it.

First, we notice that scientific laws and theories can be argued to implicitly and partly define the meanings of the predicates which occur in them; hence they also affect the nature (or at least the conceptualization) of the properties designated by the predicates of these theories (see Ch. V of Tuomela [1973] concerning this). Therefore it is natural to accept scientific theories and explanations to guide our views as to "what there is".

Secondly, in the case of structured singular events such as $\langle\langle x, t\rangle, P\rangle$ it is natural to think that for the (atomic or complex) positive property P there is a directly corresponding linguistic expression (cf. our earlier remarks). However, if we want to build an intensional semantics to account for e.g. negative predicates problems arise. For in my view there are no *negative* singular events (e.g. events with ultimately negative aspect properties). A predicate '$\neg P$' may correspond in its exemplifications to a property constellation Q in the case of singular event e but to a different constellation Q' in the case of event e'. Thus, consider two conditional composite predicates '$P \rightarrow Q$' and '$\neg Q \rightarrow \neg P$'. They may be considered logically equivalent as the latter represents a contraposition of the former. Still they are not generally intersubstitutable within our model of explanation without change in the validity of explanation (see property (e) of Theorem 2). This fits well together with the view that negative predicates correspond to different positive properties on different occasions. Thus while '$P \rightarrow Q$' may be taken to always represent the same property, '$\neg Q \rightarrow \neg P$' may correspond to different properties, and hence to different events, depending on the situation. Accepting this, (L) is not a valid principle, and the above proof of Theorem 1 becomes blocked also for this reason. Other related arguments for showing the invalidity of (L) can be developed e.g. on the basis of defining new more complex predicates which are logically equivalent to the original one but which, through yielding different constitutive statements, describe different events.[5]

Before going on I want to emphasize one additional feature, which I cannot pursue further here. This is the role of so called theoretical concepts within explanation. My *DEL*-model (in contrast to the above *weak DEL-model*) explicitly requires that theoretical concepts occur in the explanans (the weak *DEL*-model neither requires nor forbids this). Much of scientific investigation just aims at discovering underlying and 'hidden' causes for manifest phenomena and events. For instance, it is not very interesting or important from the point of view of causal explanation to know for instance that the emergence of measles-speckles is nomically, and perhaps in some sense causally, preceded by fever in patients. The important thing is to discover the underlying virus which causally accounts not only for the occurrence of the measles-speckles but the symptomatic fever as well.

IV

It may seem now that the logico-linguistic subsumptivistic approach to causality might be made to work. But matters are not that easy. First, we have not yet gone much more towards a resolution of the paradox expressed by Theorem 1 than blocked its proof, a critic may argue. Something more constructive is required, he says. Secondly, somebody may argue that the use of definite descriptions in the proof of Theorem 1 is the source of the paradox. To show that these remarks are not well taken we shall look the matter from a related but slightly different angle. We plainly discuss the requirement of constant conjunction in cases where events explicitly have some structure. This leads us to a paradox closely similar to our earlier one. This new problem shows that the ontological troubles involved are not really unique to problems of causality but present always when some kind of subsumption of event is involved. The problem is completely analogous to a difficulty Kim is struggling with within his set-theoretical framework (see Kim [1973], p. 222ff.).

Let us concentrate on the requirement of constant conjunction and ignore for a while the other aspects of causality. As we mentioned in the beginning of this paper constant conjunction seems to make sense only in the case of generic events. (This feature is of course implicit in our conditions (C) and (C*). But we may try to make it more explicit now.) Constant conjunction may conceivably be construed like this. If two singular events are causally related there must be at least two lawfully connected generic events under which they respectively fall. Suppose now we are dealing with two singular events named by the singular terms a and b. Let A and B be the generic events (or predicates, rather) which a and b respectively instantiate. These generic events are correlated by a law, which thus connects two different types of events. But instead of using our previous law $(x)\,(A(x) \rightarrow B(x)$, with the individual variables x running over singular entities (objects or events) we now, following a suggestion by Kim [1973], use $(x)\,(y)\,(A(x) \wedge P(x, y) \rightarrow B(y))$ to give the form of a causal law. (We let the variables run over singular events.) Here '$P(x, y)$' stands for a universal pairing relation supposed to correctly pair the instances of A and B (as well as of any other causally connected generic events). The pairing relation is supposed to accomplish the pairing in cases like this. If two adjacent guns are fired simultaneously and if these two events cause two deaths, how can the causes and effects be paired so that a cause becomes correlated with *its* effect and an effect with *its* cause? It seems plausible to require of the pairing relation that it makes causes and effects unique (*vis-á-vis* the other element in the pairing relation), but we shall not here pursue our investigation in that direction.

Our simple paradigm law is not perhaps the best explicate for the form of a causal law but it suffices for our present purposes. We may now try to define constant conjunction analogously with (C) as follows:

(CC) Two singular events named by a and b are said to be *constantly conjoined* (with a as antecedent and b as consequent) if and only if

$$(EA)\,(EB)\,[A(a) \wedge B(b) \wedge P(a, b) \wedge (x)\,(y)\,(A(x) \wedge P(x, y) \rightarrow \rightarrow B(y))].$$

But now we almost immediately run into a trouble analogous to the one discussed by Theorem 1. The question is about which property combinations define generic events and which do not. We assume, for simplicity, that the existence condition discussed in connection with Theorem 1 holds. We then get:

Theorem 3. If any combination of properties is allowed to define a generic complex event then every singular event satisfies (CC) with respect to any event that satisfies it with respect to at least one event.

To prove this theorem for our simple situation we assume that we have the law $(x)(y)(A(x) \wedge P(x, y) \rightarrow B(y))$ and that we have $A(a)$, $P(a, b)$ and $B(b)$. Next we assume there is a binary relation R (e.g. 'being above') such that $R(c, a)$ holds for some singular event c. Then we define a new complex predicate D by

$$(z)(D(z) \equiv (\mathrm{E}x)(R(z, x) \wedge A(x))).$$

From our assumptions it follows that $D(c)$. But then the arbitrary complex event c is seen to be constantly conjoined (with D substituted for A) with the event b according to (CC). Kim presents a similar argument within his set-theoretical framework (see Kim [1973], p. 227). He argues that the result is unacceptable. We agree with that interpretation. But the above argument about the existence of R is to be understood correctly. For we may take the underlying general claim concerning R to be either of the following two:

(i) $(\mathrm{E}R)(x)(y)(A(x) \wedge (P(x, y) \rightarrow B(y)) \rightarrow (\mathrm{E}z)R(z, x))$

(ii) $(x)(y)(\mathrm{E}R)(A(x) \wedge (P(x, y) \rightarrow B(y)) \rightarrow (\mathrm{E}z)R(z, x))$

Of these interpretations (i), contrary to (ii), says that one single R can effect the trivialization. While such a non-trivial R may perhaps be found (e.g. 'above', 'earlier', etc.) presumably some additional requirements (e.g. concerning spatiotemporal location) could still be used to deny the z's, for which $D(z)$, becoming causes instead of the x's.

On the other hand, the piecemeal trivialization effected by (ii) seems more serious, as (ii) seems easier to satisfy non-trivially than (i). Notice that in the case of (ii) the subsumption may often be effected by means of different laws.

In any case, Theorem 3 seems to us to represent a telling argument against the subsumptivistic logico-linguistic approach unless some qualifications are made concerning complex predicates permissible in the explanans. The problem of defining such permissible generic events seems to amount to giving criteria for predicates which represent natural kinds, which is one of the hardest problems in philosophy. We cannot really try to say too much in the way of a solution here.

Granting that plain constant conjunction has its defects, one should, on one's way to defining causality, rather require constant conjunctions with *explanatory* force. For is it not just the presence of predicates standing for natural kinds that guarantees explanatory power to a generalization and helps to make it a law? Granting that, our requirement of explanation gives at least a necessary condition for the presence of genuine generic events (or natural kinds) in the law. Thus, even in the case of constant conjoinedness we may require (weak) *DEL*-explanation instead of mere subsumption, analogously with our remedy to Theorem 1. For instance, our initial example above in the proof of Theorem 3 would be an explanation in this sense. But here it is central to emphasize our view that the distinction between predicates which express natural kinds and those which do not cannot be effected by formal means but in the last analysis only in pragmatic terms (see e.g. Tuomela [1973], Ch. V).

Still it is worth noticing that in the present case one can do something even by means of purely formal considerations. Namely, in our example the subsumption with the antecedent predicate A has the required formal features of an explanation. That is, the explanans $\{(x)\,(y)\,(A(x) \wedge P(x, y) \rightarrow B(y)), A(a), P(a, b)\}$ explanains $B(b)$ as far as the deductive-inferential aspects of explanation are concerned. If we substitute D for the predicate A it so happens that the conditions of our model of explanation become satisfied. However, the new explanation may remain only a potential one as it does not always satisfy the further 'material' requirement that the covering generalization $(x)\,(y)\,(D(z) \wedge P(x, y) \rightarrow B(y))$ really be a *law*. Which predicates $D(z)$ yield laws as well as philosophically and scientifically acceptable explanations depends, among other things, on the interpretation of $R(z, x)$. Hence, we may conclude, Theorem 3 does not constitute a counterargument specifically against the *explanatory-nomological* (or explanatory-subsumptivistic) idea of causation but rather it represents a challenge for further investigation of scientific explanation.

BIBLIOGRAPHY

Ackermann, R.: [1965], 'Deductive Scientific Explanation', *Philosophy of Science*, **32**, pp. 155–67.

Ackermann, R. and Stenner, A.: [1966], 'A Corrected Model of Explanation', *Philosophy of Science*, **33**, pp. 168–71.

Berofsky, B.: [1971], *Determinism*, Princeton University Press, Princeton.

Davidson, D.: [1967], 'Causal Relations', *The Journal of Philosophy*, **LXIV**, pp. 691–703.

Davidson, D.: [1969], 'The Individuation of Events', in N. Rescher *et al.* (eds.) *Essays in Honor of Carl G. Hempel*, Reidel, Dordrecht, pp. 216–234.

Goldman, A.: [1970], *A Theory of Human Action*, Prentice-Hall, Englewood Cliffs.

Kim, J.: [1969], 'Events and Their Descriptions: Some Considerations', in N. Rescher *et al. op. cit.*, pp. 198–215.

Kim, J.: [1973], 'Causation, Nomic Subsumption, and the Concept of Event', in *The Journal of Philosophy*, **LXX**, pp. 217–236.

Omer, I.: [1970], 'On the *D-N* Model of Scientific Explanation' *Philosophy of Science*, **37**, pp. 417–433.

Quine, W.: [1966], *The Ways of Paradox*, Random House, New York.

Tuomela, R.: [1972], 'Deductive Explanation of Scientific Laws' *Journal of Philosophical Logic*, **1**, pp. 369–392.

Tuomela, R.: [1973], *Theoretical Concepts*, Library of Exact Philosophy, Springer-Verlag, Wien.

Tuomela, R.: [1974a], *Human Action and Its Explanation*, Reports from the Institute of Philosophy, University of Helsinki, No. 2. Forthcoming, in 1977, in revised and expanded form, as *Human Actions and Their Explanation*, Reidel, Dordrecht and Boston.

Tuomela, R.: [1974b], 'Causes and Deductive Explanation', forthcoming in the proceedings of the Fourth Biennial Meeting of the Philosophy of Science Association held in Notre Dame, Illinois, in 1974.

Tuomela, R.: [1976], 'Morgan on Deductive Explanation: A Rejoinder', forthcoming in *Journal of Philosophical Logic*.

REFERENCES

[1] The hard question in discussing event identity is how to characterize the identity of aspect properties. I am willing to assume the possibility of synthetic identities holding between them. I cannot, however, in this paper discuss at any length the difficult problems involved here.

[2] Note here that unless deductive explanation is given a stringent formulation all kinds of paradoxes ensue already on the linguistic level of statements, quite independently of any ontological and semantical assumptions (such as (I) or (L)). For instance, even within the original Hempel-Oppenheim model it is possible to explain practically any singular statement by means of almost any explanans. This same remark applies to all attempts to clarify philosophical notions by means of some kind of subsumption. For instance, of the definitions of *determinism* at least the one by Berofsky, but also that by Goldman, are affected and trivialized by exactly this paradox (see Goldman (1970), p. 172 and Berofsky (1971), p. 168). Similarly this linguistic deducibility paradox seems to affect Goldman's definition of level-generation of action in Goldman [1970], p. 43.

[3] Speaking in general philosophical terms, I here understand explanation simply as an *argument* providing reasons for the truth of the explanandum (see Tuomela [1973], Ch. VII).

[4] Thus, for instance, both of the following two arguments are valid *potential* explanations in our (weak) *DEL*-model:

(i) $\dfrac{(x)(y)(F(x,y) \to G(x,y)) \quad F(a,b)}{G(a,b)}$ (ii) $\dfrac{(y)(F(a,y) \to G(a,y)) \quad F(a,b)}{G(a,b)}$

Note that in the proof of Theorem 1 '*b* has *K*' obviously does not explain '*a* has B', given the low, within our model of explanation.

[5] It should be noticed here also that the validity of principle (I) becomes rather questionable. For in general it is not possible to preserve the validity of a (weak) *DEL*-model-explanation by changing the explanans *Tc* into another *Tc*' by replacing a component of *Tc* by a co-referential component. (Note that coreferentiality is a linguistic notion.) As we just saw, one may argue that this property of the (weak) *DEL*-model just indicates that the new explanans *Tc*' does not describe the same event (or events) as *Tc*.

Note, however, that if $E(L, Tc)$ and if L contains a singular term a naming, say, a certain event and if $a = b$ then $E(L', Tc')$, where Tc' and L' have been obtained from Tc and L by substituting the singular term b for a allover.

PAUL WEINGARTNER

ON THE INTRODUCTION OF INTENSIONS INTO SET THEORY

I. INTRODUCTORY REMARKS

1.1. Texts of philosophers and logicians (within the time from Aristotle to De Morgan) suggest that the concept of intension is mostly introduced by and dependent on the principle of the opposite relation between extension and intension. This principle is clearly formulated by De Morgan thus: "The logicians who have recently introduced the distrinction of extension and comprehension, have altogether missed this opposition of the quantities... . In this the logicians have abandoned both Aristotle and the laws of thought from which he drew the few clear words of his dictum: 'the genus is said to be part of the species; but in another point of view the species is part of the genus'. All animal is in man, notion in notion: all man is in animal, class in class."[1] It seems that De Morgan had a very clear and adequate understanding of Aristotle who treats these two kinds of inclusion in several of his works.[2] Similar passages on the two kinds of inclusion can be found in Boethius,[3] Thomas Aquinas,[4] Cajetanus,[5] Leibniz[6] and Kant.[7]

1.2. There has never been a difficulty in interpreting the type of inclusion which De Morgan describes as "class in class" in terms of modern logic. Since this type of inclusion is unambigiously understood as the standard class-inclusion $\mathrm{a} \subseteq \mathrm{b}$ defined as $(z)(z \in a \rightarrow z \in b)$. Moreover, this interpretation can be shown to be in agreement with doctrines of traditional logic and especially with the philosophers and logicians cited in notes 2 to 7.

There was and is however a difficulty how to interpret the other type of inclusion described by De Morgan as "notion in notion".

The purpose of this paper is to give a positive answer to the following two questions:

Question 1. Is it possible to define a kind of inclusion which

(1) can be used as an interpretation of "notion in notion"
(2) satisfies the principle of the opposite relation between extension and intension: the class *a* is included in the class *b* iff the notion (set of properties, characteristics) of *b* is included in the notion (set of properties, characteristics) of *a*.
(3) can be easely introduced into the theory of classes and into set theory?

Question 2. Is it possible to construct a logic of extension and intension which

(1) satisfies the conditions (1) and (2) of Question 1
(2) can be easely incorporated into the standard theory of classes and into set theory by adding a few definitions
(3) needs only first-order predicate logic for the derivation of the theorems of this calculus?

1.3. Notation: '*a*', '*b*', '*c*', '*x*', '*y*', '*z*', '*X*', '*Y*', '*Z*' are class-variables or set-variables. Atomic sentences are formed with the usual membership-relation. Sentential connectives $\neg$, $\vee$, $\wedge$, $\rightarrow$, $\leftrightarrow$, are used in the usual way. Well-formed formulas are defined in the usual way. '*P*' for 'Propositional calculus'. '$\mathrel{\overline{\overline{\cdot}}}$' stands for extensional identity as defined in chapter 2, D1 and chapter 3, T4.1. '$\doteq$' stands for intensional identity as defined in chapter 5.

II. DEFINITIONS

D1	$a \mathrel{\overline{\overline{\cdot}}} b \leftrightarrow (z)(z \in a \leftrightarrow z \in b)$	extensional identity
D2	$a \subseteq b \leftrightarrow (z)(z \in a \rightarrow z \in b)$	extensional inclusion
D3	$a \sqsubseteq b \leftrightarrow (Z)(a \subseteq Z \rightarrow b \subseteq Z)$	intensional inclusion
D4	$a \cap b \mathrel{\overline{\overline{\cdot}}} c \leftrightarrow (x)[(x \in a \wedge x \in b) \leftrightarrow x \in c]$	
D5	$a \cup b \mathrel{\overline{\overline{\cdot}}} c \leftrightarrow (x)[(x \in a \vee x \in b) \leftrightarrow x \in c]$	
D6	$a \sqcap b \mathrel{\overline{\overline{\cdot}}} c \leftrightarrow (Z)[(a \subseteq Z \wedge b \subseteq Z) \leftrightarrow c \subseteq Z]$	
D7	$a \sqcup b \mathrel{\overline{\overline{\cdot}}} c \leftrightarrow (Z)[(a \subseteq Z \vee b \subseteq Z) \leftrightarrow c \subseteq Z]$	

D1, D2, D4 and D5 are well-known definitions of the standard theory of classes and of set theory. The new definitions are D3 (*intensional inclusion*), D6 (*intensional intersection*) and D7 (*intensional union*). With the help of the additional definitions D3, D6 and D7, a *theory of extension and intension* can be introduced into the theory of classes and into set theory. For the proof of the theorems of the theory of extension and intension only the principles of firstorder predicate logic, are needed, nothing else; i.e. no special axiom of set theory.

III. THEOREMS

T1 $a \subseteq b \rightarrow b \sqsubseteq a$

Proof. $(Z)[a \subseteq b \rightarrow (b \subseteq Z \rightarrow a \subseteq Z)]$ is a theorem (transitivity of implication). Therefore $a \subseteq b \rightarrow (Z)(b \subseteq Z \rightarrow a \subseteq Z)$ and so by D3 T1.

T2 $b \sqsubseteq a \rightarrow a \subseteq b$

Proof: $b \sqsubseteq a \rightarrow (Z)(b \subseteq Z \rightarrow a \subseteq Z)$ from D3; by universal instantiation:
$b \sqsubseteq a \rightarrow (b \subseteq b \rightarrow a \subseteq b)$ and therefore
$b \subseteq b \rightarrow (b \sqsubseteq a \rightarrow a \subseteq b)$ by P. Since $b \subseteq b$ is a theorem, T2.

T3 $a \subseteq b \leftrightarrow b \sqsubseteq a$

Proof: From T1 and T2.

T3 is the principle of the opposite relation between extension and intension. It says: *a is extensionally included in b* iff *b is intensionally included in a*. Or: class *a* is included in class *b* iff notion *b* is included in notion *a*.
Or: the class *a* is extensionally included in the class *b* iff the set of properties (characteristics) of the class *b* is intensionally included in the set of properties (characteristics) of the class *a*.

Examples: The class of men is extensionally included in the class of animals iff the set of characteristics (properties) of animals is intensionally included in the set of characteristics of men. The class of cardinal numbers is extensionally included in the class of ordinal numbers iff the set of characteristics of ordinal numbers is intensionally included in the set of characteristics of cardinal numbers.[8]

Another example raises a new question: Can we say that the class of natural numbers is included in the class of rational numbers iff the set of properties of rational numbers is intensionally included in the set of properties of natural numbers? Since there is a one-one-correspondence, this seems dubious. On the other hand, there are rational numbers which are not natural numbers. An unambigious formulation of the relation between natural and rational numbers, according to the principle T3, seems to be this: The class A of (classes of) numbers which are described by an axiom-system for natural numbers is extensionally included in the class B of (classes of) numbers which are described by an axiom--system for rational numbers, iff the set of properties which is attributed to B by the axiom-system of rational numbers is intensionally included in the set of properties which is attributed to A by the axiom-system of natural numbers.

Finally the following two questions have to be distinguished here: (1) *Is* T3 *valid*? (2) *Is* T3 *a good interpretation of the principle of the opposite relation between extension and intension*? The answer to the first question is clearly: Yes, if we accept the usual theory of classes and First-Order Predicate Logic, T3 is simply a theorem of the theory of classes, if one replaces '$b \sqsubseteq a$' by its definiens '$(Z)(b \subseteq Z \rightarrow a \subseteq Z)$'. The answer to the second question seems to me to be 'Yes' with some restrictions. T3 seems to be a quite good interpretation of the traditional principle of the opposite relation between extension and intension. The reason is that in the relevant philosophical and logical contexts this principle is almost exclusively applied to *genus* and *species* but not generally to classes (in the modern sense of the word). This is a strong restriction if one realizes that *species*, in the proper sense, for Aritotle is a class of substances in the proper sense (i.e. real existing individuals) or a first order predicate of individuals with the additional restriction that it expresses an essence. Thus animals are a *species* (there is an essential difference between plants and animals), but lions are not a *species*. If on the other hand, T3 is understood as the principle of the opposite relation between extension and intension which is generally applicable to classes (in the modern sense) one may have to choose certain unambigious (and therefore often circumstantial) formulations and to exclude other formulations in order to recognize the old principle in the new instance at all.

T4 $(a \subseteq b \wedge b \subseteq a) \leftrightarrow (b \sqsubseteq a \wedge a \sqsubseteq b)$

Proof: From T3 by P.

T4.1 $a \doteqdot b \leftrightarrow (Z)(a \sqsubseteq Z \leftrightarrow b \sqsubseteq Z)$

Proof: From D1, D2, D3 and T4.

T5 $x \in (a \cap b) \leftrightarrow x \in a \wedge x \in b$

Proof: Replace 'c' by '$a \cap b$' in D4.

T6 $(a \sqcap b) \sqsubseteq Z \leftrightarrow a \sqsubseteq Z \wedge b \sqsubseteq Z$

Proof: Replace 'c' by '$a \sqcap b$' in D6.

T7 $a \subseteq (b \cap c) \leftrightarrow (x)[x \in a \rightarrow (x \in b \wedge x \in c)]$

Proof: Use D2 and T5.

T8 $a \sqsubseteq (b \sqcap c) \leftrightarrow (Z)[a \sqsubseteq Z \rightarrow (b \sqsubseteq Z \wedge c \sqsubseteq Z)$

Proof: Use D3 and T6.

T9 $x \in (a \cup b) \leftrightarrow (x \in a \vee x \in b)$

Proof: Replace 'c' by '$a \cup b$' in D5.

T10 $(a \sqcup b) \sqsubseteq Z \leftrightarrow a \sqsubseteq Z \vee b \sqsubseteq Z$

Proof: Replace 'c' by '$a \sqcup b$' in D7.

T11 $a \subseteq (b \cup c) \leftrightarrow (x)[x \in a \rightarrow (x \in b \vee x \in c)]$

Proof: Use D2 and T9.

T12 $a \sqsubseteq (b \sqcup c) \leftrightarrow (Z)[a \sqsubseteq Z \rightarrow (b \sqsubseteq Z \vee c \sqsubseteq Z)]$

Proof: Use D3 and T10.

T13 $(a \cap b) \subseteq c \leftrightarrow (x)[(x \in a \wedge x \in b) \rightarrow x \in c]$

Proof: Use D2 and T5.

T14 $(a \sqcap b) \sqsubseteq c \leftrightarrow (Z)[(a \sqsubseteq Z \wedge b \sqsubseteq Z) \rightarrow c \sqsubseteq Z]$

Proof: Use D3 and T6.

T15 $(a \cup b) \subseteq c \leftrightarrow (x)[(x \in a \vee x \in b) \rightarrow x \in c]$

Proof: Use D2 and T9.

T16 $(a \sqcup b) \sqsubseteq c \leftrightarrow (Z)[(a \sqsubseteq Z \vee b \sqsubseteq Z) \rightarrow c \sqsubseteq Z]$

Proof: Use D3 and T10.

T17 $a \sqcap b \doteqdot a \cup b$

Proof: $[(a \sqcap b) \sqsubseteq Z] \leftrightarrow [(a \cup b) \sqsubseteq Z]$ by T4.1; transform the left part of the equivalence by T6, D2, P and T15 into $(a \cup b) \sqsubseteq Z$.

T18 $a \sqcup b \sqsubseteq a \cap b$

Proof: Use D3, transform by T10 and T13 into:
$(Z)[[(x)(x \in a \rightarrow x \in Z) \vee (x)(x \in b \rightarrow x \in Z)] \rightarrow (x)[(x \in a \rightarrow$
$\rightarrow x \in Z) \vee (x \in b \rightarrow x \in Z)]]$

T19 $a \cap b \subseteq a \sqcup b$
Proof: Use T18 and T3.

The theorems T17, T18, and T19 are—besides the theorems T3 and T4.1—the most important of the theory. They explain the relations between intersection and union. T17 says: *the intensional intersection of a and b is (extensionally) identical with the extensional union of a and b.* T18 says: *The intensional union of a and b is intensionally included in the extensional intersection of a and b.* T19 says: *The extensional intersection of a and b is extensionally included in the intensional union of a and b.*

Examples: T17: The intensional intersection of the sets of characteristics of animals and men is extensional identical with the extensional union of the sets of animals and men. T18: The intensional union of the sets of characteristics of living things and of things without senses [this is just the set of common characteristics of things without senses] is intensionally included in the set of characteristics of the extensional intersection of living things and things without senses [this intersection is just the class of plants]. T19: The extensional intersection of living things and things without senses (i.e. the class of plants) is extensionally included in the class of things which have as their characteristics the intensional union of the characteristics of living things and things without senses.

T20 Analogous to the theorems of the logic of classes, the following are theorems:
$(a \sqcap b) \sqsubseteq a$, $a \sqsubseteq (b \sqcup a)$, $a \sqsubseteq (b \sqcap a) \leftrightarrow a \sqsubseteq b$,
$(a \sqcup b) \sqsubseteq a \leftrightarrow b \sqsubseteq a$, $a \sqcap b \doteqdot b \sqcap a$, $a \sqcup b \doteqdot b \sqcup a$,
$[(a \sqcap b) \sqcap c] \doteqdot [a \sqcap (b \sqcap c)]$, $[(a \sqcup b) \sqcup c] \doteqdot [a \sqcup (b \sqcup c)]$,
$a \sqcap a \doteqdot a \sqcup a \doteqdot a$.

T21 $a \cap \Lambda \doteqdot \Lambda$ $\quad a \cap V \doteqdot a$
Proof: Replace '*b*' and '*c*' by 'Λ' or 'V' in D4.

T22 $a \sqcap \Lambda \doteqdot a$ $\quad a \sqcap V \doteqdot V$

Proof: Replace 'b' by 'Λ' and 'c' by 'a' or 'b' and 'c' by 'V' in D6.

T23 $a \cup \Lambda \doteq a \qquad a \cup V \doteq V$

Proof: Replace 'b' by 'Λ' and 'c' by 'a' or 'b' and 'c' by 'V' in D5.

T24 $a \sqcup \Lambda \doteq \Lambda \qquad a \sqcup V \doteq a$

Proof: Replace 'b' and 'c' by 'Λ' or 'b' by 'V' and 'c' by 'a' in D7.

T25 The distribution laws are analogous to those of the theory of classes:

$(a \sqcup b) \sqcap c \doteq (a \sqcap c) \sqcup (b \sqcap c)$ *Proof*: Use T6 and T10.

$(a \sqcap b) \sqcup c \doteq (a \sqcup c) \sqcap (b \sqcup c)$ *Proof*: Use T10 and T6.

T26 The following laws are analogous to those of the theory of classes:

$a \sqsubseteq (b \sqcap c) \rightarrow a \sqsubseteq (b \sqcup c)$ *Proof*: Use T8 and T12.

$(a \sqcup b) \sqsubseteq c \rightarrow (a \sqcap b) \sqsubseteq c$ *Proof*: Use T14 and T16.

T27 The following three theorems state interrelations between extensional and intensional functors:

$a \sqsubseteq (b \sqcap c) \leftrightarrow (b \cup c) \subseteq a$ *Proof*: Use T3 and T17.

$(a \sqcap b) \sqsubseteq c \leftrightarrow c \subseteq (a \cup b)$ *Proof*: Use T3 and T17.

$(a \sqcup b) \sqsubseteq c \leftrightarrow c \subseteq (a \cap b)$ *Proof*: Use T16, T3, D3, T7.

T28 $(a \sqcup b) \subseteq c \rightarrow (a \cap b) \subseteq c$ *Proof*: Use T18 and D2.

$a \sqsubseteq (b \sqcup c) \rightarrow (b \cap c) \subseteq a$ *Proof*: Use T3, T28.

IV. DEFINITION OF EXTENSION AND INTENSION

The following definitions may give a better understanding of the basic concepts that underly the calculus presented in chapters 2 and 3. They however are not needed to derive the theorems of chapter 3.

The definition of extension used in the following definition is well--known and frequently used. On the other hand, the definition D12 of intension is one of many which are possible. Another definition of intension is that with the help of the '$\in$'-*relation*: $\text{Int}(a) = \{z: a \in z\}$ which is stronger than that used here. But the definition D12 is—like D8 is for extensions—implicitly incorporated in the calculus given in chapters 2 and 3; and, moreover, definition D12 leads to the intuitively and traditionally accepted principle T3 of the opposite relation between extensions and intensions.

4.1. *Definition of Extension.*

D8 $\mathrm{Ext}(a) \mathrel{\overset{\cdot}{=}} \{x : x \in a\}$

D9 $a \cap b \mathrel{\overset{\cdot}{=}} \{x : x \in a \wedge x \in b\}$

D10 $a \cup b \mathrel{\overset{\cdot}{=}} \{x : x \in a \vee x \in b\}$

D11 a is a class (set) $\leftrightarrow a = \mathrm{Ext}(a)$

If it is wanted that the extension is taken of linguistic expressions rather than of concepts these definitions become more complicated and turn into conditional definitions like this:

If u denotes a then $\mathrm{Ext}(u) = \{x : x \in a\}$.
(Similarly for the other definitions).

4.2. *Definition of Intension.*

D12 $\mathrm{Int}(a) \mathrel{\overset{\cdot}{=}} \{Z : a \subseteq Z\}$

D13 $a \sqcap b \mathrel{\overset{\cdot}{=}} \{Z : a \subseteq Z \wedge b \subseteq Z\}$

D14 $a \sqcup b \mathrel{\overset{\cdot}{=}} \{Z : a \subseteq Z \vee b \subseteq Z\}$

D15 a is a set of characteristics (properties) $\leftrightarrow (\exists b)\,[a \mathrel{\underset{\cdot}{=}} \mathrm{Int}\,(b)]$

V. INTENSIONAL IDENTITY

Although no kind of identity other than extensional identity (as defined by D1 in chapter 2) is required for the calculus of extension and intension given in chapters 2 and 3 the question of a stronger kind of identity may be worth of investigation. Whether the identity defined by D16 is stronger than that of D1 depends on the underlying system.

D16 $a \doteq b \rightarrow (Z)(a \in Z \leftrightarrow b \in Z)$

If the underlying system is a system of set theory with the axiom of extensionality then the right parts (the definientia) of D1 and D16 are logically equivalent by that very axiom. But when the axiom of extensionality is dropped, D16 may serve as a stronger kind of identity so that the following theorem (but not its converse) holds:

5.1. $a \doteq b \rightarrow a \mathrel{\overset{\cdot}{=}} b$

This principle is a theorem of a finite matrix calculus which includes a logic of extensions and intensions and which is a finite model of set theory.[9]

REFERENCES

[1] De Morgan, 'Syllabus of a Proposed System of Logic', in: De Morgan, *On the Syllogism and Other Logical Writings* (ed. P. Heath) London 1966, p. 147–207, p. 201, note 2.

[2] Cf. Aristotle, *Met.* 1023b 17–21. *Phys.* 210a 17–20. *Top.* 121b. *Kat.* 3b 21–23.

[3] Cf. Boethius; 'In Porphyrium Commentariorum', in: Migne J.P. (ed.) *Patrologia Cursus Completus*, Series Latina, Lib. V, p. 142.

[4] Cf. Thomas Aquinas, *Commentary on Aristotle's Physics IV*, **4** (435); *Commentary on Aristotle's Metaphysics V*, **21** (1094; 1097). *Summa Theologica I*, **85**, 3 ad 2.

[5] Cf. Cajetanus (Cardinalis, Thomas de Vio), *Commentaria in Porphyrii Isagogen ad Praedicamenta Aristotelis* (ed. P. Insardus—M. Marega) Rome 1934, Cap. I, p. 25. In this text the expressions 'extensive' and 'intensive' were used for the first time.

[6] Cf. Leibniz: 'De formae logicae comprobatione per linearum ductus', in: *Opuscules et fragments inédits de Leibniz* (ed. L. Couturat) Paris 1903, p. 292–321, p. 300. This passage is translated in J. M. Bocheński, *History of Formal Logic*, Notre Dame **19**, 36.09.

[7] Cf. Kant, Logik A 147, 148, § 7.

Passages of the text to which the notes 2–7 refer have been given in my paper 'Die Fraglichkeit der Extensionalitätsthese und die Probleme einer intensionalen Logik', in: *Jenseits von Sein und Nichtsein, Beiträge zur Meinong-Forschung* (ed. R. Haller) Akad. Verlagsanstalt Graz 1972, p. 127–178, notes 2, 5, 7, 9, 12 and 14. and in: 'Bemerkungen zum Intensionsbegriff in der Geschichte der Logik', in: *Zeitschrift für Philosophische Forschung* **30** (1976), pp. 51–68, and in my book *Wissenschaftstheorie*, II, 1, *Grundlagenprobleme der Logik und Mathematik*, Stuttgart-Frommann 1976, especially chapter 3.41. It seems to me that these historical studies confirm that the gist of the traditional doctrine of extension and intension consists of the following simple thesis: Any entity of higher type than 0 (i.e. of higher type than individuals) can be viewed in two ways: (1) By "looking down" to the "individuals" relative to that entity (predicate) or to the elements relative to that class .This is the extensional view, describing the entity by its lower-type entities. It is expressed by definition D8 in chapter 4. (2) By "looking up" to the "properties" or higher classes relative to that entity. This is the intensional view, describing the entity by its higher-type entities. It is expressed by definition D12 in chapter 4 or by the following definition: $\text{Int}(a) = \{z: a \in z\}$.

[8] Several objections against the principle interpreted by T3 have been made by philosophers: Cf. F. Brentano, *Die Lehre vom richtigen Urteil*, p. 80, B. Bolzano,

Wissenschaftslehre I, § 120 (also referred to in Kneale W.—Kneale M., *Development of Logic*, p. 364 f.). These objections show that without a precise definition of "intensional inclusion" the interpretation of T3 leads to difficulties. The main-source for the construction of counterexamples seems to depend on the question of what an "additional property" is, when sets of properties (characteristics) are compared with one another. It seems that all "counterexamples" to the principle make use of properties which are, on one hand, already included in the original set of properties (by normal use of their meaning), but are, on the other hand, used as "additional properties".

[9] This calculus is described by the author in 'A Predicate Calculus for Intensional Logic', *Journal of Philosophical Logic* **2** (1973) p. 220–303. A theory of extension and intension is presented in: Weingartner, *Wissenschaftstheorie* II, 1 (cf. note 7), chapters 3.44, 3.45 and 3.46.

KLEMENS SZANIAWSKI

TYPES OF INFORMATION AND THEIR ROLE IN THE METHODOLOGY OF SCIENCE

0. INTRODUCTION

Information is often said to be one of the aims of science. In a somewhat aphoristic form: science is an information seeking process. If this is true then the concept of information becomes of crucial importance for the methodology of science, because the use of a method in science would then have to be justified in terms of its efficiency in obtaining information.

Attempts to do so are known in the literature. I would like to mention in this connection the well known book by I. Levi [2] and J. Hintikka's study [1]. In the present paper, certain results obtained so far will be systematized by means of the concept of pragmatic information (cf. Szaniawski's [5] and [6]; see also Nauta's [3]).

Since the main dividing line will be drawn between *apragmatic* and *pragmatic* varieties of information, I shall first restate briefly the principles on which they are based, then go on to discuss their interconnections in the analysis of scientific procedures.

I. APRAGMATIC INFORMATION

As the terminology suggests, apragmatic information is independent of relations that might exist between the language and its user. It is defined exclusively in terms of probabilities. In its basic form, *apragmatic information* is simply a decreasing function of probability. Out of the variety of such functions, two have been singled out because of certain desirable properties they possess (see [1] for details):

(1) $$inf(t) = -\log p(t)$$

(2) $$cont(t) = 1 - p(t),$$

where t is any statement.

A number of derived concepts can be defined, in terms of either *inf* or *cont*. The most important probably is what Hintikka calls *transmitted information*, i.e. "the amount of information that x conveys (contains) concerning the subject matter of s". If will be denoted by *transinf* (x, s):

$$(3) \qquad \mathit{transinf}(x, s) = \mathit{inf}(s) - \mathit{inf}(s|x).$$

In the above expression, $\mathit{inf}(s|x)$ is obtained from (1) by replacing the absolute probability of s by the conditional probability $p(s|x)$.

Let s and x in (3) range over the sets X and S, respectively. We assume that $S(X)$ is such that the exclusive disjunction of all s in S (all x in X) is true. For notational convenience, both sets are assumed to be finite. Under these conditions, *transinf* can be averaged. If we denote the resulting expression by $I(X, S)$ we have

$$(4) \qquad I(X, S) = E_{p(s)}\mathit{inf}(s) - E_{p(x)}E_{p(s|x)}\mathit{inf}(s|x),$$

where E_p is, of course, the operator of taking expected value with respect to the measure p.

$I(X, S)$ is the famous *Shannon expression*, i.e. the difference between the absolute entropy of S and the average conditional entropy of S, given x in X. It represents the average decrease of indeterminacy of S, due to the knowledge of x; in other words, the average amount of transmitted information (from X to S or vice versa: the expression is symmetric).

An expression analogous to I can be obtained if *inf* is replaced by *cont*.

II. PRAGMATIC INFORMATION

The numerical value of the expression $I(X, S)$ is entirely determined by the joint probability distribution function $p(x, s)$, where the sentential variables x, s range over the sets X, S, respectively. All the absolute and conditional probability distributions in (4) follow from $p(x, s)$.

Let it be remarked, in passing, that an important special case is obtained if $p(x, s)$ satisfies the condition

$$(5) \qquad \bigwedge_x \bigvee_s [p(s|x) = 1]$$

Then each x determines uniquely the answer to the question "*which s in S*?". This fact justifies calling such (transmitted) information *perfect.*

Now, in order to be able to speak of pragmatic information, we must additionally introduce something else: we have to suppose that the consequences of a person's decision depend upon the question, which s in S is true. Let, therefore, A be a (finite) set of behavioral alternatives ('actions', 'strategies') the subject is about to choose from. And let u be a real function defined on $A \times S$, representing the subject's evaluation (his 'utility') of all possible pairs (a, s). The triple $U = \langle A, S, u \rangle$ constitutes what is usually called *decision problem* in its normal form.

By *pragmatic information of X on S, relative to the decision problem* U, we shall mean the (average) improvement of decision in U, due to X. The definition is straightforward. If the decision maker does not know which x in X is true he simply maximizes u, averaged by means of $p(s)$. If he does know the true x he maximizes u, relative to the conditional measure $p(s|x)$; and what he obtains in this way must be averaged over all x in X. It follows that the definition of pragmatic information, denote it by C, is

$$C(X, S|A, u) = E_{p(x)} Max_a E_{p(s|x)} u(a, s) - Max_a E_{p(s)} u(a, s). \tag{6}$$

By introducing a convenient shorthand

$$v(a, x) = E_{p(s|x)} u(a, s) \tag{7}$$

we can rewrite (6), more symmetrically, as

$$C(X, S|A, u) = E_{p(x)} Max_a v(a, x) - Max_a E_{p(x)} v(a, x). \tag{8}$$

For some purposes, still another representation will be more convenient. If we assume, as we did, that the sets X, S, and A are finite then both $u(a, s)$ and the joint probability distribution function $p(x, s)$ are really matrices. Let us denote their product by w (in order to be able to perform the multiplication, we must first transpose the probability matrix):

$$[w(a, x)] = [u(a, s)] \cdot tr[p(x, s)] \tag{9}$$

Then pragmatic information is easily seen to be

$$C(X, S|A, u) = \Sigma_x Max_a w(a, x) - Max_a \Sigma_x w(a, x). \tag{10}$$

Equations (9) and (10) suggest an easy way to study the relation between pragmatic information C and the Shannon expression I. In order to be able to compare the two, we must somehow make the utility function vanish, since it does not enter the definition of $I(X, S)$: this last expression was defined entirely in terms of $p(x, s)$. But if we assume the utility matrix to be identity matrix then w becomes equal to the transpose of p and we have C depending entirely on the joint probability distribution.

The meaning of the assumption just made is that the utility matrix is square and has 1 on the main diagonal, 0 elsewhere. This establishes the one-one correspondence between the sets A and S, hence A may be identified with S and we have the utility function (call it u_1) defined as follows

$$(11) \qquad u_1(s, t) = \begin{cases} 1 \text{ if } s = t \\ 0 \text{ if } s \neq t \end{cases}$$

A natural interpretation of such a decision problem is that, for any s and t in S, 'action' s consists in assuming s to be true, hence $u_1(s, t) = 1$ just in case the decision maker is right, while $u_1(s, t) = 0$ in case he is wrong. The function u_1 is thus seen to be a special case of what we will call epistemic utility, following an already well established usage. In the next chapters we will discuss this concept more systematically.

For the moment, let me remark that $C(X, S|S, u_1)$ exhibits a strong similarity to $I(X, S)$; see [5] for details.

III. EPISTEMIC DECISION PROBLEM

By suitably interpreting the set A and the valuation function u in U, we obtain the model of situation often found in science: the decision to be made consists in choosing a statement for acceptance. If the basic set of statements is S then the elements of A are truth-functional compounds, most conveniently: disjunctions, of some s in S (here and below I am following closely Levi's [2]).

If we identify, for simplicity of notation, a disjunction with the set of its elements we may say that in an epistemic decision problem A is a family of subsets of S:

$$(12) \qquad \emptyset \neq A \subseteq 2^S,$$

satisfying certain conditions. It is not, however, postulated that A be deductively closed. This does not mean that we reject the postulate of deductive closure of the rational subject's body of beliefs: having accepted an a in A he is bound by this postulate to accept any a' in A such that $a \subset a'$. But insofar as his decision problem goes, a is the strongest statement that he is prepared to accept. To put it shortly, adopting a means: (i) the disjunction of all s in a is accepted as true; (ii) the same goes for all the supersets of a; (iii) if $a' \subsetneq a$ then the disjunction of all s in a' is not accepted as true.

So much for the interpretation of elements of A. As to conditions which A is assumed to satisfy they are fairly obvious:

(13.1) *If* $a \in A$ *then* $a \neq \emptyset$

(13.2) $\bigvee_{a, a' \in A} a \neq a'$

(13.3) $\bigcup_{a \in A} a = S$

In other words, some statement is accepted, the subject has a choice, and at least one a in A is true, whatever the actual s.

The exact character of A is, in each case, determined by the subject. Thus, he may want to know whether a particular s is true or not (verification of a simple hypothesis); in which case his A consists of two elements: $\{s\}$ and $S - \{s\}$. There are, of course, other possibilities.

In what is going to follow we shall limit our attention to two extreme cases:

(14.1) $A_1 = 2^S - \{\emptyset\}$

(14.2) $A_2 = \{s\}_{s \in S}$

The second case was already mentioned above: it was presupposed in (11). The first case allows any disjunction to be accepted as strongest, including S itself, the truth of which has been initially assumed. This is the case discussed in Levi's book (he interprets acceptance of S as the suspension of judgment). If A is defined by (14.1) it will be called *unrestricted*; and *restricted* if it is defined by (14.2).

Given the above interpretation of A, what shape is the function u to have in an epistemic decision problem? The answer depends on the goal the subject wants to achieve. Insofar as the epistemic decision problem is a model of scientific procedure, the utility function u may be said to define the aims of science (thus Levi considers his book "an essay on induction and the aims of science").

We are here interested in science as purely cognitive activity, thereby excluding all cases of applied research from the scope of our inquiry. Even so, the characterization of the aims of scientific research remains a fairly complex task, because: (a) there is a variety of possible aims, depending on the context; (b) in some cases, the research is supposed to achieve two conflicting aims simultaneously; (c) quite often it is not possible to go beyond a verbal and vague description of what the scientist wants to achieve.

I will try below to discuss the epistemic utility in more detail. It seems, however, that there is one postulate any epistemic utility function ought to satisfy. Namely, that every error be lower valued than any non-error. Obviously, for a given s, the adoption of an a is an error iff $s \notin a$. Hence, the postulate on u has the following form:

$$(15) \qquad \bigwedge_{s} \bigwedge_{a,a'} [\textit{if } s \in a \textit{ and } s \notin a' \textit{ then } u(a, s) > u(a', s)]$$

One simple way to ensure that (15) is satisfied can be described as follows. Suppose that the elements of A are evaluated by means of a non-negative function h, irrespective of whether they are erroneous or not. Then the following definition of u in terms of h will make u satisfy (15):

$$(16) \qquad u(a, s) = \begin{cases} h(a) & \textit{if} \quad s \in a \\ h(a) - Max_a h(a) & \textit{if} \quad s \notin a \end{cases}$$

We then have $h(a) > 0 > h(a) - Max_a h(a)$, with the possible exception of the boundary case.

The postulate (15) is a formal expression of the fact that truth takes precedence over any other epistemic value. We now come to the discussion of those 'other values'.

IV. GLOBAL THEORIZING

The distinction between global and local theorizing is due to Hintikka [1]. In his own formulation: "One of the most important distinctions here is between, on one hand, a case in which we are predominantly interested in a particular body of observations e which we want to explain by means of a suitable hypothesis h, and on the other hand a case in which we have no particular interest in our evidence e but rather want to use it as a stepping stone to some general theory h which is designed to apply to other matters, too, besides e. We might label these two situations as cases of *local* and of *global theorizing*, respectively. Often the difference in question can also be characterized as a difference between explanation and generalization, respectively." (p. 321)

We are now going to try to formalize the case of global theorizing, in the sense of defining a utility function which would express this particular goal of research. It seems, however, that global theorizing has at least two conflicting goals. One is *truth*, the other is *information*.

Suppose first that truth is the only aim. This means that all cases of accepting a true statement are treated alike, similarly all cases of committing an error, and the first are preferred to second, in accordance with (15). The utility function is, therefore, a generalization of (11):

$$(17) \qquad u_1(a, s) = \begin{cases} 1 & \text{if} \quad s \in a \\ 0 & \text{if} \quad s \notin a \end{cases}$$

If A is restricted in the sense of (14.2) pragmatic information of X on S, relative to u_1, is:

$$(18) \qquad C(X, S|A_2, u_1) = E_{p(x)} Max_s p(s|x) - Max_s p(s).$$

As indicated above, the expression behaves in a very similar way to $I(X, S)$.

On the other hand, if A is unrestricted, i.e. defined by (14.1), C becomes equal to 0:

$$(19) \qquad C(X, S|A_1, u_1) = 0$$

The explanation is fairly simple. If truth is the only aim in the sense of (17) then the maximal disjunction of all elements of S dominates any other a in A_1, i.e. is rated higher by u_1, under any s. Acceptance of a tautology (equivalently, suspension of judgment) is the best absolutely, hence any information becomes superfluous.

Another obvious aim of global theorizing is information. For reasons which have been sketched by Hintikka, *cont* is in this context a more appropriate measure than *inf*. Suppose, therefore, that elements of A are appraised in terms of *cont*. If the new utility function, say u_2, is to satisfy the postulate (15) then the function $h(a)$ in (16) will simply become $cont(a)$ and we have the following definition of u_2:

$$(20) \qquad u_2(a, s) = \begin{cases} cont(a) & \text{if } s \in a \\ cont(a)-1 & \text{if } s \notin a \end{cases}$$

which gives the following expression for the information of X on S, relative to u_2:

$$(21) \qquad C(X, S|A, u_2) = E_{p(x)} Max_a[p(a|x)-p(a)].$$

In the above formula, $p(a|x)$ is, of course, a shorthand:

$$(22) \qquad p(a|x) = \Sigma_{s\in a} p(s|x),$$

and similarly for $p(a)$.

The expression in the square brackets, to be maximized by a suitable choice of a, has been extensively discussed in the literature (cf. Hintikka [1], p. 325–329). Interpreted as the increase in the probability of a, due to the evidence x, it forms the basis of many measures of evidential support. Formula (21) shows that its maximal value, averaged over all x's in X, is identical with pragmatic information of X on S, if global theorizing's aim is defined as maximization of *cont*, subject to the condition (15).

If the joint probability distribution satisfies (5), i.e. if the information of X on S is perfect, the expression (21) depends on the marginal distribution of s:

$$(23) \qquad C(X, S|A, u_2) = 1-E_{p(s)} p(s),$$

which becomes maximal in the case of equidistribution:

$$(24) \qquad Max_{p(s)} C(X, S|A, u_2) = 1-\frac{1}{n},$$

where n is the number of elements in S.

The two aims (search for truth, search for information) are in obvious conflict. From the first point of view, an element of A is the more highly valued, the more probable it is. Whereas *cont* is inversely related to probability. In actual practice some kind of compromise between these two conflicting goals is established. Its exact nature depends upon the subject's (the scientist's) decision. In particular, upon his willingness to risk (cf. Levi's expression: 'gambling with truth'). Let us, therefore, form a 'mixed' utility, say u_3, by means of Levi's parameter q, satisfying the inequality $0 \leqslant q \leqslant 1$.

$$(25) \qquad u_3 = (1-q) \cdot u_1 + q \cdot u_2.$$

As the result we obtain

$$(26)\qquad u_3(a, s) = \begin{cases} 1 - q \cdot p(a) & \text{if} \quad s \in a \\ -q \cdot p(a) & \text{if} \quad s \notin a \end{cases}$$

The interpretation of q is obvious: the higher its value, the more willing the scientist is to risk an error in order to win information in the sense of *cont*.

The generalized expressions for pragmatic information are, in the case of unrestricted and restricted A, respectively, as follows.

$$(27)\qquad C(X, S|A_1, u_3) = E_{p(x)} Max_a [p(a|x) - q \cdot p(a)] - (1-q)$$

$$(28)\qquad C(X, S|A_2, u_3) = E_{p(x)} Max_s [p(s|x) - q \cdot p(s)] - (1-q) \cdot \cdot Max_s p(s).$$

If p satisfies (5) the above reduce to

$$(29)\qquad C(X, S|A_1, u_3) = q \cdot [1 - E_{p(s)} p(s)]$$

$$(30)\qquad C(X, S|A_2, u_3) = 1 - q \cdot E_{p(s)} p(s) - (1-q) \cdot Max_s p(s).$$

It is easily seen that Levi's epistemic utility function (cf. [2], p. 81) is a special case of u_3, obtained by means of postulated equidistribution on S. If we denote such a measure on S (used by Levi only to define *cont*) by m, and the resulting utility function by u_4, we have

$$(31)\qquad u_4(a, s) = \begin{cases} 1 - q \cdot m(a) & \textit{if} \quad s \in a \\ -q \cdot m(a) & \textit{if} \quad s \notin a \end{cases}$$

In the case of restricted A, $m(a)$ becomes constant in a, which makes u_4 identical with u_1, up to linear transformation. In other words, if the subject decides in advance to have one of the logically strongest statements accepted, he behaves, in terms of u_4, as if his only aim were truth (no matter what the value of q is). The genuine compromise between truth and *cont* (defined in terms of m) is arrived at if A is unrestricted. This is the case investigated by Levi.

Pragmatic information is then equal to

$$(32)\qquad C(X, S|A_1, u_4) = E_{p(x)} Max_a [p(a|x) - q \cdot m(a)] - - Max_a [p(a) - q \cdot m(a)].$$

To close this chapter, a remark concerning the epistemic utility u_2, which played an essential role in the above analysis. Let us compare it with the utility defined by Hintikka in an analogous context ([1], p. 325).

Hintikka describes the aim of global theorizing as follows (I am changing the symbols in an inessential way): "If a is true, we gain the (substantive) information $cont(a)$ by adopting a. If a is false, we lose the information which we could have gained by opting for $\sim a$, rather than a. According to what was just said, our net utility in this case is, therefore, $-cont(\sim a)$."

Now, it is easy to see that: $-cont(\sim a) = cont(a)-1$. It thus turns out that Hintikka's utility is identical with u_2, which has been obtained above in another way.

V. LOCAL THEORIZING

The starting point is now different. We assume a statement, call it e, as given and the problem is, how to explain e by a suitable choice of an s in S (the discussion will be limited to restricted A). Following Hintikka, we associate the meaning of the word 'explain' with the notion of transmitted information: e is the better explained by s, the more information s conveys concerning the subject matter of e.

Thus, *transinf* (s, e) is a means of defining the purpose of *local* theorizing. It seems, however, that simply to identify this purpose with maximization of *transinf* would not be an adequate description of what actually takes place. From the scientist's point of view it is not immaterial whether the intended explanation is true or not. If, therefore, a utility function u_e is to define the purpose: to explain e, it ought to satisfy the postulate (15). In order to assure this, we apply (16) once more[1], this time making h equal to *transinf*. In view of

$$(33)\qquad Max_s transinf\,(s, e) = Max_s \log\frac{p(e|s)}{p(e)} = -\log p(e)$$

we have

$$(34)\qquad u_e(s, t) = \begin{cases} \log p(e|s)/p(e) & \textit{if} \quad s = t \\ \log p(e|s) & \text{if} \quad s \neq t \end{cases}$$

Averaged over all t in S, this gives

$$(35)\qquad E_{p(t|e)}u_e(s, t) = \log p(e|s) - p(s|e)\log p(e).$$

If $p(s)$, i.e. the prior distribution of s, is uniform then maximization of the above expression is equivalent to maximization of the likelihood of e, i.e. of $p(e|s)$. In other words, the maximum likelihood principle guides the choice of explanation if all explanations are a priori considered equiprobable[2].

Obviously, another interpretation of local theorizing is also possible. It may be thought that the process of explanation is first carried out without considering the logical value of the explanans. Maximization of *transinf* takes place and after it has led to the selection of an s, an independent test of its truth is made. According to this view, maximum likelihood principle leads to the selection of the explanans, because it is equivalent to maximization of transmitted information (cf. Hintikka, [1], p. 322). Then, however, maximum likelihood principle could not be called a rule of acceptance, contrary to its current interpretation, unless we are prepared to accept statements regardless of their logical value.

VI. CONCLUDING REMARKS

On the preceding pages we tried to show how the various goals of scientific research can be defined in terms of information, in its many variants. In the case of global theorizing, the (average) value of the data, i.e. of the elements of the set X, can be computed, with respect to such a goal. This is what we proposed to call *pragmatic information*, relative to different kinds of epistemic decision problem.

It was presupposed throughout that all the probabilities are given, i.e. the joint probability distribution function of the relevant variables is known. This assumption has been, and still is, the subject of controversy. It has consistently been maintained by some writers that the only probabilities actually given are the conditional ones of the type $p(x|s)$. In other words, that we ought to dispense with the marginal distribution $p(s)$. What would be the consequences of this view?

Obviously, certain concepts would have to go, viz. those which are defined in terms of $p(s)$. It follows that the goals of research would have to be redefined. The concept of pragmatic information could be retained in a different form, relativized to a criterion of decision making, such as *maximin, minimax loss, etc.* The way to do this is sketched in [5].

BIBLIOGRAPHY

[1] J. Hintikka: 'The Varieties of Information and Scientific Explanation', in: *Logic, Methodology and Philosophy of Sciences* III, van Rotselaar and Staal ed., North Holland, 1968.

[2] I. Levi: *Gambling with Truth. An Essay on Induction and the Aims of Science*, Alfred A. Knopf, 1967.

[3] D. Nauta, Jr.: *The Meaning of Information*, Mouton, 1972.

[4] K. Szaniawski: 'The Maximum Likelihood Principle: An Evaluation'. (in Polish), in: *Rozprawy Logiczne. Essays in Honour of K. Ajdukiewicz*. PWN, Warsaw, 1964.

[5] K. Szaniawski: 'Two Concepts of Information'. *Theory and Decision*, 5, 1974.

[6] K. Szaniawski: 'Information and Decision as Tools of Philosophy of Science', *Danish Yearbook of Philosophy*, Vol. 10, 1973.

REFERENCES

[1] We are here making the rather natural assumption that only such statements are possible candidates for explaining e which are positively associated with it. It follows that *transinf* (s, e) is positive for any s in S.

[2] The connection between the maximum likelihood principle and the assumption of equiprobability of hypotheses has been investigated in [4].

B. G. MIRKIN, L. B. CHORNY

CLASSIFICATION AND RANKING MODELS IN THE DISCRETE DATA ANALYSIS
(A Survey)

I. SOME INFORMAL PATTERNS OF DATA ANALYSIS

Different kinds of empirical data obtained in scientific research may be presented in the form of a 'unit-variable' table on (i, j)th—position of which is the value of a variable j for a unit i ($i = 1, \ldots, N, j = 1, \ldots, n$). For example, these data may describe a set of individuals i characterized by questions j of a sociological questionnaire. At the same time some tables may be given of the unit-unit and variable-variable type characterizing intercorrelations between units and variables, respectively. Those tables may be obtained as a result of independent empirical research (it is in the case of sociometric matrices) but more frequently they are obtained from the analysis of a unit-variable matrix on the basis of a formalized perception of the resemblance of units or correlation of variables.

Real data may be of more complex nature: they may be given as two sets of units each characterized by a set of variables and besides there may occur some correlations between units. Such, for example, are the data obtained as results of experts evaluations of manufactured articles. Here one set includes the articles characterized by different quality indices the other is the set of experts characterized by socio-psychological variables; and the correlations between elements of the sets are the experts' preferences concerning the set of manufactured articles. Because of complexity of the general notion and since formal and non-formal analysis schemes here are, on the whole, combinations of 'unit-variable' type schemes we shall restrict ourselves in this paper to the discussion of only one set of units.

An important place in the unit-variable table is held by so called *qualitative variables*. The information of qualitative (rank or nominal) variable is completely set by the partition of the set of units with fixed or not fixed ranking of classes, respectively) into classes (groups) of such units which have the same values of a given variable. So the values of the nominal variable are not ranked (and those of the rank variable are ranked) names of the classes of a corresponding partition. A ranked partition is named a *ranking*. Therefore a *classification* or *ranking of units* in some system of variables is in fact construction of a new nominal or ranked, variable i.e. condensation of $N \times n$ 'unit--variable' table into $N \times 1$ table of units characterized by the only qualitative variable. We shall now describe the main schemes of non-formal and formal data analysis in terms of classifying or ranking units.

A. *A classification (or ranking) of units is given a priori.* The problem is to find out a subsystem of variables determing this classification (or ranking) in some natural sense. So far as a given classification (ranking) is the qualitative variable itself we are in fact speaking about 'significance', 'informativity' 'substantivity' coefficient of variables from the point of view of 'output'. Analysis of correlations between variables is closely related to it so the more 'information' variables carry about each other, the higher are correlations. For example, if the partition of some set of people according to their 'migration mobility' is given, it is possible to elucidate an 'influence' of different characteristics of a man on this index.

B. *A classification (or ranking) is not given a priori.* It is necessary to construct it and then to analyse an influence of different variables on this classification. Such is the case when it is necessary to rank individuals according to their 'potential mobility' by the variables characterizing their 'migrational mobility'.

C. *A comparison of different classifications and of the variables which generated them.* For example, it is possible to classify individuals according to variables characterizing 'migrational mobility' and according to variables characterizing 'the mode of life' and then to compare those classifications; it is possible to consider one group of variables as 'input' and the other as 'output', for example, it may turn out that the mode of life is highly informative for 'migrational mobility' (output group of variables); it is also possible to find variables 'determining' classifications, and then to analyse correlations of variables determining' classifications, and then to analyse correlations of variables determining one classification and to compare them with correlations of variables determining the other classification. It may happen that some of the classifications to be compared are given *a priori.*

The scheme C may be considered as an application of the schemes A and B to the situation when a partition of the whole set of variables into groups (for example, into 'mode of life' and 'migrational behaviour' groups) is given. The next scheme is a natural complication of the previous ones.

D. *Neither a classification of units, nor a classification of variables is given.* It is necessary to pick up some groups of variables according to which classifications (rankings) must be constructed followed by further study of correlations between variables. In the course of picking up groups of variables with the help of formal methods the problem of identification of a 'latent factor' which is characterized by the given group arises. However, a similar difficulty is connected also with 'naming' of the classes of obtained classifications.

E. Still more sophisticated analysis is connected with a transfer to the obtained classifications classes. All A through D type schemes may be applied to each of these classes. For example, we may classify each 'first level' class separately and thus construct the 'second level' classes and so on. In such a way we shall obtain a taxonomical tree with 'vertices' connected by 'determining' variables. It is possible also to analyse intercorrelations of variables in every class—it often leads to fairly interesting conclusions, especially in the case when patterns of correlations of variables in separate classes are different from those obtained in a whole set.

As we may see now, successive application of A through D type schemes in different subclasses of units and variables provides a tremendous variety of possible qualitative data analysis schemes.

II. DATA ANALYSIS METHODS

Now we shall describe some methods applied to the considered data analysis schemes.

A. One of the main formal problems arising here is the description of the *a priori classification* (*output variable*) in terms of variables which characterize units. Such a description makes it possible to fix variables really connected with a given classification by output variables and at the same time to predict 'behaviour' of other units which did not fall into the considered sample. To obtain such a description use is often made of *coefficients of informativity* of variables.

There are many methods of the informativity evaluation. Some of them evaluate the proximity of the classification determined by the variable to the output classification (Mirkin, 1970), others more sophisticated but at the same time requiring more computation, are based on evaluation of the number of entries of the given variable into different sets of variables determining completely values of the output variable (Zhuravliov, 1968). In this case we may try to find out for every class (or 'pattern') combinations values of variables determining the class (for example, Bongard, 1967).

As for informativity evaluations based on the evaluation of the proximity of classifications, they are made in terms of different coefficients of correlation between qualitative variables. Correlation coeficients evaluate in one or another way the possibility of prognosis of one variable values by those of another. There is a considerable number variety of correlation coefficients (Goodman and Kruskall, 1966, Kendall and Stuart, 1966). Many of them may be formulated in geometrical terms of the following space of qualitative variables.

Let A be a set of N units. For a given variable R we shall designate by iRj the fact that units i and j are correlated by variable R so that for a nominal variable iRj iff i and j correspond to the same value of the variable (are in the same class of R) and for a ranked variable iRj iff the value of R for i is not lower than for j $(i, j = 1, \ldots, N)$.

Let us construct for each R a Boolean $N \times N$ matrix $r = \|r_{ij}\|$, where $r_{ij} = 1 \leftrightarrow iRj$. We may notice that it is possible through such matrices to present more complex qualitative variables than rankings and partitions, for example, results of pairwise comparisons of units by some experts.

We shall denote by $\delta(R, S)$ for variables R and S *relative Hemming distance* of corresponding to them Boolean matrices r and s: $\delta(R, S) = \frac{1}{N^2} \sum_{i,j} |r_{ij} - s_{ij}|$. The value $\delta(R, S)$ means the probability of the fact that for arbitrarily taken two units i and j, $r_{rj} \neq s_{ij}$ or $r_{ij} \neq s_{ji}$.

We shall denote by U the variable which has the same value for all units; then $\delta(R) = \delta(R, U)$ corresponds to the notion of dispersion and, what is more, evidently $\delta(R, S) = 2\delta(R, S) - \delta(R) - \delta(S)$, where RS is the intersection of the variables R and S, in other words, it is the variable, the values of which are all possible combinations of variables R and S. If the variable R is nominal, then $\delta(R) = 1 - \sum_k p_k^2$, where p_k is the probability of value k of the variable R, so $\delta(R)$ is the average error of the proportional piediction of the values of the variable R. This means that $\delta(R, S)$ also expresses the possibility of the prognosis of the variables R and S. It is interesting that the dispersion of a ranking is precisely two times less than that of a corresponding nominal variable (it has the same values but their ranking is not taken into consideration).

Many well known correlation coefficients, for example, Kendall's, Goodman and Kruscall's coefficients (Kendall and Stuart, 1967), etc. may be expressed in terms of $\delta(R, S)$. This coefficients is very similar to entropy but is quite different from indices based on χ^2 value (Kendall and Stuart, 1966).

B. A classification construction according to a system of variables may be performed as well in terms of variables and their correlations as in terms of units and their distances.

A classification of units in terms of variables is something which we may name an *analogue of numerical factor analysis.* As in the case of 'quantitative' analysis a factor (the searched classification or ranking) may be looked for with the purpose of the best approximation of values of the initial variables (the principal components model) or, and it is quite different, with the purpose of the best explanation of intervariable correlations we have already obtained (the factor analysis model). In the case of discrete variables those purposes are realized in the method of approximation in qualitative variables space and in the latent structure analysis (Lazarsfeld and Henry, 1968).

The main idea of the *latent structure analysis method* is the following: the factor which explains correlations between variables has values α with probability V^α and besides for the units from a fixed class all the initial variables are independent so that $p_{ij\ldots k}^\alpha = p_i^\alpha p_j^\alpha \ldots p_k^\alpha$, where $i, j, \ldots, k$ are the values of different variables and $p_{ij\ldots k}^\alpha$ is the probability of their simultaneous realization. The quantities $V^\alpha, p_i^\alpha, p_j^\alpha, \ldots, p_k^\alpha$ may be found out from the equations with known left-hand sides:

$$p_i = \sum_\alpha V^\alpha p_i^\alpha$$

$$p_{ij} = \sum_\alpha p_i^\alpha p_j^\alpha$$

. . . .

We must consider as a substantial shortcoming of this method the fact that very often solutions of those equations have no probabilistic meaning so that latent structure model is not suitable for the analysis of such data.

The approximation method model is as follows. Let us fix some set E of the Boolean $N \times N$ matrices (for example, the set of matrices representing partitions on five or less classes). If $r^1, \ldots, r^n$ are the matrices of initial variables, then we define the unknown factor as the variable r which has the minimal sum $\sum_i \delta(r^i, r)$ for all $r \in E$.

In (Mirkin, 1971) it is shown that four natural conditions on the rule of choice of the factor sharply narrow the class E of feasible r up to the class of all possible intersections (cross-classifications) of different sets of variables. In this case the factor is sought for in the class of all cross-classifications which in high degree corresponds to the combinational groupings procedure.

Further development of the cross-classifications methods are the tree--type classifications which we get if every class of the obtained grouping will be partitioned with respect to its own variable which is the best for this particular class.

Tree-type and cross-classifications are very convenient as their classes may be characterized in the unique way by corresponding values of variables.

Let us note that criteria of choice of the system of variables with respect to which we group may be changed in wide limits.

Far deeper is the development of the classification and ranking in terms of units and distance between them (or the indices of the closeness or connection) (Ayzerman, Braverman and Rozonoer, 1970, Dorofeiuk 1971, Zagoruiko, 1972, Bonner, 1964, Rose, 1964, Constantinescu, 1964).

It is senseless even to try to describe the whole variety of proposed methods of classification. However, they may be divided purely logically into two groups of methods: those applying explicit notion of a 'class' and those working without it but directly in terms of partitions. A different approach to classification of taxonomical methods considers the question whether in some method an explicit elimination of weak connections is used or whether those weak connections are taken into consideration implicitly, by the way of optimization of some integral characteristic for classification.

For example, there are methods using the following notion: the class is a set of units whose distances from each other are less than their distances from the rest of the set. Other methods (Zagoruiko, 1972, Bonner, 1964) construct classification not using such an explicit notion of 'block'. Weak connections are usually eliminated explicitly through choosing some *threshold* quantity α and throwing off all connections smaller than α (or the sum of thrown off connections must not be greater than α). After this procedure a graph of connections is constructed and considered together with its partition into components and bicomponents.

Usually application of optimization methods of classification brings the best practical results though there are no exact algorithms of such type because of great combinatorial complexity of considered problems. *Wroclaw taxonomy method* is the nice exception from this rule—the method has a simple exact solution, however, unfortunately, if often produces unstable and poorly interpreted results.

The problem of approximation of variables in space is equivalent to the problem of such type of optimization of classification.

Let us denote by a_{ij} the number of variables R^k such that iR^*j. Then the optimal partition $R = \{R_1, \ldots, R_m\}$ $(\bigcup_i R_i = A)$ must maximize the sum of interior connections $a_{ij} - \frac{n}{2}$:

$$f(R) = \sum_s \sum_{i,j \in R_s} \left(a_{ij} - \frac{n}{2}\right) = \sum_s \sum_{i,j \in R_s} a_{ij} - \frac{n}{2} \sum_s |R_s|^2$$

As for optimal ranking, it is obtained as a solution of the problem of 'over-diagonal' connections maximization. The problem consists in obtaining the permutation $(i_1, \ldots, i_N)$ of units which maximizes

$$f(i_1, i_2, \ldots, i_N) = \sum_{k<l} a_{i_k i_l}.$$

We must notice now that those approaches reveal in the matrix of connections $||a_{ij}||$ structures of such type given beforehand: partitions or rankings. But the real structure may be represented in a more complex way, for example as a tree-type or a cycle-type graph. So it is only natural to try to discover the structure which is really 'contained' in the given system of connections. In this case we must consider as the feasible set E all possible partitions $R = \{R_1, \ldots, R_m\}$ with connections between classes given by an arbitrary relation $\varkappa$ on the set of indices of classes $\{1, 2, \ldots, m\}$. Then the problem of approximation results in the functional:

$$f(R, \varkappa) = \sum_{(s,t)\in\varkappa} \sum_{j\in R_s} \sum_{j\in R_t} (a_{ij} - a).$$

Maximizing the functional with respect to all R and $\varkappa$ we can obtain an unknown structure. Such an approach is relatively new. It was applied to the data taken from different fields of economics, genetics, biology and has given well interpreted results (those studies were done by W. L. Coopershtoch and W. A. Trofimov).

C. This scheme, as it has already been stated, is a combination of A and B type schemes.

D. Classification of variables is usually effected according to the matrix of the evaluations of proximity between variables. For the *proximity evaluation* we may use different coefficients based on the distance or entropy. Variables are grouped by methods of classification. After that the analysis of each group is carried but separately.

In such short account it is obviously impossible not to overlook some details or even essential moments. We tried mainly to emphasize an idealogical basis of classification and ranking methods. Our references are essentially far from completeness. For more complete reference list see (Dorofeiuk, 1971).

BIBLIOGRAPHY

[1] Bonner, P., 'On Some Clustering Techniques', *IBM. J. Res. and Develop.*, 1964, No. 1.

[2] Constantinescu, P., 'The Classification of A Set of Elements with Respect to a Set of Properties', *Computer Journal*, 1966, No. 4.

[3] Goodman, L. and Kruskall, W., 'Measures of Association for Cross Classifications', *J. of Am. Statistical Association*, 1954, No. 6.

[4] Kendall, M. G. and Stuart, A., *The Advanced Theory of Statistics*, v. 2, London, 1966.

[5] Lazarsfeld, P. Henry, N., *Latent Structure Analysis.* Boston, 1968.

[6] Mirkin, B. G., *An Approach to the Analysis of Primary Sociological Information.* Novosibirsk, 1970.

[7] Rose, M., Classification of a Set of Elements', *Computer Journal*, 1964, No. 3.

[8] Айзерман М. А., Браверман Э. М., Розоноэр Л. И., *Метод потенциальных функций в теории обучения машин.* М., Наука, 1970.

[9] Бонгард М. М., *Проблема узнавания.* М., Наука, 1967.

[10] Вальтух К. К. (peg.), *Межотраслевой баланс производственных мощностей.* М., Экономика, 1973.

[11] Дорофеюк А. А., *Алгоритмы автоматической классификации* (обзор)', *Сборник трудов Института проблем управления*, вып. 1, 1971.

[12] Журавлев Ю. И., Ш. Е. Туляганов, 'Измерение важности признака': *Вопросы кибернетики*, вып. 30.

[13] Загоруйко Н. Г., 'Методы распознавания и их применение'. *Советское радио*, 1972.

[14] Миркин Б. Г., *Об аксиоматических подходах к проблеме согласования классификаций.* (доклад на конференции ,,Логика и методология'', Вроцлав, 1971), Новосибирск, 1971.

G. TORALDO DI FRANCIA

WHAT HAVE PHYSICISTS LEARNED FROM EXPERIENCE ABOUT INDUCTIVE INFERENCE?

I

The problem of inductive inference or the problem of passing from knowledge of the *observed* to prediction of the *non-observed*, can be viewed from two different standpoints. One is a philosophical standpoint (*quid iuris*?), while the other is, broadly speaking, the standpoint of natural science (*quid facti*?).

The two questions to be answered can be boiled down to:[1]

Question 1: Why is science successful?

Question 2: What has natural science (or different natural sciences) found out about induction?

respectively. I do not believe that question 1 has ever been answered in a satisfactory way.

The popular answer that *nature is regular* is apparently too vague, but can be made more precise. If we take sensory stimuli as our information source, we discover that such a source does not have maximum entropy.[2] In other words, a given collection of sensory stimuli is not unconditionally compatible with any other collection of sensory stimuli. The *real* sensory world is a member of a proper subset of the set of all *possible* sensory worlds. *Defining this subset is the task of science.*

However, all this is merely *a posteriori* speculation and does not answer the question *iuris*. All we can *at best* say is that *so far* nature *has been* regular. We cannot by any logical means conclude that nature *is* (and consequently *will be*) regular. The legitimacy of prediction is still unproved.[3]

A sensible attitude to take may be the pragmatic one (see ref. [1]). If nature *is* regular, the strategy of inductive inference can be successful. If nature *is not* regular, *no* strategy can be successful. No one can predict the next element of a completely random sequence. As a consequence, we have better to assume that nature *is* regular.

A mere denial of the legitimacy of *any* process of induction does not help very much in analysing the foundations of science, as long as we recognize objectivity or at least *intersubjectivity* to science.

It is widely admitted that an observational report cannot serve as a basis for science, unless it is *intersubjective* (dreams or hallucinations do not help to understand the external world). However as soon as I assume that something is intersubjective, I have made full use of inductive inference. I believe that my observational report: "This paper is white" has an intersubjective validity. This is because in the (finite) number of cases when I have reported 'white' I have received the assent of a (finite) number of people. From this I jump to the conclusion that in *all* cases *all* people will agree with my statement. This is the meaning of the word 'intersubjective'.

I am a physicist and have no hope (or ambition) to be able to give a *full answer* to the philosophical question 1. But I believe to have something to say about question 2. And this, in turn, will shed some light on question 1.

As W. V. Quine has rightly pointed out [2], epistemological questions can arise only *after* science and *within* science. They can be understood and possibly answered only by making use of science.

I want to discuss how the problem of inductive inference can be viewed within the framework of modern physical science.

It will be shown that what physicists believe to know about induction is the result of long experience rather than the product of merely abstract speculation.

II

Physicists have a long record of application of inductive inference. It is well known that by making experiments and applying inductive inference, they have learned a lot about the physical world. However it is not very widely realized that during the course of this application *they have learned something about inductive inference* too! It stands to reason that while opening bottles, one should learn something about the corkscrew.

What I will say now has definitely a Popperian flavor [3, 4]. Physicists have not learned from physics what kind of inductive procedure is *valid*. This would be a *petitio principii*, since it would involve induction. But they have learned what procedures are *wrong*, by stumbling upon *counterexamples*.[4]

One point must be emphasized before discussing this subject. Physics, by its own nature, strives to arrive at rigorous and fully reliable statements. It is only on this level that physicists have been *forced* to give up some very common forms of inductive inference. But they can very well make use of such forms of inference on the *heuristic* level. For instance, in any trial and error procedure, the trials are suggested by some sort of induction, otherwise they could not make any sense.

In this paper I will only refer to *reliable* forms of induction, defined as those forms of induction that can lead to statements accepted by the physics community beyond reasonable doubt.

As well known, some people will question the existence of such statements. But this is, more often than not, the result of a serious misunderstanding. Modern physicists are perfectly aware that any law or theory is valid only *within a certain domain D* (*domain of validity*), specified by the class of phenomena to which it applies, by the range of the parameters involved, and last but not least, by the *precisions* of the instruments used by the experimenter [5, 6]. Physicists are well aware that *extrapolation* is an unsound procedure and that only experience can tell us whether or not the theory is valid outside the boundaries of D, as are known at present. But within D, they are *sure* that the theory leads to perfectly reliable statements.

III

The most important rule of inference that physicists have been forced to give up is the *many-to-all* rule.

Let $\{P_i\}$ represent a *finite* set of monadic predicates such that $P_i a$ can be verified by a finite set of experimental operations to be carried out on the physical object a and A a physical 'property' defined by the standard normal form[5]

$$(1) \qquad A =_{\mathrm{df}} (P_m \wedge P_{m+1} \ldots) \vee (P_n \wedge P_{n+1} \ldots) \vee (P_l \wedge P_{l+1} \ldots) \ldots$$

where some of the indices may be identical. Let B represent a different property defined in a similar way and consider the general statement

(2) $\forall x(Ax \rightarrow Bx)$

This statement may be a logical consequence of the definitions (for instance when the definition of B contains all the disjunctions of A, plus some other disjunctions: if x is an *electron*, then x is a *lepton*). Otherwise, in order to 'prove' (2) we must have recourse to experiment. But experiment can prove (2) only in the case when x ranges over a *finite* set of individuals, and when all of them are tested. In that case (2) turns out either false or trivial.

In the interesting case when the number of the individuals is, at least virtually,[6] infinite, some people believe that experimental evidence can at least *confirm* (2) and that the more individuals a are found, such that $Aa \wedge Ba$, the better is (2) confirmed (Nicod's criterion [7, 8]). In other words, *positive instantiation* renders the general statement (2) more 'probable', in one or other of the different senses that the concept of *probability* can have [9].

Now physicists have known for a long time that such a criterion is absolutely *unsound*. The reason is that they have learned at the cost of repeated failures that, no matter how many positive instances of (2) have been found, the world where

(3) $\exists x(Ax \wedge \neg Bx)$

is *not less probable* than the world where (2) is valid. This is because modern physics has taught us that, contrary to common belief, *very rare events are very common*!

Let us give an example to clarify this point. Suppose that we can observe individual uranium atoms, one at a time. Let A represent the property 'is an uranium U^{238} atom' and B the property 'does not decay within an hour'. We can observe one thousand or even one million such atoms and still be (virtually) certain that all of them satisfy condition B. This is because less than one out of ten thousand billion atoms decay within an hour.[7] *But if we conclude that* (2) *is valid we are wrong.*

Still, one might be tempted to think that, after all, we are only *very little* wrong in assuming the validity of (2) and that we are just missing a negligible detail of the physical world. Nothing could be farther from the truth. We would instead be missing a fact of tremendous importance in nuclear theory and in cosmology!

Now we are in a position to answer the question: Would a physicist, after observing one million uranium atoms, be justified in saying that (2) is *probable*? By no means, for the case of the uranium nucleus does not represent a rare exception in modern physics; it is rather the rule. Any potential barrier can be tunnelled through and will be tunnelled through in the long run. In particle physics very rare modes of decay are always to be expected. Even the electronic wavefunctions of atoms, though decaying exponentially, extend to infinity, so that an electron of an atom in the ground state can be found anywhere (and be given by the observation the energy it needs in order not to return to the nucleus).

For these reasons, *a physicist is not inclined to think that a world where* (3) *is valid is less probable than a world where* (2) *is valid,* independent of how many positive instances of $A \wedge B$ he has observed.

We will not dwell on the well known paradox of confirmation [10], since such paradox disappears as soon as we recognize that the criterion for asserting (2) is unsound.

IV

Have then physicists followed Popper's advice and given up all attempt of applying induction? The answer is no. However there is *only one case* when they apply a rule of inductive inference *with absolute confidence.* This is when they apply the principle of *space-time invariance* of physical laws.

In very simple words this principle says that nature acts script in the same way as it acts *there then*, both in the past and in the future.

I have argued elswhere [6] that since the truth of this principle is a *necessary* condition for us to carry out everyday life, it has become (either phylogenetically or ontogenetically) a necessary structure of human thought.[8]

In order to formalize this principle, we must distinguish between classical macrophysics and microphysics.

Let s and s' represent any two finite regions of space-time that can be superimposed on one another by means of a translation in space-

-time and a rotation in space. We will say that s and s′ are *congruent* and satisfy the relation $C(s, s')$. Let IB_s represent the assertion that a specified set of initial and boundary conditions *I, B* are verified in *s*, and F_s the assertion that a certain physical phenomenon *F* takes place in *s*. Our principle says that if IB_s and F_s are true, then an observation of $IB_{s'}$, (the *I, B* conditions being now specified with respect to *s′* in the same way as they were with respect to *s*), will entail the truth of $F_{s'}$. We shall write

$$(4) \qquad \forall s' \{\exists s [C(s, s') \wedge IB_s \wedge F_s] \rightarrow [IB_{s'} \rightarrow F_{s'}]\}$$

No counterexamples have been found to this principle in classical physics. *One single* experiment is sufficient to derive a universal statement. If experiments with *identical conditions* are sometimes repeated it is only in order to increase the precision of measurement or to avoid mistakes. It would be ridiculous to repeat the experiment several times in order to verify space-time invariance, which is taken for granted.

Now the rule of induction (4) cannot be applied *sic et simpliciter* to microphysics. As well known, one must introduce repeated experiments and probabilities.

Probability is an elusive concept which can have many different interpretations (classical, frequentist, subjectivist, logicistic, and so forth [1, 9]). However, the dispute between supporters of different interpretations seems sometimes to be unjustified. As a rough analogy to see the point, imagine a dispute about the concept of derivative, the physicist claiming it to be a velocity, while the geometrician wants it to be the inclination of a tangent and the economist the rate of growth of capital. They would simply confuse the concept of derivative with its applications! A similar confusion is very often likely to arise when people talk of probability.

Now it is undeniable that physicists have used probability to calculate *relative frequencies of occurrence*. For this reason they have started to say that probability *is* a relative frequency of occurrence. There is no harm in this, after all, provided that one remains within the domain of physical science and does not want to speak, say, of the probability that God exists.

Let Σ represent the equivalence class of all space-time regions that are congruent with a given *s* and $S, S' \subset \Sigma$ two *finite* subclasses of Σ.

We will say that S and S' satisfy the relation $C(S, S')$. If we carry out an experiment under identical conditions in each of the n regions belonging to S, we will find that in a fraction r of the cases a certain phenomenon F takes place. We will say that r represents the result of a *measurement* of the probability $p(F, S)$ that F occurs in S. Such a measurement is made with a precision ε,[9] which becomes better and better with increasing n.[10] This will be expressed by $p(F, S) = r \pm \varepsilon$. Let us now repeat the same experiment in each of the n' regions belonging to S'. We will find a value $p(F, S') = r' \pm \varepsilon'$ for the probability that F occurs in S'.

The generalization of (4) to microphysics will now be written in the form

$$\text{(5)} \qquad \forall S' \{\exists S[C(S, S') \wedge p(F, S) = r \pm \varepsilon] \rightarrow [p(F, S') = \\ = r' \pm \varepsilon' \rightarrow r' = r \pm (\varepsilon + \varepsilon')]\}$$

This formula expresses the *space-time invariance of probability.* Within the precision of measurement, the probability of F is the same in S as in S'. No counterexample has so far been found to this principle.

It will be noted, of course, that the meaning of the initial and boundary conditions is now a little different from what we had in classical physics. We will say that we are making experiments under identical conditions when we carry out each time a *maximum* observation on the system in the sense of quantum mechanics and find the same results (up to the precision allowed by the instruments). To verify F will mean to make successively a certain measurement and find a certain result.

Quantum mechanics shows us that probabilities become δ-functions for *macroscopic experiments,* so that (4) becomes a consequence of (5).

A principle similar to (5) is found to hold even in macrophysics, when the fixed conditions that were assumed in the application of (4), are replaced by a (possibly *incomplete*) set of specifications taken for each s of S or S' *at random* in a certain domain. In this case there may sometimes arise a difficulty in the definition of *randomness.* This difficulty is well known in the classical theory of probability and we will not attempt to solve it here. However there is no denying that in most cases the principle can be applied in a rather simple and reliable way (e.g. throwing dice, shooting a gun, and so on). Note that (5) is valid *a fortiori* when we are working in *microphysics* with *random* conditions.

The principle of space-time invariance of probability is the most general form of inductive inference in which a modern physicist has complete confidence.

All other forms of induction may be used (and are used) for heuristic purposes. But they are known to be unreliable and to admit of counter-examples.

V

Let us now show briefly how the principle of space-time invariance of probability can be applied to develop physical science.

We will first consider the case of simple laws and later discuss the case of general theories.

Case a. Suppose that we are making experiments on a *macroscopic* phenomenon with *fixed* I, B conditions. Probabilities are δ-functions, and we can apply (4). *One single* experiment is needed to learn once and for all what phenomenon F will take place as a consequence of the I, B conditions. We have a reliable *one-to-all* generalization.

Case b. Consider now an experiment in microphysics with *fixed* or *random* conditions, or an experiment in macrophysics with random conditions (within a specified range). We can apply (5). One single *set* of experiments is needed in order to determine the probability that F will hold. Of course, the more numerous are the experiments of the set, the better is the precision with which probability is determined.

The application of the rules of case a or case b is very often straightforward. However, there are some cases that deserve special attention.

Suppose we are investigating a physical object of a class for which physics has a special name, such as: star, comet, vortex, crystal, oxygen atom, electron, neutron. In order to state that x is one of these objects we must verify that x has a property such as A, defined by (1). If we observe altogether n cases, where the conditions not specified by A are taken at random, and find that in all of them the object shows also property B, we do not start to assert (2), a conclusion that we know to be unsound. We apply instead the principle of space-time invariance of probability and say that *the probability that an object specified by A will not have property B has a value of the order of* $1/n$ *or less.* This is the correct assertion that in modern physics has replaced all the statements of the form (2) as well as all the statements about the probability of (2) being true. Of course, sometimes n is very large. In this case people can assert (2) as an *abbreviation*; what they mean is only that $1/n$ is very small.

Let us now turn to the very important case of the quantitative laws of physics, expressed by equations of the type $f(q_1 \dots q_m) = 0$, where $q_1 \dots q_m$ represent the values of m physical quantities. An equation of this kind has, of course, a domain of validity, specified by the class of phenomena to which it applies, by the allowed range of each quantity and by the precision with which each quantity is measured. For this reason the correct form of the equation can only be

$$(6) \qquad f(q_1 \dots q_m) = \pm\varepsilon$$

where ε depends on the precisions [5].

When one asserts the physical law (6), one apparently implies that (6) should be true for *any* m-tuple of values for the $q_1, \dots q_m$, selected within the allowed ranges. This cannot be verified experimentally, for it would require an infinite number of experiments. When we assert (6), we assert something about *unobserved* cases that are not even space--time congruent with any *observed* case. Is this a sound procedure? Of course, it is not.

Spectral lines, resonances in particle physics and related phenomena have taught us that experimental plots cannot be smoothed out without risk. One can come across a fine spike anywhere.[11]

The correct way to describe the procedure is this. One measures the relevant quantities in n cases, selecting the conditions at random, and calculated the expression $f(q_1 \dots q_m)$. If in at least one case, $|f|$ is found to exceed ε, the law (6) is not valid and one should look for a different function. If instead (6) is found to be verified in all n cases,[12] one must conclude that *the probability that the result of a new measurement does not satisfy* (6) *is of the order of* $1/n$ *or less.*

In the case of the classical and well known laws of physics, n is so large that people tend to disregarded $1/n$ and simply say that (6) is *verified.* There is no harm in this. But one must be aware that this locution is just a matter of convention.

VI

We pass now from the case of a simple law (6) to the case of a general physical theory. A theory T will consist of a set of axioms (used in conjunction with the rules of deduction of classical logic), from which one can derive one or more laws of the form (6). Obviously, one or more such laws may directly belong to the axioms.

Physicists assert that T is valid within a domain D, consisting of all the phenomena that can be correctly predicted by T. Strictly speaking, D should include only those phenomena that are space-time congruent with cases already observed. However, it is customary to include the whole range of phenomena described by a law such as (6), provided that the number n of the points tested for validity of the law is very large and consequently the probability of falsification is very small.

Anyway, *extrapolation* is not admissible, if not for heuristic purposes.

The specification of D has necessarily a historical meaning. Physicists are always striving to make experiments in regions lying *outside* of D (by improving precisions or by enlarging the class of phenomena) and to compare the results with the predictions of T. When there is agreement, they have succeeded in enlarging the boundaries of D in that direction. But it may happen that one finds an experiment E whose result is contradicted by T.

It is curious to note that many writers tend to describe this occurrence in a false light. They would tell the story of a physical community stricken with dismay by the unexpected and *unpleasant anomaly*. Of course, the opposite is true. The dearest dream of the experimental physicist is to succeed in making an experiment that contradicts some of the oldest and most general laws of physics. And the theoretical physicist welcomes with excitement a result that gives him the opportunity to elaborate a new theory!

The new theory T' which is sought must predict all the experiments of D plus the new experiment E. In other words, its domain of validity D' must be *at least* $D+E$.

The chronological succession described, which may appear to be necessary, is often inverted. First T' is born, and next E is carried out. If the result of E agrees with the prediction of T', we say that T' has been *confirmed* by E. Otherwise a new theory must be looked for.

The actual chronological succession of E and T', which is of importance for history, is virtually irrelevant for a logical reconstruction of the development of physics.[13]

The development of physics can be described by means of a *historical ladder* of theories $T_1, T_2, T_3, \ldots$, which are valid in larger and larger domains $D_1 \subset D_2 \subset D_3 \ldots$ Obviously, I am talking of an *internal history* or of a *rational* reconstruction that does not necessarily coincide with actual history.[14]

There are two *extreme* cases and all intermediate cases are possible.

Case I. At a given time physicists have a theory T_n that agrees with all the evidence $E_n^1, E_n^2, E_n^3 \ldots$ of a given domain D_n. One or more new experiments $E_{n+1}^1, E_{n+1}^2 \ldots$ are made whose results do not agree with the predictions of T_n. Someone suggests a new law or *ad hoc* hypotheses H, which can account for the new evidence without contradicting that of D_n. Then the new theory $T_{n+1} = T_n \wedge \mathrm{H}$ is accepted, and its domain of validity is $D_{n+1} = D_n + E_{n+1}^1 + E_{n+1}^2 + \ldots$.

As an extremely simplified example of case I, let us take the situation of electricity and magnetism in 1820, when T_n comprised the laws of classical mechanics, Coulomb's law, the creation of electric current by chemical means, and so on. Oersted makes some experiments whose results are not predicted by T_n. They are in agreement with the hypothesis H_1 that a current should give rise to a magnetic field according to a convenient quantitative law (e.g. Biot and Savart's).
As a consequence the theory $T_{n+1} = T_n \wedge H_1$ becomes accepted. Its domain D_{n+1}, includes the effects of a current on a magnet. Next, in 1831 Faraday discovers electromagnetic induction, which can be expressed by a well known law H_2. One then adopts the theory $T_{n+2} = T_{n+1} \wedge H_2$ and so on.

Case II. Physicists have a theory $T_n = T_m \wedge H_1 \wedge \mathrm{H}_2 \wedge \ldots\ldots$ that includes a number of hypothesis $H_1, H_2 \ldots$, completely independent of one another and of T, like those of the foregoing examples. Someone elaborates a theory T_n', which is still valid within the domain D_n of T_n, but contains a smaller number of independent assertions.[15] This T_n' might be preferred to T_n for the sake of unity and economy. However people tend to be conservative, unless T_n' can predict some phenomena $E_{n+1}^1, E_{n+1}^2 \ldots$ outside D_n, the predictions being later verified by experiment. In this case T'_n is confirmed and becomes the new theory T_{n+1}, with $D_{n+1} = D_n + E^1_{n+1} + E^2_{n+1} + \ldots\ldots$.

As an example for case II, we can start again with electromagnetic theory. Before Maxwell the theory is represented by a number of independent laws. Maxwell with his equations unifies almost all of them. The equations predict the existence of electromagnetic waves, and when Hertz finds such waves the theory is confirmed. There remains the *ad hoc* hypothesis concerning the force experienced by an electric charge moving through a magnetic field. Einstein unifies everything with his theory of special relativity.[16] Again this theory predicts some new phenomena, such as time dilatation, mass-energy conversion and so on. A verification of these predictions confirms the theory. Finally, there remains the *ad hoc* hypothesis of the equivalence of inertial and gravitational masses. General relativity eliminates the *adhocness* of such hypothesis and predicts new phenomena. When these predictions are verified, the theory is confirmed.

As we have noticed, all the intermediate cases are possible between the extreme cases I and II. Obviously, case II is the one that physicists prefer, while the extreme case I with its *ad hoc* hypotheses is generally accepted with some uneasiness and is considered as a provisional stage.

However, I believe that such an uneasiness is largely due to psychological rather than essential reasons. The *adhocness* of a hypothesis can merely depend on that chronological succession which in my opinion has very little conceptual meaning. For an independent hypothesis H is said to be *ad hoc*, when its only purpose is to explain new experimental evidence. But sometimes H *precedes* in time the new evidence! Can it then be termed *ad hoc*?

Among the many and well-known examples of hypothesis that were independent of all preexisting science, one may mention the law of universal gravitation (born from the desire to *unify* the motion of falling bodies and the motions of the Moon and planets), the displacement current (born from an analogy and a desire for symmetry), de Broglie's hypothesis of the wave nature of matter (motivated also by analogy and symmetry). If the new facts they predict (planetary perturbations, electromagnetic waves, electron diffraction) had been discovered before, such hypotheses could have been termed *ad hoc*.

The unification of all science under one theory with the minimum number of independent hypotheses is an excellent methodological norm. However, the essential result is the construction of the historical ladder of theories.

BIBLIOGRAPHY

[1] Salmon, W. C., 'The Foundations of Scientific Inference', in *Mind and Cosmos*. Pittsburgh, 1966.

[2] Quine, W. V., *La teoria e l'osservazione*, Seminario di storia e filosofia della scienza dell'Universitá di Firenze, April 1974.

[3] Popper, K., *The Logic of Scientific Discovery* (London, 1959).

[4] Popper, K., *Objective Knowledge*, Oxford, 1973.

[5] Dalla Chiara Scabia, M. L. and Toraldo di Francia, G., *Rivista del Nuovo Cimento*, **3**, 1 (1973).

[6] Toraldo di Francia, G., *Rivista del Nuovo Cimento*, **4**, 144 (1974).

[7] Hempel, C. G., 'Recent Problems of Induction', in *Mind and Cosmos*, Pittsburgh, 1966, p. 120.

[8] Nicod, J., *Foundations of Geometry and Induction* (New York, 1930), p. 219.

[9] Costantini, D., *Fondamenti di calcolo delle probabilitá* (Milano, 1970).

[10] Hempel, C. G., Theoria, **3**, 206 (1937); *Mind*, 54, **1**, 97 (1945); *Journ. of Symbolic Logic*, **8** 122 (1943).

[11] Popper, K., 'Truth Rationality and the Growth of Scientific Knowledge', in *Conjectures and Refutations* (New York, 1965), p. 247.

[12] Lakatos, I., 'History of Science and its Rational Reconstruction', *Boston Studics in the Philosophy of Science*, **viii** (1971), p. 91.

REFERENCES

[1] I take the view expressed several times by several people that learning (even by the newborn baby) in everyday life is just a part of science and that no discontinuity occurs when passing to the self-conscious speculations of full-fledged science.

[2] I use this expression in the sense of information theory. The source would have maximum entropy if all random collections of sensory stimuli had the same probability of occurring.

[3] The *genetic* explanation represents no proof either. One starts from the assumption that nature *is* regular and then proceeds to show how nature has phylogenetically or ontogenetically exploited this property to the benefit of the species or of the individual.

[4] The theory that inductive procedures cannot be *verified* but can be *falsified* by experimental evidence might perhaps be termed a meta-(Popper theory)!

[5] An individual variable x will be omitted throughout, for brevity.

[6] For instance the number of the elementary particles in the universe may be finite, but in the present context is practically infinite.
[7] The lifetime of U^{238} is 4,5 billion years.
[8] Let us stress once more that this does not *explain* anything. It just *describes* the way we think about nature.
[9] The situation is exactly the same as for any other measurement in physics (See ref. [5]).
[10] Probability calculus teaches us that $\varepsilon = O(1/n^{\frac{1}{2}})$
[11] We are referring to the case when one is plotting a *completely* unknown phenomenon. In other cases one may know by other evidence or law that spikes finer than a certain quantity are not allowed.
[12] It will be emphasized that the form of the function f must be selected *before* carrying out the measurements for, *a posteriori*, one could choose any wild function that passes through the n points measured. This would obviously be a wrong procedure.
[13] Popper [11], arguing with Keynes, expresses the opposite opinion. I agree with Keynes.
[14] Some historians criticize this *unfaithfulness* to history. I believe that a purely *external* history is an illusion. External history can only be made under the guidance of an internal history, possibly made subconsciously. See ref. [12].
[15] I am aware that this is only a rough description of the situation. It may sometimes be difficult to assess why one theory appears to be more *unitary* than another one.
[16] As well known, the Michelson-Morley experiment had very little influence on Einstein.

PART B

(Papers presented by title)

JOSEPH AGASSI

VERISIMILITUDE: POPPER, MILLER, AND HATTIANGADI

Though the truth is one, it may be approached by different routes, perhaps. If we expect scientific theory to be successful in different ways, then we can measure its success in different ways, and each success is, in a way, an approach to the truth. Nevertheless, there is a great appeal to the idea that the consequences of a theory can be divided into true and false and a measure of the truth-content and the falsity-content be defined. Yet, even were this possible, and even were it possible to measure the truth-content of a given theory, this will not enable us, as Miller has ventured to argue against Popper, to declare Einstein nearer to the truth than Newton in the sense which Popper has suggested — a sense which captures one of our intuitions of the situation.

The important question any theory of verisimilitude should answer is, how does science progress? I wish to argue that this question is still open, and that anyway progress can only be decided when taking science as a whole: there is no point in speaking of the verisimilitude of a fraction of science.

I. BACKGROUND

It is probably an ancient Babylonian idea that science has to increase the accuracy of its observations and predictions beyond the commonly accepted level. Yet the theory of it is still far from being satisfactory, or even clear. The principle in question is an involved affair, as any student of science discovers if he goes about it the right way. (Alternatively, our educational system being what it is, a researcher may discover it in his first laboratory work in which he designs his own instruments rather than runs it on ready-made ones.) Thus, the simplest Babylonian observations, i.e., of planetary motions, involve observations of angles and of time, these leading at once to the earliest scientific theories of material strength and of flow of water, the one being significant for measuring angles, the other time. Moreover, each refined measurement has to be correlated to other measurements — and this will turn out to be crucial in our discussion. Thus, to stay with our first example, measuring time correlates to measuring positions of planets vis-a-vis the whole firmament, i.e., the fixed stars, but not too well because the idea that the whole firmament rotates with uniform diurnal cum annual motion must be rectified, e.g. by taking account of the precession of the equinox. And, as Jerome Ravets has shown, this small matter hides in its logical recesses the whole Copernican Revolution, no less. Thus, the increase of precision is a matter of the increase of approximations to the results, and this is done in a sense locally, in a sense globally. It can easily be noted that the more advanced a science is the more global the adjustments tend to become — both in the sense that the corrections tend to become increasingly theory laden as we theorize about the instruments we correct, and in the sense that theories merged into more universal ones, thus making the global domain of application ever wider. Moreover, the choice of different theories may lead us to the choice of different instruments and thus to different measurements. Thus, alternatives are not two different theories, but at times two different sets of theories and instruments and data. This matter was first noticed by Pierre Duhem who saw each scientific change a change in the whole of science. He therefore required that the changes be small so as not to lose us our bearings. But Einstein has changed all that and shown us that even great revolutions need not lose us the connexion between old theories and new, where results of the older ones are approximations to those of the newer.

It is thus intuitively easy to grasp that the matter of approximation to the truth is far from being a matter of mere experiment or mere small change of theory but is a matter of the most general view of science which we may evolve, no less. It makes sense, then, to speak of theories as approximating each other if we take into consideration a theory about the world as a whole and measurement techniques and results of measurements and anything else we can think of in that connexion, and tie them up in a grand philosophy of science and of nature.

And so we should not be surprised to find the idea of approximation to the truth — of increased verisimilitude, to use Popper's term — though ancient, is still quite intriguing. Indeed, its career is very checkered: in the Age of Reason it was put into cold storage. Perhaps it was because it is hard to see Ptolemaism as approximate Copernicanism, or even of Brahe's view as such an approximation. Kepler has devised the mathematical method by which to do so—actually following an idea of Brahe—namely of freely transforming coordinates. Kepler's own successive systems likewise show series of approximations.[1] Newton showed that both Kepler's and Galileo's theories are excellent approximations to his. In addition he invented his celebrated perturbation theory which enabled him to begin with Kepler's views and in stages arrive at increasingly more Newtonian results in those places where Kepler's were unsatisfactory. Without this technique, or some alternative to it, Newtonianism would have been utterly paralyzed as long as the Newtonian many-body problem remained insoluble—which it still is. This technique is quite universally applied when we decide, from the viewpoint of the new theory, the domain of applicability of the old one. Thus, though we declare the old theory false, we do not discard it, either in science or in technology, but freely employ it when the new theory permits us to. Nevertheless, in the Newtonian era there was surprisingly little mention of approximations and no discussion of them, no competing theories about them, etc.

Indispensable as approximations were to Newtonianism and to successive developments in physics, there was little or no interest in the theory of approximations because the idea of science as successive approximations to the truth was philosophically suppressed by the Newtonians. This was done for the simple reason that Newton put all the weight of his immense authority on the claim that his theory is true and not a mere approximation to a better theory of the future. The most celebrated Newtonian ever, and the best theoretician of approximation, is probably Laplace. Laplace was very worried by the possibility that the inverse square law is an approximation to an exceedingly similar law with the number 2 replaced by the number $2+\varepsilon$ where ε is exceedingly small. Laplace felt, quite stupidly, incidentally, that such an idea threatens the very existence of science and is a monstrosity and an absurdity. (His argument was meant to apply likewise to additions of factors of higher orders of the inverse of the distance. He made no allusion to this better possibility of modifying Newtonianism, though he must have heard of Boscovich.) Indeed, without discussing Laplace's tremendously important theory of approximations and its contribution to other studies, such as Gauss's, and indeed to the whole growth of the theory of probability, we can say the peculiar feature of Laplace's theory is that it pertained to precision of measurements with no reference to theories at all.

Einstein has changed all that. He argued that Newtonian kinematics is an approximation to relativistic kinematics, that Newtonian kinetic energy approximates the relativistic one, that general relativity yields Newtonian gravity as an approximation. Approximations now became respectable also on the most global level.

Approximations now could hardly avoid being also a more sophisticated matter. The advocate of a new theory had now to show the old theory to be a special case as well as an approximation. There may be an exception: Fresnel's optics, suitably simplified (here it is ambiguous, but let us go on), is a special case, not a merely approximate special case, of Maxwell's electrodynamics; Maxwell's equations, suitably streamlined, are a special case of special relativity, though of course, only approximate to the covariant electromagnetic equations of general relativity. Similarly, any non-relativistic wave-equation for the electron may be nothing more than an approximation to a relativistic one, i.e., not a special case. The ideal remains: the old may well be both a special case, and, even then, a mere approximation. Thus, in general relativity only a stationary gravitational field is in any sense such that results from the Newtonian theory of gravity are approximate to it.

Here Popper's theory of science as conjecture and refutations comes in quite handy: when a new theory yields an old one as an approximation, this very fact invites a crucial experiment, i.e., a test—at times by suggesting a new measuring technique to increase the precision of our measurements in a very special manner. Also, and while we are busy improving our measuring techniques, the new theory tells us what factor to vary so as to make the error of the old boldly manifest itself.

A simple example, now that we can look backward, is the crucial test between Newtonian and Galilean gravity, which can be performed by moving a grandfather clock on sea level from south to north and seeing it accelerate however slightly over a long stretch of time. An amazingly sophisticated example, as if made to measure, is the Lee and Young mode of reasoning which led to the overthrow of the principle of con servation of parity. Step one is irrelevant here though crucial for that study and for other aspects of the philosophy of science—which interest me more but must be ignored here: there was a puzzle: two particles looked suspiciously similar and the quation was either, why? or are they really different? The reason people were forced to declare them different in spite of their great similarities is the principle of conservation of parity. Hence, obviously, rather than attack the problem why are they so similar, it may be advisable to ask, are they really different, to wit, is not the principle of conservation of parity false?

The situation was simple: there was a wealth of evidence conforming to the principle: all effort to overthrow it had ended in failure. Clearly, Lee and Young saw only one way to repudiate the principle, and that was by overthrowing it by empirical means. They reasoned thus. Assume the principle false. Assume, further, that it is true for some domain, approximate for another, and divergent from still another. The question was, which domains? To answer this they had to outline the dimmest notion of what should replace the principle. Working on this notion with the existing data in hand, they concluded that under certain conditions, as yet unexamined, the violation of the principle should be easily perceived. This turned out to be true. Once this was established, incredibly smaller deviations from the principle could be detected in more ordinary circumstances though with great ingenuity (since it is almost impossible to detect a neutrino).

So much for the present situation. Enter David Miller, former student and disciple of Sir Karl Popper.

II. VERISIMILITUDE

The theory that scientific theories are successive approximations to the truth, to repeat, is quite traditional. It was explicitly stated by a few classical scientists like Laplace who, however, insisted that the series of approximations should be finite with the trut as the end-point. Duhem postponed the end-point to the very distant future while declaring all science to evolve as one and so while demanding all changes to be small so that yesterday's science, as a whole, should approximate today's science. Einstein was the first who offered the same theory without any demand for continuity and without the end-point as a necessary condition: at times he expressed hopes regarding it, at times doubts; but never more than mere hopes. Popper has left the question of end-point utterly open on a point of principle (of the validity of Hume's critique of induction).

Popper offered a formal, or semi-formal description of the idea of series of approximations, which is very intuitive and very seductive: one later theory is a better approximation (in a strong sense) to the truth than another earlier theory, if and only if, all the true consequences of the earlier theory are also consequences of the later theory and all the false consequences of the later theory are consequences also of the earlier theory (and they are different). It is hardly necessary to explain the seductiveness of this description, or, to nail it down, this *Popper's theory of verisimilitude* (in the strong sense).

Miller's theorem pertains to the Popperian concept of verisimilitude. It says, if one theory is more verisimilar than another (in the strong sense), then it is true. Miller and others (myself included) have published successive proofs of his theorem. All these proofs fade as compared with Hempel's indirect proof which J. Hattiangadi tells us, Hempel has invented (prior to Miller) on the spur of the moment during Hattiangadi's oral doctoral examination in Princeton. Here is the proof as reproduced by Hattiangadi:

1. A is false (given)
2. B is false (given)
3. All true consequences of B are consequences of A
4. All false consequences of A are consequences of B
5. At least one true consequence of A is not a consequence of B

(3–5: Definition of verisimilitude)

6. $A \supset B$ is true (since A is false)
7. $B \vdash A \supset B$ (Theorem of propositional calculus)
8. $A \vdash A \supset B$ (Because $A \supset B$ is true by (6) and follows from B by (7), by (3) it must follow from A)
9. $A \vdash B$ (simplifying (8))
10. $A \vdash A$ (Theorem of propositional calculus)

11. $B \vdash A$ (Since A is false, by (1), and it follows from A by (10) it must by (4) follow from B)
12. $B \dashv\vdash A$ (from (9) and (11))

(5) Contradicts (12), and so we see that one of the assumptions (1)–(5) must be mistaken.

So much for the quotation, verbatim, from a note by Hattiangadi. What are we to do now? Should we, as Hattiangadi suggests, give up verisimilitude altogether as the fundamental characteristic of science and suggest alternative views? Hattiangadi has taken such a bold step, and has thus offered a revolutionary new theory of science. In his view science breeds problems and the series of scientific theories is the series of ever growing clusters of problems. This, like many surprisingly novel ideas, may sound to the untutored rather banal: no one has ever denied that science breeds problems; for example, even Thomas S. Kuhn's old-fashioned reactionary theory of science includes the view that, science breeds problems. Also, according to Kuhn, to stay a bit with this example, the value of scientific problems is that they are small and manageable (he invented a name for such problems; "puzzles" he calls them, and he calls research "puzzle solving"); whereas Hattiangadi thinks the growth of science is the growth of more and bigger and harder problems. But, more philosophically, admitting that the growth of science is both more problems and more verisimilitude, we can still differ as to which is more fundamental of these two characteristics. And to say, with Hattiangadi, that the growth of problems is more fundamental than the discovery of some truths, is bold and revolutionary indeed.

For my part, I wish to opt for verisimilitude a bit longer: I find it still a challenge to rectify Popper's theory and to see why our intuition of verisimilitude survives Miller's onslaught on Popper's theory. Of course, this move of mine may betray sluggishness of thought, or even dogmatism. Perhaps, however, it means that I am fascinated by the fact that our intuition of verisimilitude can be captured in ways other than Popper's.

The simplest suggestion to try out (it does not work, we shall see) should be to relativize Popper's idea—relative to given experiences, of course. Is this move possible? It should be noticed that Popper's theory of verisimilitude makes no reference to experience. This is both attractive and suspect. It is attractive because any reference to our experience in deciding verisimilitude may be question-begging: experience supports (a) the preference of Einstein's theory of gravity over Newton's and experience supports (b) the theory that Einstein's theory of gravity is more verisimilar than Newton's. If we want (b) to be different from (a), if we want (b) to tell us something (a) does not already tell us, then we may want (b) to be based on a different kind of reasoning, whether a different source of experience or some a priori argument, and we do not have additional experience. Yet the very a priori character of the argument is disquieting: is it synthetic a priori and thus otiose or is it analytic and so we can prove that science progresses towards the truth?

We can conceive of series of conjectures and refutations not converging to the truth: Max Black has designed a model for this. Hence, the a priori claim cannot be analytic, and we reject on a priori grounds all claim for the status of synthetic a priori knowledge. This, in itself, need not be troublesome since the claim that we do, that progress in science is increased verisimilitude, need not be knowledge: it may be a conjecture. Yet, is it backed by any argument? We said, we have no extra empirical evidence to support it, and we have no analytic evidence as such evidence will amount to a proof. Perhaps not to a complete proof, but to probability. Clearly, Popper will have to reject this rescue move out of hand. What, then, is the very status of Popper's theory of verisimilitude? How can we modify if before we decide on its status? Are our theories of verisimilitude become increasingly more verisimilar? Will there be a case where we have an increased verisimilitude by one theory of verisimilitude but not by its successor or *vice versa*?

Here we can see why Popper's verisimilitude is so attractive, and thus why Miller's theorem is such a blow. Popper's theory is so very attractive, so excellent, because it offers a criterion of increased verisimilitude which is utterly objective and has no reference to knowledge at all. It says, given that one theory contains all the truths contained in the other, and the other contains all the truths contained in the one, then, and only then, is the one more verisimilar to the other. Since each theory has infinitely many true consequences and infinitely many false ones, not to mention the fact that some of these are undecidable, it follows that *we can never know whether one theory is more verisimilar than another* (*in Popper's sense*)! The crucial experiment only excludes the claim that the early theory is more verisimilar than its successor, and so, as long as we accept its verdict we cannot prefer the old to the new. But this is far from telling us that we are sure we have progressed in the sense that we are sure another crucial experiment is impossible whose verdict will go the other way! This can hardly be otherwise, or else (a) above will become synonymous with (b)!

III. THE PROBLEM

The situation now is very tricky. We may have cases where the same evidence condemns an accused, a scientific theory, by one theory of verisimilitude, and not by another. We do not condone actual court misjudgments based on erroneous theories—on the supposition that science progresses, that later theories are better approximations to the truth than others. But perhaps this supposition itself is revisable and decisions based on it are reversible! This sounds a bit too much! We would all hate it if we could have a court case reversed not on the strength of a new scientific theory but on the strength of a new theory of verisimilitude.

Without passing any judgment, I think we all feel, as a matter of fact, that there is too much skepticism in the denial that Einstein is an advancement over Newton, though by no means that it is too much to doubt the truth of Einstein's own theory. Let us take this for a fact, and without defending or attacking it let us try to explain it. The explanation may lead us to suspect that we are too confident, or to reinforce our confidence. There is no telling.

The simple fact is that there was a crucial experiment between Newton and Einstein and Einstein won. In order to make our case more general, let us imagine more than one crucial experiment, and let us imagine that some battle was won by one theory, some other by its competitor. What can we do then?

We can dismiss the evidence that goes one way and retain the evidence that goes the other way. Such things are done regularly in the history of science, and with some effort, ingenuity, and luck one can force Mother Nature to reverse her verdict. How much the revision is forced, and how much of its success is due to nature herself is, again, an open question. Often, indeed, such questions are not resolved since a revolution sweeps the field and alters all problem-situations within it. Examples from the old quantum theory and the vector quantum theory spring to mind, but even older cases are still undecided. Historians of science defend the wave against the particle theory of light, yet, clearly, the wave theory was a theory of longitudinal waves in an elastic medium, highly penetrable yet so hard!

But at times the complementary characteristics of two theories may persist. It is not clear to me whether the wave versus corpuscular quantum theories qualify, since these are rather two readings, two interpretations, of the same theory. But we do not need examples, really. If two theories cover the same ground, are subject to series of crucial tests, and the tests do not all go the same way, then, obviously, we want a third theory to cover both!

What this indicates, first and foremost, is that we want new theories to avoid the errors of the old theory that have been empirically exposed. Of course, were this not the case we would not be troubling ourselves to test our theories in the first place. Query: is this all there is to it?

We have noted that Popper's theory of verisimilitude, when applied to a concrete case, says much more about it than that experiment decided one way rather than the other: it declares—as a conjecture—that all future experiment will go the same way, of course. Now, as Miller has claimed, this amounts to saying more than Popper has intended to say, namely that the later theory is true. Query: *Can we have the one without the other? Can we have a theory which says, the latter theory is not true, yet all crucial experiments, past, present and future, between the old and the new favor the new? Is such a theory consistent?*

It seems to me that Miller has asked this question, but I cannot say for sure. What I can say for sure, is that he seems to have asked it, and that he definitely answers it in the negative if he does. (In which case I am in agreement with him). How can he prove this? Notice that we no longer accept Popper's definition about verisimilitude as pertaining to all the consequences of the competing theories, but we try tentatively (and unsuccessfully, it will turn out) to apply it only to the experimentally decidable consequences of our theories. As long as Miller has no theory of empirical decidability—and no one does, of course—it is hard to see how he can tackle the question in all its generality. Fortunately, in this case the problem can be solved by a simple empirical refutation, as we shall see. But first let me explain at length the difference between all consequences and all empirically decidable consequences of a theory.

Popper's classical *Logik der Forschung* of 1935 is the locus classicus of this distinction, the distinction he makes between content and empirical content. If we declare the content of all mathematical theories to be zero (for many obvious and good reasons) then we can say, logical positivism identifies content with empirical content, and Popper opposed this identification. His theory of verisimilitude applies to all content; can we qualify it to empirical content only?

IV. POPPER ON TRUTH

I must apologize to the reader now: this section will be an exegetic digression on Popper now, even though I have decided many times to avoid Popperian exegesis. The popularity of Popper's philosophy these days much relates to its being a theory of the growth of knowledge, and I am afraid I detect some confusion within his doctrine on this point. The confusion, I think, stems from Popper's misinterpretation of his change of opinion about truth—which seems to me to make the change look much smaller than it is. The question, does the method of conjectures and refutation bring about the growth of knowledge? is now for Popper the same as, does this method increase verisimilitude? His theory of verisimilitude, that is, comes to bridge between his old views and his newer views. And without exegesis I think we shall miss the importance of Miller's theorem, an importance that lends more credibility than anything else I know to Hattiangadi's revolutionary idea which opens as a new theory of the growth of science.

Well, then.

Popper's theory of verisimilitude belongs to the late fifties and early sixties. His theory of science as a series of approximations to the truth belongs to the mid-fifties, to his classic essay on Berkeley as a precursor of Mach of 1954 and his celebrated and monumental essay "Three Views Concerning Human Knowledge" of 1955 (both in his *Conjectures and Refutations*, 1962). His *Logik der Forschung* of 1935 has nothing about truth. Indeed, in its final pages the author declares that methodology can do away altogether with the concept of truth and rest science on the concepts of deducibility (and thus content) and of contradictions (as falsehood) plus the requirements to test theories within science and eliminate contradictions, whether in its theories, between its diverse theories, or between any given theory and the empirical outcome of a test to it.

Thus, Popper's *Logik der Forschung* and his 'Three Views' are incompatible. The inconsistency was removed in Popper's *Logic of Scientific Discovery* of 1959 which contains his own translation of his *Logik der Forschung*, comments on it, and other matters. Popper now says in comment on his earlier disclaimer that we do need the concept of truth. He says he had rejected the concept of truth in 1935 because he was wary of it: had he known Tarski's definition of it he would not have rejected it. This is very unsatisfactory: where is the mistake in the claim that we can do without it? Why do we need it on top of the requirements to test and to eliminate contradictions? Popper does not answer the question there. He discusses it in his *Conjectures and Refutations*, Chapter 10. But in his *Logic of Scientific Discovery* his interest lies elsewhere: there he asserts, and emphatically, that he was always an adherent to the correspondence theory of the truth: as other parts of his great book show, he always was driven by a realist impulse; but the absence of a definite view of truth made him hesitate—regrettably but understandably.

All this, I wish to say, is contestable. An author has the right, as anyone else, to interpret his own work; but he has no monopoly over the task, and his interpretation can be contested: there is no privileged access, not even to one's own previous meanings.

Further, I wish to contest Popper's reading of his own earlier text. Let me say, first, Popper's confession of a realist impulse is neither here nor there, as he himself says in his great book. All writers have a realist or an objectivist impulse, anyway, even the most ardent defenders of phenomenalism and/or subjectivism. Also, Popper's tenor in the whole volume is to resolve all metaphysical disputes by refusing to take sides, a la Kant and a la Carnap. He does so in a few places, and it is quite possible that since contradiction is rejected by all theories of truth (except Hegel's, which, of course, runs contrary to logic and so does not count), his view was deliberately presented as neutral to the controversy over truth, just as it was, for example, neutral to the controversy over the question, should observation reports be couched in terms of our perception of the facts or in terms of the perceived facts? Clearly, the two neutralities correlate!

In any case, what is lost in Popper's reading of his change of mind is the problem of the growth of knowledge. For, in the *Logik der Forschung* Popper commends the method of conjectures and refutations on account of its fruitfulness, and he says no more about it (except that he means it in the intellectual, not pragmatic, sense of fruitfulness). He raises the problem of the growth of knowledge in his 'Three Views' just because what he says in *Logik der Forschung* is no solution to it: we can imagine a series of conjectures and refutations which leads nowhere, and every prospector who came up with nothing may feel that this is not enough. Indeed, looking for a needle in a haystack can offer no approximation, but either success or frustration. Hence Popper's theory of conjectures, corroboration, and refutations, as a modification of his theory of conjectures and refutations, with corroborations as guarantees of progress. At the same time he offered his theory of verisimilitude which refers neither to refutations nor to corroborations as guarantors of progress. And so Popper has two guarantees for progress, in both cases the progress being towards the truth.

Both guarantees, as guarantees, are question-begging, of course. The question need not be, what guarantee have we that our series of theories approximate the truth? The question need not even be, why do we think, or for what reason do we think, that we approximate the truth? We can ask, why do we not feel so frustrated when engaged scientifically even without finding the final answer as the one who seeks a needle in a haystack and who is only satisfied with the final answer?

The answer is, clearly, that in the course of scientific change we feel that we gain a new depth of understanding—of old facts, old theories, old questions, old techniques even. Are these mere self-deceptions?

I shall not enter techniques here, since, reather than elaborate a theory of techniques in general, I should observe that scientific theories have to explain relevant techniques and their applications. So let me stay a bit with increase of depth of understanding of facts, theories and questions.

V. POPPER ON PROGRESS

The crucial fact about Popper's theory of experience is that he denies that scientific empirical information is always final; on the contrary, he says, being scientific, empirical information is testable, i.e. refutable; and so at times it does get refuted.

This leads to the question, can we make it a policy to overthrow the empirical information which clashes with our theory and stick to our theory? This, says Popper, amounts to the policy to rescue our theory dogmatically. This, says Popper, will make the theory in question lose the scientific status, since it will deprive it of its refutability!

And now to the question that has plagued Popper's critics for decades and which he has tried to answer with his theory of verisimilitude. Suppose our theory is refuted by given empirical evidence. Suppose we have consequently given up our theory and invested great efforts in seeking an alternative to it. Suppose then the evidence is refuted and we have been forced to return to the old theory. Is this not stagnation?

The question can be put in more elaborate ways, with the aid of obviously more elaborate examples. It all comes to the same criticism of Popper's system; it does not insure progress. Oh, we know that we cannot insure the constant imput of new ideas and new experiences; this we accept from Popper. But if his system permits stagnation in the face of new imputs, then this may be its downfall!

For example, what will we do if a new crucial experiment will go the way of the old theory? We shall rush to recheck the old crucial experiment. And if it will reverse itself?

This is, we know, quite impossible. The question is, does Popper tell us that it is impossible or does he allow for its possibility?

We have two conflicting answers to this yes-or-no question. In his *Logik der Forschung* Popper appeals to the fruitfulness of science as the reason for continuing the scientific research (the Forschung of the title). The fruitfulness is an empirical fact, unexplained. Is it an empirical fact used as a transcendental argument the way Kant uses the empirical fact of scientific certitude in his transcendental argument? I think Popper is sensitive to this question, as he answers it, and in the negative: unlike certitude, which is present once and for ever or never, fruitfulness can cease to be present: as long as the game of science is fruitful, Popper says, let us play it. And when the well dries, he implies, we may reexamine the whole situation. Thus, the question, can science stagnate? is answered by Popper in his *Logik der Forschung* with, yes. His 'Three Views' still says, yes, but a weaker yes. Put the question thus: given that scientific imput can dry and that stagnation is possible, what if the imput goes on; can there still be stagnation in science? *Logik der Forschung* seems still to say, yes; 'Three Views' now says, no. This is the import of the theory of verisimilitude, now defunct.

VI. SCIENTIFIC PROGRESS

Let me repeat: we all think that science progresses, that both theory and observation together increase our knowledge of the world in many ways, not least of which is precision. We intuitively feel that empirical information can be replaced by better empirical information, theory by better theory, and so newer facts may take us further into the future, not pull us back into our past. Query: why? how? by what virtue?

Let us note that a theory of approximation to the truth is already embedded in the previous paragraph. The task is to do more than what is accomplished in the previous paragraph. But for this we may have to examine carefully what it says and at least eliminate some errors which it permits. Here is one.

Though we all feel that we cannot reverse the wheels of history, least of all where the history of science is concerned, we are not so determined about the matter as to forbid an occasional revival of a hypothesis universally declared dead. The prime example is the corpuscularian theory of light. But the matter is contestable here: the old corpuscularians were Newtonians. Hence only some of their statements were revived.

This raises the question, what constitutes a theory? A theory is a set of statements taken together. But how large is the set? Should it be, as Duhem says, all the statements accepted at a given moment? Accepted by whom?

The question is not easy. We all feel that there is some truth in phlogistonism, that something, some release, is common to all combustion, acidulation, putrifaction, etc. Indeed, this something, this release, is not of matter, perhaps, but of chemical energy; perhaps it is even a transfer of a piece of matter—of an electron. It is hard to pin down matters here.

A better example is Prout's theory of matter as composed of hydrogen atoms, first refuted by the surprisingly improved measurements of atomic weights of J. S. Stas, then revived by the Bohr-Rutherford model and verified by Aston's mass-spectrography and the discovery of isotopy (which refute Stas' results, of course), later refuted again by the discovery of the neutron, and later revived again as a by-product of the nuclear-spin theory which views the neutron as a merely excited proton (i.e. hydrogen atom) (and by similar considerations that do away with the mezon as a particle proper while in the nucleus).

We can say it is very unlikely that one part of old views will be revived. The portion of any period's scientific progress which is a revival of a once refuted theory is, thus, also very small. And the smallness of that portion makes it no threat for progress. So be it; yet the first paragraph of our present section now must be understood in this spirit. And so, for example, the modification of Popper's theory of verisimilitude proposed above, as related to empirical content rather than content, is empirically refuted, unless we take the two theories to be very broad in content!

This says no more than that scientific theories are improbable, and so it is increasingly improbable that increasingly precise tests to them will yield corroborations rather than refutations. It is therefore ever more unlikely that a refuted theory will come back to life than that an unrefuted theory will remain unrefuted.

Hence the critics who fear that Popper allows for stagnation via the revival of theories have misplaced fears. More likely he allows for stagnation by allowing infinite variations on the same hypothesis.

We now have a modern variant of Laplace's fears. Suppose Newton's theory is refuted. Surely, we can make small variations in the theory and keep refuting them! It was, indeed, Einstein who said so, and who said he felt such a procedure is futile. Popper agrees. He feels that such modifications are ad hoc, and that ad hoc hypotheses are excluded by his system on account of their being irrefutable. On this he is plainly in error. Without reducing the simplicity of a theory by Popper criteria of simplicity we can modify Newton's theory in many ways and try to refute each of them. The result will be stagnation.

And so, the truth of the matter is that Popper's theory fails to legislate as improper certain moves known as improper because they lead to stagnation.

Here, I think, we can see the error of Popper's theory of verisimilitude. We have here a model, a very simple one indeed, of series of quasi-Newtonian false theories, such that in each step new possibilities are explored, and explained falsely, yet which offer better answers to questions which older theories offer. The better answers need not be true, and at times the older theory may be easier to elicit the same replies. The question is, can we say that in a sense the newer theory, which answers falsely a question not asked by the old theory, is nearer to the truth?

Clearly, we feel we want to say yes. Indeed, space technology largely uses this model, and undoubtedly the increased precision of reliable prediction is their major object. And so the increased verisimilitude of their admittedly false theories is of a different sort form that of Einstein!

Nor is this sufficient. Often an old theory presents a question differently from a new theory—in connexion with the metaphysical assumption each rests upon—and here we have, again, a dimension of proximity to the truth, which has to do with our choice of a metaphysical framework for a scientific theory.

To conclude, the problem is, is the avoidance of past errors a sufficient cause for progress towards the truth? The answer seems to be, not always, but at times this leads to a better understanding of the world, to having more questions answered, to having a better comprehension of old answers, to having a better unifying metaphysical presupposition within which to couch our science.

Why then the stress on the avoidance of past errors? Answer: This is the empirical elements in science. And, as long as we stress this element as opposed to other elements that science may share with non-scientific activity, we also stress the element of the avoidance of past errors. But at times researches pay attention to other characteristics of science, and worry about the empirical component later. When they do so, whether they are right or not much depends on their problem-situations. I do not know why and to what extent science is empirical; nor whether the answer to that question speaks in favor of science or not. Intuitively we all agree that we can be, and often are, too much or too little empirical! And so, we do want some degree of empiricalness, of avoiding refuted mistakes.

Finally, my view shares a lot with Hattiangadi's concern with questions, yet my stress is more on answers to them than his, especially on the verisimilitude of these answers, though not in Popper's sense of verisimilitude but in a modified and more pluralistic sense.

F. M. BORODKIN

ON A GENERAL SCHEME OF CAUSAL ANALYSIS[1]

The aim of this paper is to formulate some principles of analysis of causal scheme structure without numerical evaluation of causality as well as to construct on the heuristic basis some index of interdependce of variables for which functional causal relation is defined.

I. ANALYSIS OF CAUSAL SCHEME STRUCTURE [2]

The functional causality is well defined by P. Suppes in [1], p. 67. I shall adduce this definition. P. Suppes wrote: "In a deterministic setting, we say that $y_{t'}$ *is the functional cause of* x_t if $t' < t$ and there is a function f such that

$$x_t = f(y_{t'})$$

Once errors are introduced, this equation becomes

$$x_t = f(y_{t'}) + \varepsilon,$$

and, now reverting to our explicit random variable notation, assumptions about the distribution of errors are made so that the function f holds for the conditional expectation of x_t, given a value of $y_{t'}$. Explicitly, we then have the following definition.

The property $y_{t'}$ *is the functional cause of* x_t if and only if
(i) $t' < t$
(ii) there is a function f such that for all real numbers y

$$E(X_t/Y_{t'} = y) = f(y)\text{"}$$

For the sake of simplicity, in the sequel, time will not be used, as a property of causality.

Usually causal relations are formulated for some definite set of variables beforehand on the basis of professional experience (see, for instance, [2], [3], [4]). I shall formalize a statement about causal relations structure by means of graph theory. The causal relations structure defined on the definite set of variables can be presented by a graph in the following way.

Suppose that the definite set of variables is given and variables are numbered by integers $i = 1, 2, \ldots, k$. Each variable i corresponds to one and only one vertex.[3] In our case we have the directed graph G, which can be represented by its incidence vertex matrix $A = ||a_{ij}||$, where

$$a_{ij} = \begin{cases} 1 & \textit{if an arrow}^4 \textit{ goes out from the i-th vertex and} \\ & \textit{goes to the j-th vertex} \\ 0 & \textit{otherwise} \end{cases}$$

In our further analysis two special types of graphs will be useful, namely minimal[5] and maximal.[5] Now I shall define these graphs.

Let some graph G with its incidence matrix A be given. I shall call this graph an *initial graph* or *IG*. The set of paths from the i-th vertex to the j-th vertes will be designated by μ_{ij}. The set μ_{ij} can correspond to every pair of vertices (i, j). Now consider two graphs $-G = (X, \Gamma)$ and $G = (X, \Gamma^*)$, where X is a set of vertices Γ is a binary relation on the elements of this set and $\Gamma^* \subseteq \Gamma$. In the graph G^* every pair of vertices can be assigned to μ^*_{ij}, the set of paths from the i-th vertex to the j-th vertex. Then the graph G^* is a permissible graph for G iff $\mu_{ij} \neq \varnothing$ involves $\mu^*_{ij} \neq \varnothing$ for every $i, j = 1, 2, \ldots, k$. Let $\mathscr{F}$ be the set of permissible graphs for some fixed *IG* and A^* the incidence matrix of some permissible graph G^*. Let then

$$\min_{G \in \mathscr{F}} \sum \sum a^*_{ij} = a_0 \qquad i, j = 1, 2, \ldots, k.$$

The *minimal graph* for some fixed *IG* is such a permissible graph $\tilde{G}$, for the incidence matrix $\tilde{A} = ||\tilde{a}_{ij}||$ of which the following condition holds:

$$\sum_i \sum_j \tilde{a}_{ij} = a_0, \qquad i, j = 1, 2, \ldots, k.$$

If the *IG* has no circles, the minimal graph for such *IG* is found in a very simple way, namely,

$$\tilde{A} = A - A \dot{\times} \sum_\lambda A^\lambda. \qquad \lambda = 2, \ldots, k.$$

where the symbol $\dot{\times}$ designates the following operation. Let

$$||c_{ij}|| = ||a_{ij}|| \dot{\times} ||b_{ij}||. \qquad i, j = 1, 2, \ldots, k$$

Then $c_{ij} = a_{ij} b_{ij}$ for every i, j. All the operations of summation and multiplication are boolean.

The minimal graph has a remarkable property. If some linkage in the *IG* can be either direct[6] or indirect,[7] then every arrow in the minimal graph represents only direct linkage (in causality sense). The binary relation of the minimal graph is the necessary condition for the existence of the fixed *IG*.

Consider an illustration from Blalock [6]. The causal scheme includes five variables: X_1-urbanization, X_2-percentage of nonwhite population, X_3-whites' income, X_4-nonwhites' level of education, X_5-nonwhites' income. The incidence matrix of *IG* for this scheme is the following:

$$A = \begin{bmatrix} 0 & 1 & 1 & 0 & 1 \\ 0 & 0 & 1 & 1 & 1 \\ 0 & 0 & 0 & 0 & 0 \\ 0 & 0 & 0 & 0 & 1 \\ 0 & 0 & 0 & 0 & 0 \end{bmatrix}$$

The minimal graph for this *IG* has the following incidence matrix:

$$\tilde{A} = \begin{bmatrix} 0 & 1 & 0 & 0 & 0 \\ 0 & 0 & 1 & 1 & 0 \\ 0 & 0 & 0 & 0 & 0 \\ 0 & 0 & 0 & 0 & 1 \\ 0 & 0 & 0 & 0 & 0 \end{bmatrix}$$

The direct linkage between X_1 and X_5 is not obligatory. It is enough that direct linkages between X_1 and X_2, X_3 and X_4, X_4 and X_5 exist. But the existence of these direct linkages is necessary if the *IG* is correct.

I should like to say some words about another property of the minimal graph, discovered through numerous experiments. Obviously, the causal scheme can be constructed only by experts. Usually, graphs formulated by experts are very diverse, but the difference among minimal graphs for the given *IG* is very small.

Now I shall define a maximal graph for *IG*. Consider some initial graph *G*. Graph *G'* is an *absorbing graph* for *G* iff the graph *G* is permissible for *G'*. Designate the set of absorbing graphs for the fixed *IG* as $\mathcal{T}$, and the incidence matrix of some absorbing graph—as $B = ||b_{ij}||$, $(i, j = 1, 2, \ldots, k)$. Let then

$$\max_{G' \in \mathscr{T}} \sum_i \sum_j b_{ij} = b_0 \qquad (i, j = 1, 2, \ldots, k).$$

The *maximal graph* for the fixed *IG* is a certain absorbing graph $\hat{G}$, for incidence matrix $B = ||\hat{b}_{ij}||$ of which the following condition is satisfied:

$$\sum_i \sum_j \hat{b}_{ij} = B_0 \qquad (i, j = 1, 2, \ldots, k).$$

The interpretation of a maximal graph is obvious: some new arrows appearing in the maximal graph duplicate paths which exist in the *IG*. The incidence matrix of the maximal graph can be found out by means of the following formula:

$$\hat{B} = \sum_\lambda A^\lambda, \qquad \lambda = 1, 2, \ldots, K.$$

In this formula all the operations are boolean. For instance, the maximal graph for Blalock's scheme has the following incidence matrix:

$$\hat{B} = \begin{bmatrix} 0 & 1 & 1 & 1 & 1 \\ 0 & 0 & 1 & 1 & 1 \\ 0 & 0 & 0 & 0 & 0 \\ 0 & 0 & 0 & 0 & 1 \\ 0 & 0 & 0 & 0 & 0 \end{bmatrix}$$

Obviously, the minimal graph for the given *IG* is unique if the *IG* has no circuits and the maximal graph is unique for any graph. The graph can be used to check up *causal scheme identifability*. If the causal scheme can be identified, the matrix $\hat{B}$ is triangulable.

Nobody can guarantee that the fixed *IG* mappes only direct linkages. Any graph between minimal and maximal graphs, that is any graph which is permissible for the maximal graph, proves to be a graph which has only direct linkages. It is easy to show that the number of such graphs

$$h(\mathscr{F}) = 2^\Theta \quad , \quad \textit{where}$$

$$\Theta = \sum_i \sum_j r_{ij} \qquad (i, j = 1, 2, \ldots, k)$$

and

$$||r_{ij}|| = \hat{B} - \tilde{A}$$

After these considerations the analysis of the minimal and maximal graphs become simpler and more definite.

Now I shall define another two structures which are connected with analysis of causal schemes: automaton and simple structure. Let some *IG* be given. The *automaton* is such subgraph of the *IG* that for every pair of its vertices at least one path from one vertex to the other exists. Thus, automaton is a strongly connected graph. If we deal with *IG* which has a great number of vertices and arrows, usually it is very difficult to find out identifiability of causal scheme before its numerical evaluation. Moreover, in some cases automatons, if they exist, can be crossed out of the scheme and the rest of scheme can be evaluated. Connected but not strongly connected graph is called a *simple structure.* A very simple algorithm of automaton discovering can be constructed on the basis of the following statement. Let us have, for some *IG*, the maximal graph $\hat{G}$ with an incidence matrix $\hat{B} = ||\hat{b}_{j}||$ $i, j = 1, 2, ..., k$. The set R of vertices together with incident arrows is an automaton if $\hat{b}_{ij}\hat{b}_{ji} = 1$ for every $i, j \in R$. The *algorithm* of the automaton discovering is the following.

1. Find out the maximal graph $\hat{G}$ with the incidence matrix $\hat{B} = ||\hat{b}_{ij}||$ $i, j = 1, 2, ..., k$.

2. In the matrix $\dot{B}$ an arbitrary row i_1 and a column $j_1 = i_1$ are fixed. Thus, we have two vectors: $\hat{b}_{i_1} = (\hat{b}_{i_1}, ..., \hat{b}_{1k})$ and $\hat{b}_{j_1} = (\hat{b}_{i_1}, ..., \hat{b}_{ki_1})$. Find the product of the two vectors on the basis of the following rule: the *product* of two vectors $X = (x_1, x_2, ..., x_i, ..., x_k)$ and $y = (y_1, y_2, ..., y_i, ..., y_k)$ is a vector $z = (z_1, z_2, ..., z_i, ..., z_k)$ for which $z_i = x_i \cdot y_i$ for every $i = 1, 2, ..., k$. Hence, we shall find a vector $b_i = (b_1, b_2, ..., b_k)$ in which $b_r = \hat{b}_{i_1r} \cdot \hat{b}_{ri_1}$, $r = 1, 2, ..., k$. Suppose that numbers of nonzero components of the vector b_{i_1}. form the set R_1. Then all the vertices with numbers $i \in R_1$ and incident arrows form one automaton.

3. Cross out of the matrix $\hat{B}$ all the rows and columns with numbers $i \in R_1$. As a result we shall have the matrix $\hat{B}$.

4. Apply the steps 2 and 3 to the matrix $\hat{B}$. The procedure ends when all the rows and columns of the matrix $\hat{B}$ are checked up.

However, we cannot be satisfied by simple crossing out the automatons. Indeed, if we only remove automatons from the *IG*, we can lose some causal ties. For instance, consider the following graph

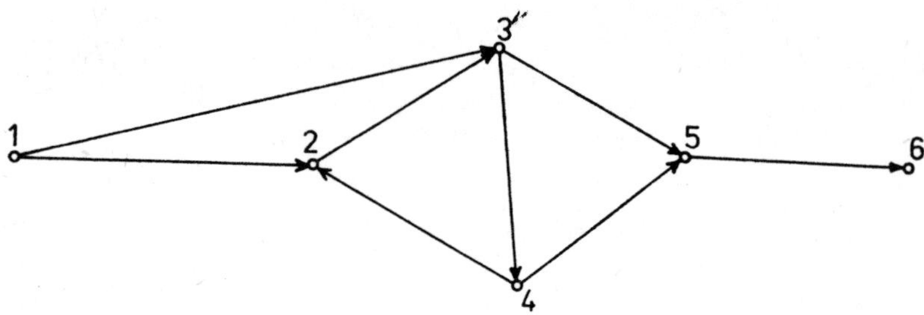

Simple removal of the automaton consisting of the 2-nd, 3-rd, 4-th vertices and incident arrows will result in the following graph

1. 5 ⟶ 6

But in the *IG* there is a path, for example $1 \to 3 \to 5$, from which it is clear that the 1-st variable effects the 5-th variable. It is logical to keep such ties. Similar necessity can arise also in the case of removing some variables from the initial causal scheme.

Now I want to present an algorithm for correct reduction of a graph preceded by some necessary definitions. Let some initial graph $G = (X, \Gamma)$ with the incidence matrix $A = ||a_{ij}||$, $i, j = 1, \ldots, k$, be given. The vertices i and j are incident if $a_{ij} + a_{ji} > 0$. Suppose that some set β of variables with incident arrows must be removed from the graph. As a result we shall have a graph $G^* = (X^*, \Gamma^*)$, where $X^* = X \setminus \beta$, $\Gamma^* \subseteq \Gamma$, with the incidence matrix $A^* = a^*_{ij}$, $i, j \in X^*$. Let the vertices from β be incident to vertices from some subset $S \subseteq X \setminus \beta$. We denote by μ_{rjt} the set of paths connecting vertices $r, t \in S$ and passing through a vertex $j \in \beta$. We shall call some graph $G^* = (X^*, \Gamma^*)$, $X^* \subseteq$ X, $\Gamma^* \subseteq \Gamma$ a *reduced graph* for the graph $G = (X, F)$, if $\mu_{rjt} \neq \emptyset$ involves $a^*_{rt} = 1$ for every $r, t \in S$ and $j \in \beta$. The set of arrows connecting vertices from S with vertices from β will be denoted by U, and an arrow directed from a vertex i to a vertex j will be designated by U_{ij}.

The problem of causal scheme reduction is solved by the following algorithm.

1. The matrix $A = ||a_{ij}||$, $(i, j = 1, 2, \ldots, k)$ is put in correspondence with the matrix $P = ||p_{ij}||$, in which

$$p_{ij} = \begin{cases} 1 & \textit{if } a_{ij} = 1, \quad i, j \in \beta \quad \textit{or } u_{ij} \in U \\ 0 & \textit{otherwise} \end{cases}$$

2. Let the number of vertices in S be equal n. Consider the set of matrices $D_n = ||d_{ij}^{(m)}||$ $(m = 1, 2, \ldots, n+2)$,

$$D_{m+1} = D_m \cdot Q_m, \textit{ where } Q_m = ||q_{ij}^{(m)}||,$$

$$q_{ij}^{(m)} = \begin{cases} 1 & \textit{if } d_{ij}^{(m)} > 0, \quad i, j \in \beta \quad \text{or } u_{ij} \in U \\ 0 & \textit{otherwise} \end{cases}$$

and $D_1 = P$.

3. Let

$$\tilde{B} = ||\tilde{b}_{ij}|| = \sum_{m=1}^{n+2} D_m$$

Then in the matrix A^*

$$a_{ij}^* = \begin{cases} 1 & \textit{if } a_{ij} + \tilde{b}_{ij} > 0, \quad i, j \in X \setminus \beta \\ 0 & \textit{if } a_{ij} + \tilde{b}_{ij} = 0 \quad \textit{or } i, j \in \beta \end{cases}$$

Now we must answer at least one more question: *can we always reduce an initial causal scheme, and if not, what reduction rule (or rules) can be formulated?*

From [2], [4], and [6] we can get some conditions for reduction. Indeed, in accordance with Boudon's terminology, there exist, for every causal scheme, some *implicit factors* "that act on the explicit variables of the causal scheme without stated explicity". ([4], p. 200). These factors must be uncorrelated if we want to get correct causal numerical evaluation. We know nothing about correlation among implicit factors, but at least we can state that the following condition can be observed: any variable which is subject to remove from the initial scheme must not influences (directly or indirectly through variables which are subject to removal) more than one variable from the considered set of variables.

This rule is realized on the incidence matrix in a simple way.

1. The matrix $A = ||a_{ij}||$ associates with the matrix $T = ||t_{ij}||$, in which

$$t_{ij} = \begin{cases} 1 & \text{if } a_{ij} = 1,\ i, j \in \beta \text{ or } i \in \beta \text{ and } j \in X \setminus \beta \\ 0 & \text{otherwise.} \end{cases}$$

2. Let the number of elements in the set β be equal n. Find out the matrix

$$B = ||b_{ij}|| = \sum_{\lambda=1}^{n} T^{\lambda}$$

Now we can state our rule. *Some vertex* $i_1 \in \beta$ *can be removed out of the scheme* iff

$$\sum_{j \in X \setminus \beta} b_{ij} = 0$$

For example, in the Blalock's structure mentioned above we cannot remove the variables x_1 and x_2 can remove x_3, x_4 and x_5.

In the 3-rd Section I shall formulate a hypothesis from which it follows that for the correct numerical evaluation of the causal connection between two variables it is necessary to take into consideration not only their common cause but also their common effect. If this hypothesis is correct, we cannot remove from of the scheme those variables which are effected by more than one variable from the rest of the scheme (from the set $X \setminus \beta$). In order to satisfy this condition we must consider the transpose of the matrix A in the previous algorithm. Suppose that in this case we have a resultant matrix $D = ||d_{ij}||$ instead of the matrix B. Then some vertex $i \in \beta$ can be removed out of the scheme iff $\sum_{j \in X \setminus \beta} b_{ij} = 0$ and $\sum_{j \in X \setminus \beta} d_{ij} = 0$.

From all the above remarks it follows that when constructing causal scheme for the analysis and evaluation, we must observe *at least* the following rules.

(1) At the first step the causal scheme must include the maximal possible number of variables, independently of measurability and their (variables) character. Any variable which is supposed to be causally connected with others under consideration must be included in the initial causal scheme.

(2) In accordance with reduction rules some variables, (unmeasurable or others), are removed out of the initial scheme.

(3) If the researcher intends to evaluate the causal scheme numerically, it is necessary to find out all the automatons and to remove them out of the scheme.

(4) The minimal and maximal graphs are defined on the basis of the new scheme. These graphs are analyzed. Then both of them together with the new initial scheme can be numerically evaluated.

II. NUMERICAL EVALUATION OF CAUSAL CONNECTIONS

Let consider a system of regression equations defined on K variables

$$Bx+u = x, \tag{1}$$

where B-traingulable matrix of regression coefficients $||\beta_{ij}||$, $\beta_{ii} = 0$, $i, j = 1, \ldots, k$, and for every $a_{ij} = 0$ $\beta_{ij} = 0$, x—vector—column consisted from k variables, u—vector-row consisted from k exogenious variables. Vectors x and u are normally distributed and $E(u_i u_j) = 0$, $i, j = 1, \ldots, k$, $(E(u_i x_j) = 0$, $i, j = 1, \ldots, k$, $i > j$, $E(u_i) = E(x_i) = 0$, $i = 1, \ldots, k$, $E(x_i^2) = 1$, $i = 1, \ldots, k$.

The system (1) can be rewrited in a following way:

$$\begin{aligned} u_1 &= x_1 \\ u_2 &= x_2 - \beta_{12} x_1 \\ u_3 &= x_3 - \beta_{13} x_1 - \beta_{23} x_2 \\ &\ldots\ldots\ldots\ldots\ldots\ldots \\ u_k &= x_k - \beta_{1k} x_1 - \beta_{2k} x_2 - \ldots - \beta_{(k-1)k} x_{k-1} \end{aligned} \tag{1'}$$

In this expression $u_1, \ldots, u_k$ are linear functions of random variables $x_1, \ldots, x_k$. The covariation matrix $||\sigma(u_p, u_q)||$ for linear functions is $||\sum_{i,j=1}^{k} \sigma_{ij} \beta_{ip} \beta_{jq}||$, $p, q = 1, \ldots, k$. Under our conditions $||\sigma_{ij}||$ is the correlation matrix $r = ||r_{ij}||$, $||\sigma(u_p u_q)|| = 0$, $\beta_{ip} = 0$ for $i \geqslant p$. It can be proved that in this case every β_{ij} is defined by the method of least squares for every equation from the system (1′) separately.

But it is very hard to interpret β_{ij}-coefficients. Now we shall build so called influance coefficient which is similar to correlation coefficient for the case of causal structure.

Let elements of B-matrix be calculated by the method of least squares. B-matrix is truangulable, and $\beta_{ii} = 0$. From

$$x = (I-B)^{-1}U = U\Big(\sum_{\lambda=1}^{k-1} B^{\lambda}+I\Big). \tag{1}$$

From this equation it follows that every component of x-vector is a linear function of u-vector components and every coefficient of this function is equal to sum y production of β-coefficients along all the paths which lead from some component of u-vector to the connected component of x-vector. In other words the j-the component of x-vector is equal to the j-th component of u-vector plus some linear function of the rest of components. The covariation of the j-th component of u-vector and the part of the j-th component of x-vector, which is the sum of the j-th component of u-vector and the addend which is a linear function of the i-th component of u-vector, we shall call the *influance coefficient.* The coefficients of the last linear function are the (i,j)-th elements of the matrix

$$\|c_{ij}\| = \sum_{\lambda=1}^{k-1} B^{\lambda}+I,\; i,j = 1, 2, \ldots, k,$$

divided by square root from production of the corresponding variances. In other words, the influance coefficient in

$$P_{ij} = \frac{c_{ij}}{\sqrt{D_j/D_i+c_{ij}^2}}$$

where $D_S = E(u_s^2)$, u_s—the s-th component of u-vector. It is obvious that $-1 \leqslant p_{ij} \leqslant 1$ and $p_{ij} = 1$ only in the case when $D_j = 0$, e.i. when the j-th component of x-vector is linear function of the rest of component.

III. SOME REASONING ABOUT USING THE ENTHROPY FUNCTION IN CAUSAL ANALYSIS

The enthropy function is very useful in causal probability analysis. It provides us with certain numerical evaluation and gives the opportunity to interpret causal relations in very natural terms of information.

Indeed, in very kind of analysis of experimental results we always deal only with information about events and/or values. Equivalence between approaches based on information (or enthropy) and probabilistic approach can be easily demonstrated.

Now I wish to remind some well known facts about the enthropy function. The *enthropy function* for some random variable V is

$$H(V) = -\int_{-\infty}^{\infty} f(v)\log f(v)\,dv, \text{ where}$$

$f(v)$ is a density function.

Suppose that two random variables X and Y are given. Then

(1) $$H(X, Y) = H(X)+H(Y/X) = H(Y)+H(X/Y)$$

where $H(Y/X)$ is a *conditional enthropy*,

$$H(Y/X) = -\iint_{-\infty}^{\infty} f(x, y)\log f(y/x)\,dx\,dy$$

where $f(y/x)$ is a conditional density function.

For normal distribution with parameters (O, σ_i^2)

(2) $$H(X_1, \ldots, X_n) = \log(2\pi e)^{2/n}\sqrt{\mathrm{Det}\,||a_{ij}||}, \quad i, j = 1, 2, \ldots, n,$$

where a_{ij} is the correlation matrix,

$$a_{ij} = \sigma_i\sigma_j r_{ij}, \quad i, j = 1, 2, \ldots, n.$$

and r_{ij} is a correlation coefficient. We shall now need some definitions.

Definition (cf. Suppes, [1], p. 61). The property $Y_{t'}$ is a (*weak*) *prima facie quadrant cause of the property* X_t if and only if

(i) $t' < t$

(ii) For all x and y if $P(Y_{t'} \geqslant y) > 0$ then $P(X_t \geqslant x/Y_{t'} \geqslant y) \geqslant P(X_t \geqslant x)$

The term *quadrant* is used because (ii) may be rewritten as

$$P(X_t \geqslant x, Y_{t'} \geqslant y) \geqslant P(X_t \geqslant x)P(Y_{t'} \geqslant y),$$

which expresses the relation between the probability of any quadrant $X_t \geqslant x$, $Y_{t'} \geqslant y$ under the given distribution and the probability of this quadrant in the case of independence".

Now I shall give one more definition. Consider a finite set of continuous random variables $X_1, \ldots, X_n$ about which it is known that the property X_n is related to a subordinate system (in sense of time) and the ordering of the rest of the variables is unknown.

Definition. The system of properties $X_1, \ldots, X_{n-1}$ is a *generalized prima facia quadrant cause of the property* X_n if and only if

(i) X_n is the subordinate (in sense of time) property in relation to the system $X_1, \ldots, X_{n-1}$;

(ii) $P(X_n \geqslant x_n / X_1 \geqslant x_1, \ldots, X_{n-1} \geqslant x_{n-1}) \neq P(X_n \geqslant x_n)$.

This is a usual condition for dependent systems. But I think that the relations among properties in these conditions are of great interest.

The condition (ii) of Definition is equivalent to the following:

$$H(X_n) - H(X_n / X_1, \ldots, X_{n-1}) > 0$$

But the other clause is more interesting:

$$(3) \qquad H(X_n) - H(X_n / X_1, \ldots, X_{n-1}) > c, \quad c > 0$$

The term c can be considered as a threshold for the process of receiving causal information. In some fields, such as psychology, this threshold can be measured experimentally. In others, nonexperimental sciences, it can be studied as, in certain sense, the evaluation of intelligibility threshold of scientific methods. Suppose that $X_1, \ldots, X_n$ are normally distributed with parameters (O, σ_i). From (3) it follows that

$$\log_a \sqrt{\frac{\operatorname{Det}||a_{pq}||}{\operatorname{Det}||a_{ij}||}} > \log_a a^c, \quad \begin{array}{l} p, q = 1, 2, \ldots, n-1 \\ i, j = 1, 2, \ldots, n \end{array}$$

But $a > 1$, $c > 0$. Therefore

$$\frac{\operatorname{Det}||a_{pq}||}{\operatorname{Det}||a_{ij}||} > a^{2c}$$

Remembering that multiple correlation coefficient

$$R^2 = 1 - \frac{\operatorname{Det}||a_{ij}||}{\operatorname{Det}||a_{pq}||}$$

we get inequality

$$(4) \qquad R^2 > 1 - \frac{1}{a^{2c}}$$

The sense of this inequality is very simple. For the stated conditions the ability of identifying the causal relation depends on measurement ability. For instance, if $a = 2$ and $c = 1$, $R^2 > 0{,}75$. It means that for $R^2 \leqslant 0{,}75$ we cannot state that the causal relation exists because the measurement error is too great. Naturally if we have only two properties X and Y, under the stated conditions

$$r_{xy}^2 > 1 - \frac{1}{a^{2c}}$$

Consider the case of three properties X, Y, Z, where Z is a subordinate property. For this system

$$1 - r_{xy}^2 > a^{2c}(1 + 2r_{xz}r_{xy}r_{yz} - r_{xz}^2 - r_{yz}^2 - r_{xy}^2) \tag{5}$$

The inequality

$$r_{xz}^2 + r_{yz} > 1 - \frac{1}{a^{2c}}$$

is the necessary condition for the independence of the properties X and Y. Indeed, suppose that

$$r_{xz}^2 + r_{yz}^2 \leqslant 1 - \frac{1}{a^{2c}}. \tag{6}$$

Then, from (5)

$$r_{xy}^2(a^{2c} - 1) > 2a^{2c}r_{xz}r_{yz}r_{xy} - a^{2c}r_{xy}^2$$

If $r_{xy} = 0$, we have the contradictory $0 > 0$.
If (6) is correct, $r_{xy} \neq 0$ and

$$|r_{xy}| > \frac{2a^{2c}}{2a^{2c} - 1}|r_{xz}r_{yz}|,$$

where $|t|$ is an absolute value of t.
The necessary condition stated above can be easily generalized for the case of arbitrary finite number of properties. Suppose that we have some set of normally distributed properties $X_1, \ldots, X_n$ and the property X_n is subordinate. Then the necessary condition for *mutual independence* of the properties $X_1, \ldots, X_{n-1}$ is the following inequality:

$$\sum_{i=1}^{n-1} r_{in}^2 > 1 - \frac{1}{a^{2c}}$$

On the basis of this analysis the following hypothesis seems to be of theoretical and empirical interest.

Hypothesis. The nonzero correlation of two random variables can be not only because of common cause, but also because of common effect.

The truth of this hypothesis is obvious for a case where the researcher deals with goal systems. In this case the whole system (or its parts) is oriented to the achievement of the set goal (or goals inherent to this system), and a need arises to coordinate the behavior of its elements. Of course, I do not mean to imply time reversibility since the goal stands out as a factor related to the past for each given effect. If, however, some special goal has already been achieved or is being continuously achieved "in part" and the researcher has at hand only a set of observed values of all variables, then it appears as though the future values of some variables determined the past values of other variables, since the values of some variables are dependent on the need to plan the future states. In an artificially constructed system it is often easy to relate the goal in the past to one of the causes for the change in the values of variables. In natural as well as poorly known structures which, moreover, have a system of goals, this representation may be difficult or even impossible. From the above statements a still more rigid hypothesis follows: the correlation between variables may be the higher, the more strict the requirement to achieve some goals and the weaker the association between certain variables and characteristics of the goal the higher the correlations between these variables may be. The possibility of this situation should make the sociologist think over the correlation between variables, not only in terms of the past but also of the future states of the system. If a goal (in the form of a system of variables describing it) can be singled out explicity and if the correlation between planned and actual states is sufficient, then, in the study, time can be reversed, and, as a consequence, we may deal with recursive systems. Otherwise, the use of the causal analysis technique is unjustifiable, and it is necessary to use a simultaneous-system approach.

BIBLIOGRAPHY

[1] Suppes, P., 'A Probabilistic Theory of Causality', *Acta Philosophica Fennica*, Fasc. **XXIV**, 1970. North-Holland Publishing Company, Amsterdam

[2] Simon, H. A., *Models of Man*, New York, 1964.

[3] Ashby, Ross, W., *Design of a Brain. The Origin of Adaptive Behaviour*. London, 1960.

[4] Boudon, R. A., 'A New Look at Correlation Analysis', in: H. M. Blalock and A. B. Blalock (eds), *Methodology in Social Research*, McGrow-Hill, New York, 1968, pp. 199–235.

[5] Ore, O., 'Theory of Graphs', *American Mathematical Society Colloquium Publications*, v. **XXXVIII**, AMS, 1962.

[6] Blalock, H. M., 'Correlation and Causality: The Multivariate Case', *Social Forces*, **39**, March, 1961, pp. 246–251.

[7] Бородкин Ф. М., 'Статистическая оценка связей экономических показателей', *Статистика*, Москва 1968, 203 стр. (The Statistical evaluation of economic variables connections).

[8] Бородкин Ф. М., Лукацкая М. Л., 'К вопросу о статистической оценке причинного влияния' — Научные труды Новосибирского государственного университета. Выпуск VIII. Математические методы решения экономических задач, Новосибирск, 1966. (About statistical evaluation of causal influence).

[9] Borodkin, F., Doncheva, S., Penkov, M., 'About Measuring Influence', in: *A system of parameters*. Paper submitted to 7-th World Congress of Sociology, Varna, September, 16p. See also: Бородкин Ф. М., С. Дончева, М. Пенков, 'Об измерении влияния в системе параметров', сб. *Социология и математика*, АН СССР, Институт экономики и организации промышленного производства, Академия Наук Народной Республики Болгарии, Институт социологии, Новосибирск 1970, стр. 32—48.

REFERENCES

[1] This work has been done in the Institute of Economics and Organization of Industrial Production, Siberian Department of the USSR Academy of Sciences, Novosibirsk.

[2] Some results of this section are stated in [7], pp. 20–46.

[3] The theoretical-graph terminology follows [5].

[4] directed edge

[5] These graphs in Ore's terminology [5] correspond to *basic* graph and *transitive closure* graph.

[6] It means that the given two variables are linked through some variables which are not included in this, causal scheme.

[7] It means that the given two variables are linked through some variables included in the studied causal scheme.

H. DISHKANT

LOGIC OF QUANTUM MECHANICS

"When we go far in the direction of the very small, quantum theory says that our forms of thought fail, so that it is questionable whether we can properly think at all". These words of Bridgman express the main problem of quantum logic.

It is a known and basic fact that a state of knowledge can change in process of cognition. Transitions between these states of knowledge are either transitive or symmetric. The first case can be best investigated in terms of intuitionistic logic, as shown by Kripke and Grzegorczyk, while the other requires application of quantum logic.

Thus, intuitionistic and quantum logics are the main nonclassical logics, and analogies between them turn out to be of importance for methodology of science.

In physics quantum logic is known as theory of the lattice of results of quantum observations. The problem of equivalence between such a logical approach and canonical formalism of quantum mechanics constitutes the most frequent subject of investigations. However, according to us, the well known in physics recipes of quantization are natural and exhaustive way of formal analysis, and thus we set forward the following

Thesis 1. The usual quantization of a classical theory C is a simple substitution of quantum logic for the classical logic on which C is based. In such a way we obtain the corresponding quantum theory C^{+}.

Being unable to prove this thesis, we will show that a classical physical quantity turns by such a substitution into a quantum observable.

Indeed, let the domain of individuals of C be the set of real numbers. Let $P(x)$ mean "a physical quantity (for instance energy) has a value x". Then we have in C:

$$\vdash \exists x P(x), \qquad (")$$

$$x \neq y \vdash P(x) \Rightarrow \neg P(y). \qquad (" \; ")$$

Here $\vdash$ is the sign of derivability in C, $\exists$, $\Rightarrow$ and $\neg$ are existential quantifier, implication and negation of the classical logic, respectively, (") denotes that our physical quantity takes some value, and (" ") means that there is only one such value.

In conformity with our thesis the corresponding quantum theory C^+ contains the same signs, but the logical signs must be considered as belonging to quantum logic. For any fixed value of x, $P(x)$ is a sentence and corresponds in quantum logic to a closed subspace of a Hilbert space. Quantum negation $\neg$ corresponds to the transition to the orthogonal complement, and $\exists x$ is the operation of taking a closed linear envelope of the set of all subspaces corresponding to any value of x. Derivability of quantum implication means that a subspace corresponding to the antecedent lies in a subspace corresponding to the consequent. Hence, (" ") means that subspaces corresponding to different values of a physical quantity are orthogonal, and (") denotes that they form a complete system in a Hilbert space. This means that P corresponds to a selfadjoint operator of a Hilbert space presented in a spectral form i.e. to a quantum observable. To enrich our analogy we put forward

Thesis 2. Intuitionistic logic is a natural basis of thermodynamics.

Indeed, the knowledge of physical quantities (energy, entropy, etc.) for large parts of a thermodynamic body can be transfered to the knowledge of these quantities for small parts of this body. Such transitions are transitive, of course. But the classical limit of exhaustive knowledge is not accessible in thermodynamics because of uncontrolled thermal fluctuations. In such situation intuitionistic logic is adequate. We hope that intuitionistic logic permits to avoid paradoxes connected with irreversibility of mechanical processes.

Some words should be said about the use of non-classical logics in classical mathematics. Here intuitionistic logic is more successful. Fitting has shown Cohen's forcing to be equivalent to intuitionistic set theory. This success is attained thanks to the existence of an imbedding of classical logic in intuitionistic logic: any formula X (not containing $\forall$) is derivable in classical logic if and only if $\neg\neg X$ is derivable in intuitionistic logic.

Using our analogy, we may expect that the use of quantum logic in classical mathematics would be possible after some imbedding of classical logic in quantum logic has been discovered. Analogous imbeddings but on the level of informal theories are well known in physics as limit transitions from a quantum theory to the corresponding classical one. It permits us pose

Thesis 3. The most important imbedding of classical logic in quantum one is the generalized version of the limit transition $h \to 0$, where h is Planck's constant.

In the limiting case $h = 0$ we get a classical theory and a classical logic.

WILHELM K. ESSLER

ON POSSIBILITIES AND LIMITS OF THE APPLICATION OF INDUCTIVE METHODS

Inductive reasoning aims at determining the degree of confirmation involved when passing from a premise (which is regarded as problematic in the context) to a conclusion (which is regarded as hypothetical). An *inductive conclusion* assigns such a degree of confirmation to a conclusion relative to the given premise, the assignment being effected in accordance with a generally described procedure (a generally characterized function). *Inductive logic* is the theory of these inductive methods; it delinaates them from other procedures or functions, it systematizes them, and it shows in which situations the application of certain inductive methods is justified.

The question whether inductive logic is useful for accruing knowledge of reality has split philosophers of science into two bitterly opposed camps. Some philosophers, especially Carnap and Hintikka, developed systems of inductive methods, supposing (implicitly or explicitly) them to be applied in the empirical sciences—or at least applicable—while others, particularly Popper and Feyerabend, claim that such systems are a waste of time, since there is no moment in the process of attaining empirical knowledge at which inductive methods could be efficently employed and science the dynamics of establishing theories takes place in an entirely different way.

When the alternatives are given in such a radical manner, only one of them can be valid. Therefore, if it can be shown that inductive logic *may be* and *is* used for advancing empirical knowledge, the other position becomes untenable, although this demonstration can in turn be compromised by arguing that at least one process of theory building *exists* in empirical science which does not involve inductive logic. In the following we hope to show that in this case, both positions have their merits. According to this conception, inductive logic can be used to determine the degree of confirmation of hypothetical conclusion with respect to accepted premises in *some* cases, but *not* in *all*. The following four types of statements (at least) may serve as hypotheses in empirical sciences:

(1) *Singular experiential data*: they describe particular observations or measurements made at some moment on some object, *b*, by applying observation or measurement techniques (eventually, *b*'s position at that moment as well as other factors must be taken into account).

(2) *Generalized observational* (including *generalized measurement statements*): they ascribe certain properties to an object, *b*, without referring to whether and when these properties have actually been found.

(3) *General theories*: they describe in which way the properties and relations of singular objects are interconnected.

(4) *Metatheories*: they state which general (i.e. *formal*) properties general theories should have in physics for instance; that they be time-invariant, i.e., valid independent of time; that they be coordinate-invariant, i.e., they don't change when a given coordinate system is replaced by another; that they summarize the general observational statements in a simple way, etc.)

Hypotheses of the 4th type belong to that field of research concerned with the metatheoretical presuppositions of empirical knowledge which describe the structure of our general theories about experience. In the following, we shall investigate the relationship of singular experiential data to generalized observational statements as well as generalized observational statements to general theories. As a consequence we can dispose of an apparent circle thought to be involved in the process of attaining empirical knowledge. It runs as follows: in empirical science, generalized observational statements are not only accepted or rejected on the basis of singular experiential data, but also as a consequence of general theories. On the other hand, since the acceptance or rejection of these general theories depends entirely on generalized observational statements, this procedure seems, at first glance, to be circular: on the basis of accepted singular experiential data we accept generalized observational statements and consequently, general theories; on the basis of accepted general theories we reject, on the other hand, generalized observational statements, even if these are supported by accepted singular experiential data, casting, therefore, doubt on the relevance of the latter. This apparent circle can nevertheless be resolved in a way dipicting not only the scientific, but also the everyday process of knowledge.

In order to keep the following arguments from becoming too abstract, they shall be illustrated by the law of falling bodies. With appropriate modification they may though be applied to other fields of experience as well.

Let R_1 be an unambiguous instruction for an experiment, to the effect that object b (heavy enough to render air resistance a neglegible factor or, alternatively, in an approximate vacuum) being dropped from a tower, the distance from its starting point being measured after one second. To determine this distance, a scale of finite length, finitely subdivided, is used (for instance, 100 m long with subdivions in mm). Let q be the number of half open intervals on this scale, marked by the terms $\pi_1, \ldots, \pi_q$. Let the measurement instruction imply that an object belongs at most to one of these concepts, so that measurements carried out at the times a_i lie only in one of these intervals, and that every object belongs to one of these concepts, i.e. the scale is long enough to cover the expected results. The measurements taken at the moments $a_1, a_2, \ldots$ $\ldots, a_k$ lead, therefore, to k results, described by the conjunction Φ_k. Using Φ_k, the distance which the body covers in one second is estimated; it is the result of adding the products of the possible length of the distance and the probabilities of these lengths with respect to the data of Φ_k. The application of such a probability function, however, always assumes the uniformity of the domain of things (in this case: realizations of experiment instructions). Depending on the degree of accuracy of observation and measurement techniques and the absence or presence of disturbing factors, we may assume a greater or lesser uniformity and employ a corresponding probability function (characterized by a probability distribution). The measurements being sufficiently accurate and all important disturbing factors being eliminated, one may consider the uniformity to be nearly maximal and choose a probability function whose values for the singular predictive inference are nearly identical with the observed relative frequency.

Using this probability function, representing the inductive method, the quality of the estimation concerning the distance which the body has covered in *one* second is calculated. Then the estimated value is subjected to a confidence interval, K_1, which decreases as the assumed degree of uniformity and coincidence of the experiential data with this uniformity increases. The confidence interval is to be such that the probability of error (the probability that the actual value does not belong to it) with respect to Φ_k is below a certain value r_ε, for example, 0,005.

Let R_2 be another measurement instruction differing from the former only in that, that the object in question be allowed to fall *two* seconds. In a certain number of applications, this leads to an estimation of the distance within a different confidence interval K_2. Applying a third instruction—R_3—the object falls for *three* seconds etc. If these intervals (which in probability theory and in inductive logic are horizontally marked) are entered vertically in a new graph (vertical line: length of falling (distance), horizontal line: time of falling), the result is a line, showing the values at these intervals:

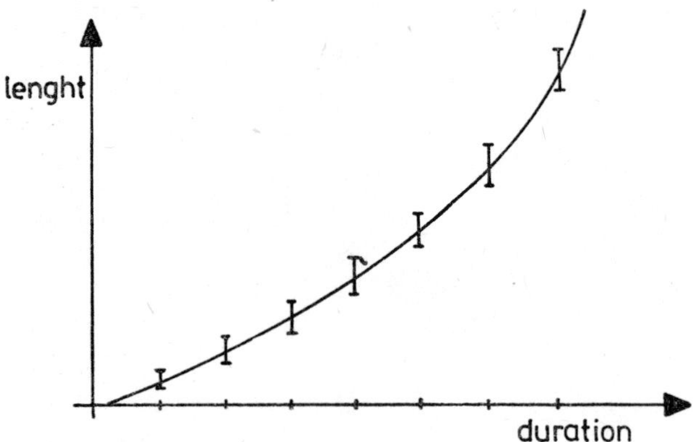

Obviously these intervals cover an infinite set of curves, some of which are wavy, unlike the one shown; each amplitude of such waves is, however, smaller than the respective interval. Even if the given confidence intervals are supplemented by finite or denumerably infinite further confidence intervals—the results of many different experimental instructions using inductive arguments of the above kind, infinitely many curves remain which run through all these intervals. Every one of these possible curves represents a possible law of nature, describable in a language containing a theory of real numbers. We choose here the *most simple*, provided we know of no other law of nature in other fields which could be connected to this one. If we do, we choose the *most simple one consistant with the interdependent laws of nature.* (It has not been decided as yet whether or not the concept of the simplicity of a law or of a system of laws is the same as that of the simplicity of a system of concepts a system of terms together with the rules for their use. Apparently the term 'simplicity' is used in different ways in different contexts, having varying intensions according to the context so that these contexts and classifications ought to be delineated).

General observational statements are deducible from such general theories. They are, of course, not deductively equivalent with statements of the following kind: "Whenever some experiment is carried out, the results obtained are within a fixed interval". For then experimental instructions which were never carried out would lead to the same difficulties as the well known ones adhering to dispositional concepts; the fact that we accept them (since we accepted the general theories) does mean, however, that we expect, with a probability $1-r_\varepsilon$, to obtain results in some class of experiments within a certain confidence interval around the estimated value, where r_ε is, for instance, the number 0,005.

Thus we need inductive methods in order to use the results of a measurement instruction for estimating the actual value of the object to be measured; not only do we need them for estimating this value, but also for calculating the quality of this estimation which leads to the confidence interval.

The value chosen from this interval in constructing a general theory (in the given example: of the laws of falling bodies) is then—provided this theory is accepted—regarded as the *actual* value (actual with respect to this theory), but it need not necessarily coincide with the *estimated* value and normally will not do so. By means of this theory, however, a value from this confidence interval is chosen, hence a value being sufficiently close to that estimated value.

The generalized observational statements do not yield any inductively justifiable conclusions for general theories because the generalized observational statements are formulated in a language which attributes or denies properties from a definite (or, idealized: denumerably infinite) class of properties to certain objects (realizations of measurement instructions), while general theories are among those judgements of a language connecting certain points of a space-time-continuum by a continuous function (or by the corresponding mathematical relation). The language in which the generalized observational statements are formulated is *interpreted* in the language to which the general theory belongs by interpreting the properties $\pi_1, \ldots, \pi_q$ as intervals of real numbers.

When selecting the general theory out of the class of possible theories in this new language neither the curve is estimated nor the quality of its estimation calculated; attention is given rather to criteria such as simplicity and continuity.

Furthermore, it will even be allowed that the curve in one case or another does *not* fail within a confidence interval, so that the normal case is *not* continuously realized.

Let the intervals (with the same curve) be of this kind:

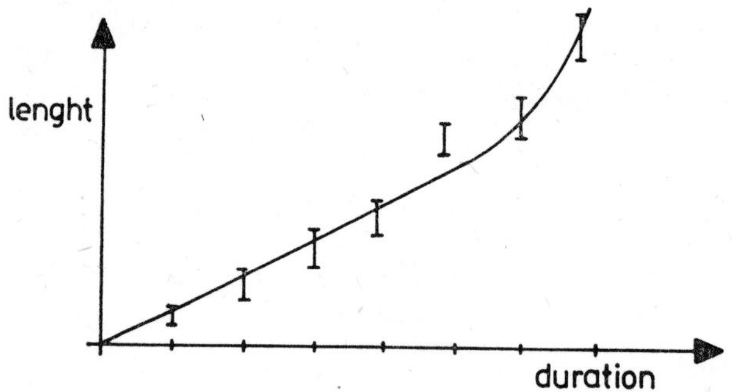

Whoever believes that the acquisition of empirical knowledge must always proceed from observational statements to elementary theories and from there to complex theories has to search for laws corresponding to these observations. Of course there are laws that satisfy these measurements, i.e. there are curves which pass through each of these intervals. Yet, only a finite set of measurements can be executed, and, consecuently, only a finite set of such intervals is available. Therefore, at least one function of real numbers (moreover: an infinite set of them) can be given, all of which pass through these intervals. This, however, is no longer a function which can be described by simple law statements of the following kind: "If an object (air resistance being neglected) being dropped close to the perigee, it covers in time t distance s, where $s = \frac{1}{2} gt^2$, g being a constant number", because in the case of $t = 5$, for example, we have to make an exception.

It would be simple, of course, to reject the generalized observational statement in the case of $t = 5$ and thus save the original general theory. With respect to generalized observational statements the importance of a general theory as compared to that of singular experiential data, increases with the decrease of such exceptions (the more seldom they are, the simpler are the laws which rescue the generalized observational statements). Furthermore, the general theory grows stronger the more it is connected with other general theories. If isolated, we are more inclined to give it up than when deductively (or even approximately) derived from other accepted general theories—theories themselves agreeing with almost all generalized observational statements. If a new theory takes such singular deviations into account, but fails to show a connection with other theories, it will be given up (and with it the generalized observational statement in question).

If, however, it should turn out that these apparent exceptions coincide with exceptions concerning other laws, their rejections will be revoked and a new overall theory will be adopted, from which a new general theory of the laws of falling bodies may be deductively or approximately inferred.

If it is decided to reject certain generalized observational statements because of a general theory, we have to deal with the question of what to do with the singular experiential data which have been accepted as the basis for the now rejected generalized observational statements. The following possibilities at least present themselves:

(1) The margins of error for the generalized observational statements may be widened, so that the old general theory again passes through the corresponding intervals.—Scientists, however, are not happy with such a solution. Since r_ε had been chosen very small already. the degree of confirmation $1-r_\varepsilon$ is not substantially increased, while on the other hand, the margins of

error have to be enlarged in unproblematic cases too, because a generalized observational statement does not tell us whether or not it is problematic without referring to the theory. But in this case, a vast host of new general theories has to be admitted, all of them having equal formal status with the accepted general theory, since they, too, pass through all intervals and are, some of them, simple as in the sameway. The more the general observational statements deviate, the more the margins of error have to be widened following this procedure the less specific becomes the content of the totality of these generalized observational statements, which means that fewer of all possible theories can be excluded.

(2) It is supposed that the world is not as simple as at first thought and therefore, more complicated laws (graphically represented by more complicated curves) have to be introduced. If the exceptions are restricted to single points in the time scale, be $t = 5$, and do not appear elsewhere in their neighbourhood, say at $t = 4{,}75$ or $t = 5{,}25$, this argument will not be accepted unless it can be explained from other sources *why* the experiment instruction with $t = 5$ *is* exceptional. Formally the experiment instructions are equal, they differ only in the time factor. The more complicated law collides with the metatheoretical conditions of continuity and simplicity for general theories without offering any explanation for doing so.

(3) It is assumed that other factors have been involved in these measurements. This may mean one of two things: (a) The realizations of the measurement instruction at $t = 5$ have been unlucky. Some additional factors (i.e. physical values, which are representable by mathematical functions or relations have interfered, but have passed unnoticed and, consequently, have not been eliminated.—This case, based on the conception that the fundamental laws of the world are non-statistical (i.e. roughly speaking, deterministic), may then be resolved by ignoring the questionable experiments and executing this experiment instruction again in such a way that the disturbing factors are, as far as possible, excluded. (b) It is not probable, but yet possible,

that each of twenty thousand successive tosses of a coin results in *Heads*; it is likewise not probable, but yet possible, that with the k experiments we all too frequently obtained improbable exceptions, which were then responsible for the estimation initially accepted together with its confidence interval.—This case can be remedied by declaring the realizations of the experiment instructions un-representative and searching for new (representative) realizations; this procedure can only be justified when we already know of other relationships (with respect to other general theories).

(4) It is argued that the measurements or observations were not carried out correctly. Thus the instructions for observing or measuring were not actually realized and no corresponding results obtained either.—This apparently suggests a solution at a different epistemological level than the other three proposed for solutions. Those proposals assumed that singular experiential data are irrevocable because of the uniqueness of the events concerned and that, at most, they can be neclegted or taken as inexplicable and isolated because they colide with other accepted sentences. In this forth case, however, a metatheoretical argument is stated: it is said that the rules for using the single expressions which occur in the experiment instructions were not fulfilled. If somebody *observes the observer* and on the basis of this *observation* (*of an observation*) makes such a judgment, one will accept it in case the person is reliable. If, however, there are no such judgments, one will be careful not to reject singular experiential data in this way. If the observation incorporates procedure innate operations, as for instance ascertaining colours, the rejection of singular experiential data of this kind implies that the observer is lying.

WŁADYSŁAW KRAJEWSKI

CORRESPONDENCE PRINCIPLE AND THE IDEALIZATION

I. CORRESPONDENCE PRINCIPLE

The *Correspondence Principle* (*CP*) established by Niels Bohr[1] is a postulate for new theories (and, in the same time, a description of the way of introducing new theories) in contemporary physics. According to it, a new theory T_2 must be more general than the old theory T_1 and must include it as a special case when some parameter p characteristic for T_2 tends to zero: $p \to 0$ (or to infinity but then we can consider $1/p$). The physicists say also that T_2 passes asymptotically into T_1 as $p \to 0$. The logicians and philosophers of science often say that T_2 is a consequence of T_1 and the assumption that $p = 0$:

(1) $\qquad T_2 \wedge p = 0 \to T_1$

The relation of T_2 to T_1 is called *Correspondence Relation* (*CR*). The classical example is the relation of Quantum Mechanics (*QM*) to Classical Mechanics (*CM*). The characteristic parameter for *QM* is Planck's constant h. This constant being assumed to be equal to zero, the equations of *QM* pass into the equations of *CM*, what can be presented as follows:

$$QM \wedge h = 0 \to CM$$

Another classical example of *CR* is the relation of *Special Relativity Theory* (*SRT*) to *CM*. Here the parameter is the light velocity c and the assumption—its infinity. In short:

$$SRT \wedge c = \infty \to CM$$

It should be explained that we designate by SRT its main laws but not all (e.g., this implication is not valid for the equation $E = mc^2$).

II. RADICAL MEANING VARIANCE CONCEPTION AND ITS REPUDATION

In the past decades CP was repudiated by some philosophers of Science (Kuhn, Feyerabend, Hanson, Toulmin) who claimed that in the course of scientific revolution the meaning of the main concepts changes, hence the old theory T_1 and the new one T_2 are incommensurable. This so called *Radical Meaning Variance* conception (*RMV*) has been criticized in papers by many authors and even in a special book.[2]

The opponents of *RMV* usually stress that if theories T_1 and T_2 divided by a scientific revolution were incommensurable they could not be incompatible, hence they could not be rival ones—they could held both; however, the adherents of *RMV* speak about the rivalization of such theories—hence, they are inconsistent. This is a good argument but it does not solve the problem of the relationship between T_1 and T_2.

The opponents of *RMV* often say that there are some changes of meaning in the course of the scientific revolution but they are usually not big and do not hinder comparing T_1 and T_2. It is, however, not clear why these 'little' changes allow the comparison. The problem must be approached otherwise.

I will not discuss here different theories of meaning, I will restrict myself to some remarks. The meaning of a concept is determined by its definition and not by the 'whole theory' as it is often claimed. When T_2 maintains the old definition of a concept *a* its meaning does not change. We have, of course, new synthetic sentences about *a* but not new analytic ones.

When in T_1 there is no explicit definition of a concept *a* we must see whether its extension changes or not. If the founders of T_2 use this concept for describing the same objects as it was done in T_1, the meaning of *a* has been unchanged.

This situation happens very often. The creation of *SRT* is a classical example of the scientific revolution. The adherents of *RMV* claim that essential shifts of meaning occur here: the length of a solid is absolute in *CM* and depends on the coordinate system in SRT, the mass is constant in *CM* and depends on the velocity in *SRT*, etc. I cannot agree with that. The mass is the measure of inertia (and of gravity) of a body. This definition is common to both *CM* and *SRT*, hence the meaning of 'mass' is unchanged. *CM* and *SRT* are incompatible because the former considers the mass as constant and the latter considers the (equally defined) mass as variable. The situation with 'length' is analogous. True, there is no simple explicit definition of 'length' (only some operational definitions can be done here) but it is doubtless that Einstein speaking about 'length' meant the same feature of a body as Newton did.

Sometimes it is said that the transition from *CM* to *SRT* changes not only the semantics but even the syntax of some concepts. Philip Frank claimed that the concept of mass changes its syntax because in *CM* it is a function of one variable $m = f(x)$ and in *SRT*—a function of two variables $m = f(x, v)$, where x means the body and v—is velocity.[3] I do not think this argument is a serious one. When biologists measure the mass of a young animal or a plant they investigate its dependence on the age t and therefore consider the mass as a function of two variables: $m = f(x, t)$. Does this mean that the meaning of 'mass' changes radically in biology?

Another example. Sometimes it is claimed that the transition from classical to quantic theory of radiation changes the meaning of 'radiation frequency' because in the former the frequency variation is continuous and in the latter it is discrete. What we have here, however, is again only the contradiction between some synthetic theses about radiation. The definition of 'frequency' has been unchanged: in both theories frequency is the number of oscillations in a unit of time.

Of course, sometimes the definition of a concept in the new theory is changed. In some cases there is only a change of the intension and the extension remains unchanged. E.g., after Robert Koch's discovery the definition of tuberculosis has been changed (into 'the illness caused by Koch's bacillus') but the same illness has been identified as tuberculosis just as it was done before Koch. I see no problem here.

In other cases the new definition changes both intension and extension, the latter can be widened or narrowed. We must, of course, take this change into account when we compare both theories. Usually it does not present difficulties. It seems after all that this situation is rare in science.

III. REAL LOGICAL DIFFICULTIES

The refutation of *RMV* does not mean that everything in the logical status of *CP* is correct. There are serious objections against the simple version of *CP* presented in Section I.

As we have noticed in the case of *QM* and *SRT*, T_2 is incompatible with T_1. In general, T_2 is incompatible with the assumption $p = 0$. This is evident in the two cases considered. The assumption $h = 0$ is incompatible with the foundations of *QM* ($h > 0$). The assumption $c = \infty$ is incompatible with the foundations of *SRT* ($c =$ const.). In other words, the two terms of the antecedent in the implication (1) are incompatible, the antecedent is false, hence the implication is vacuously true and has no value for science!

This strange situation has been noticed by some authors (although not by many!) but is rarely analysed. Some physicists say that the assumptions $h = 0$ and $c = \infty$ are not to be understood literally what is, of course, no solution. Recently there has been a vivid discussion among Polish philosophers of science on this issue. Different solutions have been suggested.[4]

One group of authors say that we can speak about an approximate implication (1) under certain conditions: in *SRT* when $v/c \approx 0$, in *QM* when h is very small in relation to other magnitudes, etc. We can call this solution an *approximate-implicative* version of *CP*.

Others say that we must abandon the implication (1) at all and interpret *CP* in the following way: T_2 explains all phenomena explained by T_1 and some additional phenomena which could not be explained by T_1. Sometimes certain conditions of the precision of explanation were added. We can call this solution an *explanative* version of *CP*.

Finally, others say that we must reexamine the whole issue taking into account the idealizational nature of physical laws (what is stressed especially by the methodological school from Mickiewicz University in Poznań). I think it is the right way (although I do not agree in some details with the philosophers from Poznań).

IV. IDEALIZATION AND FACTUALIZATION

Theoretical physics as well as other advanced sciences deal with ideal models. The main physical laws are idealizational ones, i.e. they are *strictly* true only in ideal models. Sometimes they are *approximately* true in real models; more often, however, in order to apply them to real objects, we must apply a procedure which we can call 'factualization' (the Poznań methodologists call it 'concretization'). It consists in abrogating the idealizing conditions. We obtain then factual laws which are true (or approximately true) for real objects.

The scheme of an *idealizational law* can be presented as follows:

$$C(x) \wedge p_1 = 0, \wedge \ldots \wedge p_n = 0 \rightarrow D(x)$$

where $D(x)$ is a functional dependence between certain parameters of x, $C(x)$—the set of factual conditions, $p_1 = 0, \ldots, p_n = 0$—idealizing conditions.

The factualization consists in the gradual abrogation of the idealizing conditions and replacement of them by the factual conditions $p_1 > 0, \ldots$ $\ldots, p_n > 0$. We must then replace the dependence $D(x)$ by other, usually more complicated, dependence $D_1(x), \ldots, D_n(x)$. The scheme of the *factual law* is as follows:

$$C(x) \rightarrow D_n(x)$$

For example, the Boyle law is true only for the ideal gas, in which the intermolecular forces a and the volumes of the molecules b vanish. We see that here $p_1 = a$, $p_2 = b$. $C(x)$ is a factual condition that a given mass of a gas is in a constant temperature, $D(x)$ is Boyle's equation

$$pV = const.$$

where p is the pressure of the gas, V—the volume of the gas.

When we abrogate the idealizing assumptions we must replace Boyle's equation with van der Waals' one:

$$\left(p + \frac{a}{V^2}\right)(V - b) = const.$$

Van der Waals law is also only approximately true for real gases but with much higher degree of precision than the Boyle one. The passing from Boyle's law to van der Waals' one is a simple example of the factualization.

Of course, when Boyle discovered his law he thought that it was a factual one. Only in the XIX century did it turn out that this is an idealizational law.

This is a common situation. A law of T_1 supposed to be factual turns out to be idealizational only in the light of a later theory T_2. For example *CM* turns out to be a *double* idealizational theory in the light of *SRT* and *QM*: the idealizing assumptions are: $c = \infty$ and $h = 0$.

Hence, we see that *CR* is a relation between an idealizational theory and a factual one (or, at least, a more factual one).

The relation between Boyle's and van der Waals' laws is, of course, also a *CR*. We see that *CP* had governed the development of physics in the XIX century, long before it was formulated by Bohr.

V. ANALYSIS OF THE CORRESPONDENCE RELATION

I think that it is possible to maintain the implicative version of *CP* in a qualified form which I shall call *sophisticated implicative* version.

A simple example will explain it. The law of composition of two identically directed velocities has in *CM* the simple form

$$v = v_1 + v_2$$

and in SRT—a more complicated form

$$v = \frac{v_1 + v_2}{1 + \frac{v_1 v_2}{c^2}}$$

In terms of SRT the former law is valid only under the idealizing condition $c = \infty$. *SRT* assumes the factual condition (incompatible with the former) $c = \mathit{const}$. We cannot simply consider a conjunction of the idealizing condition and the *SRT* law because we would obtained a contradiction. However, we can choose a more 'sophisticated' way. We "suspend" the factual condition in the SRT law and then obtain an 'abstractionized' law which is, strictly speaking, not a law of any physical theory but only an interpreted mathematical equation. Now we may make a conjunction of the idealizing condition and the 'abstractionized' *SRT* law and deduce from it the *CM* law.

We designate T_1 reinterpreted in terms of T_2 by T_1' and the 'abstractionized' T_2 by T_2^*. We can then set up the following table:

T_1	(*CM*)	$v = v_1 + v_2$
T'_1	(*CM* in terms of *SRT*)	$c = \infty \rightarrow \quad v = v_1 + v_2$
T_2	(*SRT*)	$C = const. \rightarrow v = \dfrac{v_1 + v_2}{1 + \dfrac{v_1 v_2}{c^2}}$
T_2^*	(*abstractionized SRT*)	$v = \dfrac{v_1 + v_2}{1 + \dfrac{v_1 v_2}{c^2}}$

The implication has then the following form:

$$T_2^* \wedge c = \infty \rightarrow T'_1$$

In other words, the reinterpreted corresponded theory is a consequence of the conjunction of the abstractionized corresponding theory and of the idealizing assumption (the need of which has been shown by the corresponding theory).

In the general form we write the idealizing assumption as $p = 0$ and the implication is as follows:

$$(2) \qquad T_2^* \wedge p = 0 \rightarrow T'_1$$

This is the general scheme of *CR*, a more 'sophisticated' one then the scheme (1). In the antecedent of (1) there is a contradiction, in the antecedent of (2)—there is none.

REFERENCES

[1] For the history of this principle see: Klaus Michael Meyer-Abich, *Korrespondenz, Individualität und Komplementarität. Eine Studie zur Geistesgeschichte der Quantentheorie in den Beiträgen Niels Bohrs*, Wiesbaden 1965.

[2] Carl R. Kordig, *The Justification of Scientific Change*, Dordrecht 1971.

[3] Cf. Jerzy Giedymin: 'Logical Comparability and Conceptual Disparity between Newtonian and Relativistic Mechanics', *The British Journal for the Philosophy of Science* **24** (1973), pp. 270–276.

[4] See: *Correspondence Principle in Physics and the Development of Science* (in Polish) (eds. W. Krajewski *et al.*), Warszawa 1974.

PAVEL MATERNA

PRAGMATIC MEANING AND TRUTH

I. $\mathscr{L}_\mu$-LANGUAGES

$\mathscr{L}_\mu$-languages are languages of applied intensional logic that are based on a modified simple theory of types (see [1], [6]). Their alphabet consists of: *variables of different types*, i.e. letters $p, q, \ldots, z$ with lower indices (natural numbers) and upper indices (greek letters and concatenations of greek letters and parentheses); *logical constants* i.e. identity signs of types $((o\alpha)\alpha)$ where α is a type, connectives (unary of the type (oo): $\sim$, negation, binary of the type ((oo)o): say, disjunction $\vee$, conjunction $\wedge$ etc.); *quantifiers* Π and Σ, both of them of the type $(o(o\alpha))$; ι-operators of the type $(\alpha(o\alpha))$; *λ-operator*; *extra-logical constants* (denumerably many): capital letters $A, B, \ldots, Z$ with lower indices (natural numbers) and upper indices (types, i.e. greek letters etc.).

We define types as usually, i.e.:

1. o, ι, μ *are types*;
2. *If* α *and* β *are types, so is* $(\alpha\beta)$;
3. *There are no other types but those defined by* 1, 2.

Our metavariables: $v_1, v_2, \ldots$ with upper type indices range over the variables; $C_1, C_2, \ldots$ range over extra-logical constants (again, they are provided with upper type indices); $E_1, E_2, \ldots$ range over the expressions; $F_1, F_2, \ldots$ range over the wffs; $\alpha, \beta, \ldots$ range over types.

Definition 1. *Well-formed expressions* (*wfes*):

1. Variables and extra-logical constants are wfes;
2. If $E_i^{(\alpha\beta)}$ and E_j^β are *wfes*, $(E_i E_j)$ is wfe of the type α; $\lambda v_i^\beta E_j^\alpha$ is *wfe* of the type $(\alpha\beta)$;
3. Any expression is *wfe* only owing to 1, 2.

Definition 2. *Well-formulas* (*wffs*):

E_i^α is a *wff* iff $\alpha = (o\mu)$ or $\alpha = 0$.

Abbreviations:
Parentheses, where omitted, are associated to left.
Instead of "$= E_i^\alpha E_j^\alpha$" etc. we write "$(E_i^\alpha = E_j^\alpha)$", "$(E_i^0 \wedge E_j^0)$" etc.
Instead of "$\Pi \lambda v_i^\beta E_j^0$" we write "$\forall v_i^\beta E_j^0$".

The ontology of any $\mathscr{L}_\mu$-language is based on three fixed sets:
U — its members are of the type ι and are called *individuals*;
T — its members are of the type o and are called *truth values*: t, f;
W — its members are of the type μ and are called *possible worlds*.

Definition 3. *Ontological objects* (*Objs*):

1. Individuals are $Objs^\iota$, truth values are $Objs^0$, possible worlds are $Objs^\mu$;
2. any function that maps the set of $Objs^\beta$ into a set of $Objs^\alpha$ is an $Obj^{\alpha\beta}$;
3. as usually.

An Obj the type of which is not $\alpha\mu$ will be called a *set-theoretical Obj* (*s-t-Obj*).

Speaking of Objs we shall use symbols $a_1, a_2, \ldots$ with upper type indices.

Definition 4. An *interpretation of the basis of* an $\mathscr{L}_\mu$-language $\mathscr{L}$ is a function $I^{\mathscr{L}}$ (we shall omit the upper index), the domain of which is th class of extra-logical constants of $\mathscr{L}$. It must satisfy the following condition: For any α and C_i^α: $I(C_i^\alpha) \in Objs^\alpha$.

Definition 5. A *valuation of variables* of an $\mathscr{L}_\mu$-language $\mathscr{L}$ is any function $V^{\mathscr{L}}$ (we shall omit the upper index) such that its domain is the class of variables of $\mathscr{L}$ and that it satisfies the condition: For any α and v_i^α: $V(v_i^\alpha) \in Objs^\alpha$.

Let $V_{(v_i^\beta / a_t^\beta)}$ be the valuation differing from V at most in giving the *Obj* a_t as the value to the variable v_i.

Definition 6. An *interpretation of an* $\mathscr{L}_\mu$*-language* $\mathscr{L}$ is a function $Int^{\mathscr{L}}$ (we shall omit the upper index) such that its domain is the class of ordered triples $\langle I, V, E_i \rangle$, where I, V are as above and E_i is a wfe of $\mathscr{L}$. The following conditions define the function *Int*:

$$Int(I, V, C_i^\alpha) = I(C_i^\alpha);$$

$$Int(I, V, v_i^\alpha) = V(v_i^\alpha);$$

$$Int(I, V, E_i^{(\alpha\beta)} E_j^\beta) = Int(I, V, E_i^{(\alpha\beta)})\big(Int(I, V, E_j^\beta)\big);$$

$Int(I, V, \lambda v_i^\beta E_j^\alpha)$ = the function that maps the set $\{a_t^\beta, t = 1, 2, ...\}$ into the set $Int(I, V_{(v_i/a_t)}, E_j^\alpha)$ $(t = 1, 2, ...)$;

$Int(I, V, (E_i^\alpha = E_j^\alpha)) = t$, if $Int(I, V, E_i^\alpha) = Int(I, V, E_j^\alpha)$, f otherwise;

$Int(I, V, \sim E_i^0) = t$, if $Int(I, V, E_i^0) = f$, f otherwise;

(Similarly for the binary connectives.)

$Int(I, V, \forall v_i E_j^0) = t$, if for all $t = 1, 2, ...$ $Int(I, V_{(v_i/a_t)}, E_j^0) = t$, f otherwise.

Definition 7. A wff F is *true in* I *for a possible world* w iff $Int(I,V,F) = t$ or $Int(I, V, F)(w) = t$ for all V.

Definition 8. A wff F is *logically true* iff F is true in all I for all w.

Let $Cn(\{F_1, ..., F_m\})$ be the set of logical consequences of the set of wffs $F_1, ..., F_m$. (This means: if $F_k \in Cn(\{F_1, ..., F_m\})$ then for any I, F_k is true in I for all such w that $(F_1, ..., F_m)$ is true in I for w.) A set of wffs $F_1, ..., F_m$ is said to be *consistent* iff there is no wff F_k such that $F_k \in Cn(\{F_1, ..., F_m\})$ and $\sim F_k \in Cn(\{F_1, ..., F_m\})$.

Definition 9. A *restricted* $\mathscr{L}_\mu$*-language* (an $\mathscr{L}_{\mu_r}$*-language*) is an ordered pair $\langle \mathscr{L}, P \rangle$, where $\mathscr{L}$ is an $\mathscr{L}_\mu$-language and P is a consistent non-empty set of wffs of $\mathscr{L}$; furthermore, P must be effectively decidable in the set of wffs of $\mathscr{L}$, and if F_i is logically true, then $F_i \notin P$. (We call P the set of meaning postulates of the $\mathscr{L}_{\mu_r}$-language).

Definition 10. Let $\mathscr{L}$ be an $\mathscr{L}_{\mu_r}$-language. An *admissible interpretation of the basis of* $\mathscr{L}$ (I_{adm}) is any interpretation of the basis of $\mathscr{L}$ satisfying the condition:

If $F_i \in P$, then F_i is true in I_{adm} for all w.

Definition 11. A wff F of an $\mathscr{L}_{\mu_r}$-language is *analytically true in* this *language* iff F is true in all I_{adm} for all possible worlds. (Clearly, F is analytically true iff $F \in Cn(P)$.) A wff F is *analytically false* iff $\sim F$ is analytically true. A wff F is *factual* (in the given language) iff it is neither analytically true nor analytically false (in this language).

Some consequences: a) Any logically true wff of an $\mathscr{L}_{\mu_r}$-language is an analytically true wff of this language.

b) There are analytically true wffs of $\mathscr{L}$ which are not logically true;
c) In every $\mathscr{L}_{\mu_r}$-language there are some factual wffs.

Definition 12. Let E_i^α be a wfe of an $\mathscr{L}_\mu$-language ($\mathscr{L}_{\mu_r}$-language). The *meaning* of E_i^α in the interpretation I (I_{adm}), for the valuation V is $Int\,(I, V, E_i^\alpha)$ $(Int(I_{adm}, V, E_i^\alpha))$

Any object of the type $\alpha\mu$ will be called a *concept*.

The objects of the type $o\mu$ are propositions;
—„— —„— $\iota\,\mu$ are individual concepts;
—„— —„— $\alpha\mu$ are properties of objects of the type α
—„— —„— $o\alpha_k\alpha_{k-1}\ldots\alpha_1\mu$ are relations-in-intension between the objects of the types $\alpha_1, \alpha_2, \ldots, \alpha_k$, respectively.

II. A PRAGMATIC EXTENSION OF $\mathscr{L}_{\mu_r}$-LANGUAGES

It appears that an $\mathscr{L}_{\mu_r}$-language is able to represent semantically such declarative sentences of a natural language that do not contain deictical elements (and that are, therefore, tenseless). A similar idea has been Montague's conception ([2], [3], [4]): as the basic language for semantic representation of natural language the language of intensional logic has been chosen by him. In order to represent declarative sentences that contain indexical expressions Montague has used the idea of conceiving the relevant aspects of 'possible contexts of use' as 'points of reference' (see D. Scott, [5]). I shall use this idea, too, but with some modifications.

For Montague, the possible worlds can be conceived as a sort of points of reference. For me, there is a principial difference between the possible worlds on one hand and the points of reference on the other, at least if we intend to let the latter play the role of (relevant aspects of) possible contexts of use. The possible worlds are not 'contextual'; typically, we need them to represent any tenseless sentence that does not contain indexical expressions. To represent the sentence

(1) *The whales are mammals*

we cannot write

(2) $\forall x(W(x) \rightarrow M(x))$,

as this would—contraintuitively—mean that (1) is an analytic sentence. (Indeed: if 'W' and 'M' denoted classes, i.e. if they were of the type o, then for any individual it would be trivial to belong or not to belong to any of these classes—either the class in question would contain this individual as its member—which would mean that '$W(x)$' ('$M(x)$') is for such a valuation trivially (necessarily) true—or it would not, and then these sentences are trivially false. For the similar arguments against the representations of (1)-like sentences by means of (2)-like forms see [6]). We write—within an $\mathscr{L}_{\mu_r}$-language—

(3) $\lambda w^{\mu} (\forall x^{\iota}(W^{o\iota\mu} w^{\mu} x^{\iota} \rightarrow M^{o\iota\mu} w^{\mu} x^{\iota}))$.

The type of (3) is no more o, it is oμ, i.e. (3) names a proposition: a function that maps the set W into the set T according to whether any individual having the property of whaleness has in the given possible world the property of mammalness or not.

On the other hand, if we analyze a sentence such as

(4) *I am hungry*

we see that its 'meaning' is dependent not only on the possible worlds: it is dependent also on the context of use. This context is such that it determines some actual-world-objects: this determination is on its part given by 'points of space' and 'points of time', i.e. by where and when (4) is uttered. So the 'actualizing factors' appear to be the members of two sets: the set, say, S_1 of points of space and the set, say, S_2 of points of time.

My proposal concerning the representation of 'pragmatic' meaning consists in extending an $\mathscr{L}_{\mu_r}$-language to a language $\mathscr{L}_{\mu\sigma_r}$ as follows:

(i) The set of types is extended by adding the types σ_1 and σ_2 to the elementary types ι, o, μ.

(ii) No expression the type of which is $\alpha\mu\sigma_i$ or $\alpha\mu\sigma_i\sigma_j$, $i, j = 1, 2$, is admitted as a wfe. (We modify, therefore, the definition of wfes as follows: 1′. Variables and extra-logical constants are wfes, iff they are not of the type $\alpha\mu\sigma_i$ or $\alpha\mu\sigma_i\sigma_j$, $i = 1, 2, j = 1, 2$. 2′. ... $\lambda v_i^{\beta} E_j^{\alpha}$ is a wfe of the type $(\alpha\beta)$, iff $\alpha \neq \gamma\mu$ or $(\beta \neq \sigma_i$ and $\beta \neq \sigma_i\sigma_j$, $i, j = 1, 2.)$

(iii) E_j^{α} is a wff iff $\alpha = 0$ or $\alpha =$ oμ or $\alpha =$ oσ_i or $\alpha =$ o$\sigma_i\sigma_k$, $i, k =$ $= 1, 2$.

(iv) In the ontology two sets, S_1 and S_2, are added to the sets U, T, W. The members of S_1 and S_2 are of the types σ_1 and σ_2, respectively.

I must explain the necessity of (ii). The objects of the type σ_1 or σ_2 or $\sigma_i \sigma_j$ help to pick up (directly) some object of the actual world. If an expression of the type, say, $\alpha\mu\sigma_1$ were a wfe, it would mean that this expression names (under an interpretation) a function that takes the points of space to concepts. The role of the points of space would then consist not in picking up some actual object but in picking up a concept, which would be inacceptable from the presented viewpoint. A wfe of the type $\alpha\sigma_i$ $(\alpha\sigma_i\sigma_j)$ must be therefore such that α is a type of some s-t-object.

(We could try to apply the principles of $\mathscr{L}_{\mu\sigma r}$-languages to the representation of (4). We suppose that our $\mathscr{L}_{\mu\sigma r}$-language contains a present-tense constant 'Q' of the type o $(o\mu)\sigma_2$; the constant 'E' corresponding to the English word 'I' would be of the type $\iota\sigma_1\sigma_2$, 'H' (for 'to be hungry') would be of the type $o\iota\mu$. Our representation of (4) would be

$$(5) \qquad \lambda s_1^{\sigma_1} \lambda s_2^{\sigma_2}(Q s_2^{\sigma_2}(\lambda w^{\mu}(H w^{\mu}(E s_2 s_1))))$$

and its meaning would be an object of the type $o\sigma_2\sigma_1$).

The definitions 1′.–6′. corresponding, for an $\mathscr{L}_{\mu\sigma r}$-language, to the definitions 1.–6. can be easily formed.

Definition 7′. A wff $F^{o\mu}$ (or F^o) is *true in I for the possible world* w iff $Int(I, V, F) = t$ or $Int(I, V, F)(w) = t$ for all V. A wff $F^{o\sigma_i}$ (or $F^{o\sigma_i\sigma_j}$) is *true in I for the point of space* (*time*) $a_k^{\sigma_i}$ (*for the pair* $\langle a_k^{\sigma_i}, a_m^{\sigma_j}\rangle$ *of points of space* (*time*)) iff $Int(I, V, F)(a_k^{\sigma_i}) = t$ for all V $((Int(I, V, F)(a_k^{\sigma_i}))(a_m^{\sigma_j}) = t$ for all V).

Definition 8′. A wff F is *logically true* iff F is true in all I for *a*) all possible worlds, or b) all $a_k^{\sigma_i}$ (all $\langle a_k^{\sigma_i}, a_m^{\sigma_i}\rangle$). Analogous modifications can be done for Definition 11 and Definition 12. To go back to our example (5), imagine that we have chosen a point of space, say the place where I am just standing (let us give this place an $\mathscr{L}_{\mu\sigma r}$-language name '$K_1$') and a point of time, say the moment at which I said the sentence (4) (let us give this moment the name 'K_2'). (Naturally, 'K_1' is of the type σ_1, 'K_2' of the type σ_2). We could show that Church's

β-conversion rule leads to the following result (I omit the upper indices):

$$((\lambda s_1 \lambda s_2(Qs_2(\lambda w(Hw(Es_2 s_1)))))K_1)K_2$$

which is convertible into the expression

$$(6) \qquad QK_2(\lambda w(Hw(EK_2 K_1))).$$

Now, $Int(I, V, EK_2 K_1) = Materna$ (say: 'M') for all V and in the given (intended) interpretation I of the basis of our language. The meaning of '$\lambda w(HwM)$' is a proposition that maps the set W into the set T linking every possible world with truth iff Materna has the property "to be hungry" in this world. Finally ,'QK_2', being of the type $o(o\mu)$, names a class of propositions—videlicet the class of those propositions that are truly uttered in the moment named by 'K_2'. If (and only if) the proposition named by '$\lambda w(HwM)$' belongs to this class, (5) is true in the intended I for the pair of the points od space and time named in the object language 'K_1', 'K_2', respectively.

BIBLIOGRAPHY

[1] Church, A., 'A Formulation of the Simple Theory of Types', *The Journal of Symbolic Logic* **5** (1940), 56–68.

[2] Montague, R., 'Pragmatics', in: R. Klibansky, ed.: *Contemporary Philosophy*, Firenze 1968, 102–122.

[3] Montague, R., 'Pragmatics and Intensional Logic', in: D. Donaldson, G. Harman, eds: *Semantics of Natural Language*, 2nd ed., Reidel 1972, 142–168.

[4] Montague, R., Schnelle, H.: *Universale Grammatik*, Braunschweig 1972.

[5] Scott, D., 'Advice on Modal Logic', In: K. Lambert, ed.: *Philosophical Problems in Logic*, Reidel 1970, 143–173.

[6] Tichy, P., 'An Approach to Intensional Analysis', *Noûs* V/3, 1971, 273–297.

TADEUSZ NADEL-TUROŃSKI

SEMANTIC COMPLEMENTARITY IN QUANTITATIVE EMPIRICAL SCIENCES

This paper is an application of Prof. Wójcicki's notion of approximative truth to some problems of inter-theory and inter-structural relations.

I. DEGREES OF PRECISION

Let us have two operational structures $\mathfrak{X}$ and $\mathfrak{X}'$ where $\mathfrak{X} = p\mathscr{X} = (X, p_1 F_1, \ldots, p_n F_n)$ and $\mathfrak{X}' = p'\mathscr{X} = (X, p'_1 F_1, \ldots, p'_n F_n)$. Let us define the relation D_k between the operational structures (read: "is more precise with respect to F_k than"):

Definition 1. $\mathfrak{X} D_k \mathfrak{X}'$ iff for every $x \in X, p_k F_k(x) \subseteq p'_k F_k(x)$.

Let us define now another relation D^k between the operational structures (read: "is strictly more precise with respect to F_k than"):

Definition 2. $\mathfrak{X} D^k \mathfrak{X}'$ iff for every $x \in X$, $p_k F_k(x) \subsetneqq p'_k F_k(x)$.

We can define now the relations of 'being more precise' and of 'being strictly more precise' between the operational structures. We shall use the following symbols for these relations: 'D' and '$\hat{D}$', respectively.

Definition 3. $\mathfrak{X} D \mathfrak{X}'$ iff for every i, $\mathfrak{X} D_i \mathfrak{X}'$.

Definition 4. $\mathfrak{X} \hat{D} \mathfrak{X}'$ iff for every i, $\mathfrak{X} D_i \mathfrak{X}'$ and for some j, $\mathfrak{X} D_j \mathfrak{X}'$. Let $K_I(\mathfrak{X})$ be the symbol of the class of all idealizations of the operational structure $\mathfrak{X}$, i.e. all the quantitative structures $\mathscr{X}$ such that for every i, $F_i \in p_i F_i$. And let $K_R(T)$ be the symbol of the class of all realizations of the theory T.

We have then:

$$\mathfrak{X} D \mathfrak{X}' \quad \text{iff} \quad K_I(\mathfrak{X}) \subseteq K_I(\mathfrak{X}')$$

and

$$\mathfrak{X} \hat{D} \mathfrak{X}' \quad \text{iff} \quad K_I(\mathfrak{X}) \subsetneqq K_I(\mathfrak{X}').$$

The notion of approximative truth of theory T in the operational

structure $\mathfrak{X}$ is defined as follows:

$$(*) \qquad T \in AVer(\mathfrak{X}) \text{ iff } K_I(X) \cap K_R(T) \neq \emptyset.$$

Hence:

For every T, $\mathfrak{X}$, $\mathfrak{X}'$, if $T \in AVer(\mathfrak{X})$ and $\mathfrak{X}D\mathfrak{X}'$, then $T \in AVer(\mathfrak{X}')$. Let $\mathfrak{X}(F)$ be the symbol of the class of operational structures with the system of quantities $\mathfrak{F} = (F_1, \ldots, F_n)$. This class, when partially ordered by the relation D, will be symbolized by $\mathfrak{F} = (\mathfrak{X}(F), D)$. We can now define the relation of 'being more precise in $\mathfrak{F}$' between two theories T and T'.

Definition 5. $TD_{\mathfrak{F}}T'$ iff for every $\mathfrak{X} \in \mathfrak{F}$, if $T' \in AVer(\mathfrak{X})$, then $T \in AVer(\mathfrak{X})$.

Let us define the relation of 'being equally precise in $\mathfrak{F}$':

Definition 6. $T \underset{\mathfrak{F}}{\approx} T'$ iff $TD_{\mathfrak{F}}T'$ and $T'D_{\mathfrak{F}}T$.

We can also define the concept of approximative equivalence of two theories in the given operational structure $\mathfrak{X}$.

Definition 7. $T \underset{\mathfrak{X}}{\approx} T'$ iff $T \in AVer(\mathfrak{X})$ and $T' \in AVer(\mathfrak{X})$.

Let Ξ be a subclass of $\mathfrak{F}$: $\Xi \subseteq \mathfrak{F}$. Practically, we can have only some Ξ—not $\mathfrak{F}$—because of the impossibility of avoiding some error of observation. We can define now the concept of 'approximative equivalence in Ξ' of two theories.

Definition 8. $T \underset{\Xi}{\approx} T'$ iff for every $\mathfrak{X} \in \Xi$, $T \underset{\mathfrak{X}}{\approx} T'$.

Invention or application of some more precise technique of measurement is equivalent to transition from some $\Xi_1 \nsubseteq \mathfrak{F}$ to another $\Xi_2 \nsubseteq \mathfrak{F}$ such that: $1° \Xi_1 \nsubseteq \Xi_2$, $2°$ for every $\mathfrak{X} \in \Xi_2 - \Xi_1$ there exists some $\mathfrak{X}' \in \Xi_1$, $\mathfrak{X}\hat{D}\mathfrak{X}'$. Of course, approximative equivalence of two theories in Ξ_1 does not imply their approximative equivalence in Ξ_2. Equal precision of two theories in $\mathfrak{F}$ does not imply their approximative equivalence in some $\Xi \nsubseteq \mathfrak{F}$, because there may exist in Ξ some operational structure $\mathfrak{X}$ such that both of them are false in $\mathfrak{X}$.

Approximative equivalence of two theories in some $\Xi \nsubseteq \mathfrak{F}$ does not imply their equal precision in $\mathfrak{F}$, because there may exist some operational structure $\mathfrak{X} \in \mathfrak{F} - \Xi$ such that one of these theories (the more precise one) is approximatively true in $\mathfrak{X}$ and the other (less precise) is false in $\mathfrak{X}$. Two theories T and T' are *logically equivalent* iff $K_R(T) = K_R(T')$. We shall denote by $K_{AR}(T)$ the class of approximative reali-

zations of the theory T, i.e. the class of operational structures in which the theory T is approximatively true.
If T and T' are defined in $\mathfrak{F}$ and $K_{AR}(T) = K_{AR}(T')$, then 1° $T \underset{\mathfrak{F}}{\approx} T'$, 2° if there exists in $\mathfrak{F}$ such non-empty class Ξ that $\Xi \subseteq \mathfrak{F} \cap K_{AR}(T)$, then $T \underset{\Xi}{\approx} T'$.

II. GENERATED THEORIES AND APPROXIMATIVE STRUCTURES

Let the language of the theory T_n include symbols of quantities $F_1, \ldots$ $\ldots, F_n$ and let the methods of measurement for each of them, $p_1, \ldots, p_n$, respectively, be given. Let us enrich this theory by adding the symbols of quantities $F_{n+1}, \ldots, F_{n+m}$ which are to be measured by methods $p_{n+1}, \ldots, p_{n+m}$, resp. We denote this enriched theory by T_{n+m}. We now assume that quantities $F_1, \ldots, F_n$ define the quantities $F_{n+1}, \ldots, F_{n+m}$, i.e. the quantities $F_{n+1}, \ldots, F_{n+m}$ are definable in T_{n+m}. Let us have another theory T^*_{n+m}. The only difference between T^*_{n+m} and T_{n+m} is this: T^*_{n+m} includes some other quantities $F^*_{n+1}, \ldots, F^*_{n+m}$ instead of $F_{n+1}, \ldots, F_{n+m}$. Quantities with stars are also definable by the quantities $F_1, \ldots, F_n$, i.e. they are definable in T^*_{n+m}. For every i, the method of empirical measurement of F^*_i is the same as the method of empirical measurement of F_i.

Definition 9. We call the theory T^*_{n+m} the *simplification of the theory* T_{n+m} iff for every i, the quantity F^*_i is definable by the same quantities (not necessarily all of them) from $F_1, \ldots, F_n$ as the quantity F_1.

Example: $F_1 = m_0$ (rest mass), $F_2 = v$ (velocity), $F_3 = c$ (speed of light) F_1, F_2, F_3 are three of the quantities $F_1, \ldots, F_n$ in T_n. Let us have T_{n+1} with the quantity $F_{n+1} = m = m_0/\sqrt{1-(v^2/c^2)}$ and T^*_{n+1} with the quantity $F^*_{n+1} = m^* = m_0$. T^*_{n+1} is the simplification of T_{n+1} (degenerated case because m^* is simply the same as m_0).

Definition 10. T^*_{n+m} is *ε-near* T_{n+m} *in the region* Ω iff there exists such an m-tuple of positive numbers $\varepsilon = (\varepsilon_{n+1}, \ldots, \varepsilon_{n+m})$ that for every k, $(n < k \leqslant n+m)$, $\varrho(F^*_k, F_k) < \varepsilon_k$, in some region of values of the quantities $F_1, \ldots, F_n$ being the subset of Re^n. $\varrho(\cdot, \cdot)$ is a function of distance in a sense of metrization of functional space.

Example the same.

Definition 11. Theory T^*_{n+m} is ε, Ω*-generated by* T_{n+m}—we denote

this fact by $T^*_{n+m} \in G^{\varepsilon}_{\Omega}(T_{n+m})$—iff T^*_{n+m} is ε-near T_{n+m} in Ω and T^*_{n+m} is a simplification of T_{n+m}.

Thus the theory T_{n+m} (and even T_n alone) determines the whole class $G^{\varepsilon}_{\Omega}(T_{n+m})$. We shall write also $G^{\varepsilon}_{\Omega}(T_n)$ or even $G^{\varepsilon}_{\Omega}(T)$, for short. Let $\mathfrak{X}$ be an element of $K_{AR}(T)$ and let $\mathfrak{X}^*$ be an element of $K_{AR}(T^*)$, where $T^* \in G^{\varepsilon}_{\Omega}(T)$. If the numbers ε_k are small enough (less than the error of measurement), we can say—after necessary reinterpretation of both T^* and T (other quantities—strictly speaking)—that both theories T^* and T are approximatively true in both operational structures $\mathfrak{X}^*$ and $\mathfrak{X}$, i.e. they are approximatively equivalent in both structures. Let $K_{AR}(G^{\varepsilon}_{\Omega}(T))$ symbolize the class of operational structures in which some theory generated by T is approximatively true. Of course we have inclusion $K_{AR}(T) \subseteq K_{AR}(G^{\varepsilon}_{\Omega}(T))$. Transition from T to some $T^* \in G^{\varepsilon}_{\Omega}(T)$ is, in most cases, very easy (some terms in some formulas are simply omitted) although *not deductive*. For this reason $K_{AR}(T)$ and $K_{AR}(G^{\varepsilon}_{\Omega}(T))$ are often thought of as identical, the assumption being, however—semantically speaking—incorrect. We call the operational structure $\mathfrak{X}^*$ an approximation of the operational structure $\mathfrak{X}$ (the former is an approximative realization of T and the latter—of T^* being generated by T).

III. SEMANTIC COMPLEMENTARITY

It often happens that two different theories T_1^* and T_2^* are known and the theory T, which generates them, is yet unknown. Sometimes T_1^* and T_2^* are thought of as describing some 'special cases' and T as being 'more general' than T_1^* and T_2^*. That is incorrect, because—as we have said above—the transition from a generating theory to a generated one is not a deduction, or—what is the same—the generated theory is not a fragment of the generating one (even enriched with some additional statements or assumptions). In these cases we represent some system of material objects by means of two different operational structures (with different quantities and different system of quantities) $\mathfrak{X}_1^*$ and $\mathfrak{X}_2^*$, in which theories T_1^* and T_2^* are approximatively true, respectively. The relation between T_1^* and T_2^* is called *semantic complementarity*. We cannot regard them as both true in the same structure, because the sum $T_1 \cup T_2$ is inconsistent if one puts $F^*_{1k} = F^*_{2k}$. This is valid also when the generating theory T is known, but then our effect is to some exten masked, because the operational structure $\mathfrak{X}$, being the approximative realization of T, seems to be the 'true representation' of the given system of material objects, operational structures X_1^* and X_2^* seem to be "essentially the same structure" being the approximations of it and the theories T_1^* and T_2^* seem to be 'special cases' of T—all this in some extralogical sense. It is often the case in science (physics) that we have some T_1^* and T_2^* (different 'models') or can obtain them from some known T and use them for different purposes.

IV. CORRESPONDENCE RULE

It is often the case, also, that for every $T^* \in G_\Omega^\varepsilon(T)$ we have—after reinterpretation as described above, i.e. after putting quantities without stars instead of quantities with stars in T—non-empty difference $K_{AR}(T) - K_{AR}(T^*)$. If we know a criterion that allows us after measurement of values of a single or few quantities to decide whether the given operational structure belongs to $K_{AR}(T)$ or to $K_{AR}(T) - K_{AR}(T^*)$—we call such a criterion *correspondence rule*. Correspondence rule tells us in what set of operational structures two theories—one of them being generating and the other generated—are approximatively equivalent (after necessary reinterpretation of some symbols in one of them).

In our *Example* we have: For every X, if $(v^2/c^2) < \varepsilon_1 < 1$, then T is approximatively equivalent to T^* in $\mathfrak{X}$. This is the correspondence rule connecting the relativistic mechanics with the classical one.

BIBLIOGRAPHY

[1] Wójcicki, R., 'Metody formalne w problematyce teoriopoznawczej', *Studia Filozoficzne*, Nr 1/1972.

[2] Wójcicki, R., *Basic Concepts of Formal Methodology of Empirical Sciences*, Warsaw 1972 (Preprint)

LESZEK NOWAK

MARX'S CONCEPT OF LAW OF SCIENCE

The present paper is to offer a reconstruction of Karl Marx's concept of laws in the social sciences, to list the main characteristics of those laws as interpreted by Marx, and to suggest an explanation of his concept. It may be supposed that the rather few, and certainly not systematized, contexts in which Marx refers to the properties of the laws of science include novel ideas, different from those which underlie the best known concepts current in the present-day philosophy of science. This is the first reason for which it is worth while to undertake the task of a logical reconstruction of the Marxian concept of laws of science. The second reason is that the sketchy formulations of the author of *Capital* seem to be a good point of departure for methodological analysis of the real function of laws in science.

The description of laws in the social sciences, as offered in this paper, will cover their logical aspect (structure of laws, their epistemological aspect (relationships between laws and facts), and their ontological aspect (properties of laws connected with the categorial properties of those facts to which the laws apply.)

I. THE STRUCTURE OF LAWS OF SCIENCE

The essential idea to be found in the works of the founders of the Marxist theory and pertaining to the structure of laws of science is the Marxian concept of 'abstraction' (*abstract law*).

A typical example of Marxian 'abstraction' is offered by the *law of value*, whose simplified form states that if demand and supply balance each other, then the market prices of the commodities correspond to their values.[1] The condition on which this correspondence depends is very peculiar in this case: it is a condition which never materializes. To quote Marx, "In reality, supply and demand never coincide, or, if they do, it is by mere accident, hence scientifically = 0, and to be regarded as not having occurred. But political economy assumes that supply and demand coincide with one another."[2] When explaining the *Marxian concept of abstraction*[3] we may say that it corresponds to that of an *idealizational statement*, i.e., a strictly general statement:

(I.1) $\textit{if } G(x) \textit{ and } p_1(x) = d_1 \textit{ and } \dots \textit{ and } p_k(x) = d_k,$
$\textit{then } F(x) = f(H_1(x), \dots, H_n(x)),$

where (a) $G(x)$ stands for a realistic assumption, that is, a condition which can be satisfied; (b) $p_i(x) = d_i$ stands for an idealizing assumption, that is, a condition which requires that a property p_i should manifest itself in its minimum degree d_i, whereas in fact that property cannot manifest itself in that degree.[4] Thus, the law of value is an idealizational statement, to be written for brevity in the following form:

(I.2) $\textit{if } x \textit{ is a commodity, and if } D(x) - S(x) = 0,$
$\textit{then } P(x) = f(V(x)),$

where D stands for demand, S for supply, P for price, V for value, and f for a linear function. It can be seen that the first condition in the antecedent of (I.2) is a realistic assumption, whereas the second, as follows from Marx's comment quoted above, is an idealizing assumption.

The peculiarity of idealizational statements consists in that they may not be directly referred to empirical situations. If we do so, we commit a methodological error, for which Marx used to blame representatives of *vulgar* economy. The error is due to the fact that the contradiction between a general law and complex actual conditions can be eliminated by finding certain intermediate stages, but not by trying directly to adjust facts to abstract concepts.[5] It is self-evident why idealizational statements in the form of (I.1) may not be directly referred to facts: to apply a law to a given case is the same as to demonstrate that that law covers the case in question, i.e., that the case in question satisfies the antecedent of that law. But the antecedent of (I.1) is not satisfied by facts since that antecedent includes idealizing assumptions. Idealizational statements may be interpreted as applicable to certain theoretical constructions that cover certain aspects of facts; such theoretical constructions (a perfect gas, a rigid body, commodity in the sense used in *Capital*, a rationally acting decision maker, etc.) thus as it were inherit certain aspects of facts (real situations), namely those which are not precluded by the idealizing assumptions adopted in a given case. Such was more

or less Engels's intention when he interpreted the law of value by stating that, in accordance with what the economists assume, only the same values are mentioned, and in the sphere of an abstract theory it is really so.[6]

When we refer to an idealizational statement to explain certain observed facts, we do so "through a number of intermediary stages, a procedure which is very different from morely including it under the law."[7] The finding of those intermediary stages, that is, the *concretization* of an idealizational statement consists, to put it briefly, in the removal of the idealizing assumption (through the adoption of a realistic assumption which is incompatible with it) and in making an appropriate correction in the consequent. Such a correction may be specified with precision or roughly. In the former case we arrive at a statement which may be called a *strict concretization* of an idealizational statement. A strict concretization of (I.1) is

(I.3) *if* $G(x)$ *and* $p_1(x) = d_1$ *and* ... *and* $p_{k-1}(x) = d_{k-1}$ and $p_k(x) \neq d_k$, *then* $F(x) =$
$= g(f(H_1(x), \ldots, H_n(x)), h(H_{n+1}(x), \ldots, H_m(x)))$,

where p_k stands for the additional factor taken into account after the removal of the idealizing assumption $p_k(x) = d_k$. Thus for instance, a strict concretization of the law of value (I.2) as formulated in *Capital* would be

(I.4) *if* x *is a commodity and if* $D(x) - S(x) \neq 0$,
then $P(x) = f(V(x)) + \beta(D(x) - S(x)))$,

where f is a linear function, and β stands for a function which depends on the fluctuations of demand and supply.[8]

In the latter case. that is, when the correction made as a result of the removal of idealizing assumptions is a rough one, we arrive at an *approximate concretization* of an idealizational statement. For (I.1) it would be

(I.5) *if* $G(x)$ *and* $p_1(x) \approx d_1$ *and* ... *and* $p_k(x) \approx d_k$,
then probably $F(x) \approx f(H_1(x), \ldots, H_n(x))$.

As can be seen, this is a *factual statement* (which includes realistic assumptions only), which says that in the case of (empirically given) objects which have the properties $p_1, \ldots, p_k$ in degrees that come close to the limiting one assumed in the antecedent of (I.1), the equation in the consequent of that formula is satisfied approximately with a specified probability. Idealizational statements are usually concretized in a strict manner with reference to a number of idealizing assumptions, and are next concretized in an approximate manner with reference to the remaining idealizing assumptions. Factual statements obtained in this way refer to empirically given events. This is the procedure which Marx used in his *Capital*.

II. LAWS AND REGULARITIES

Marx used the term *law* both in the sense of a law of science and in the sense of that to which a law of science applies. To avoid misunderstandings we shall use the term *law* in the former sense, and the term *regularity* in the latter. Thus a regularity may be characterized as follows: a law is an "inner and necessary connection between two seeming contradictions";[9] hence a regularity is an (1) inner, (2) necessary connection (3) between seemingly contradictory facts. By referring to the various contexts in which Marx used the said terms we shall try to reconstruct and next to explain the meaning of that description of a regularity.

That a regularity is a connection may be explained by reference to (I.1). Thus the regularity described by that formula would be the relationship f that holds between the factor F, on the one hand, and the factors $H_1, \ldots, H_n$, on the other.[10] This regularity holds in the sense that F is an f-transformation of the factors $H_1, \ldots, H_n$, but not conversely. In other words, (I.1) refers to a regularity that consists in one factor (a dependent variable) being a function of other factors (independent variables).

Regularity as a connection between most essential factors. What is characteristic of *Capital* is the incessant stressing of the distinction to be made between essential and adventitious factors, between an inner connection and its manifestation, between that which is real and that which is apparent. That is not merely a *facon de parler* nor a display of Hegelian phraseology. For Marx it is the issue of carrying out the fundamental duty of science: "it is a work of science to resolve the visible, merely external movement into the true intrinsic movement";[11] "all science would be superfluous if the outward appearance and the essence of things directly coincided."[12] This is why "vulgar economy feels particularly at home in the estranged outward appearances of economic relations" which "seem the more self-evident the more their internal relationships are concealed (...)".[13] Marx also makes a clear distinction between two approaches to research: "One of these conceptions fathoms the inner connections, the physiology, so to speak, of the bourgeois system, whereas the other takes the external phenomena of life, as they seem and appear, and merely describes, catalogues, recounts and arranges them under formal definitions."[14] The latter approach is characteristic of the early stages of science, whereas mature theoretical studies are associated with the former, and it is then only that science carries out its true task. The connection between the method of revealing the essence of facts with that of idealization becomes thus evident: the method of proceeding from the abstract to the concrete (i.e., the idealization method in Marx's terminology) serves to reveal the essence of the facts covered by the investigations.

That connection can also be seen in Marx's analyses of economic issues. Consider, for instance, the law of value (I.2). It is based on the idealizing assumption that demand equates supply. Now, as Marx says, "Whenever else this formula (that demand equates supply—*L.N.*) is resorted to (and this is then practically correct), it serves as a formula to find the fundamental rule (...) which is independent of and rather determines, competition; notably as a formula for those who are held captive by the practice of competition (...) to arrive at what is again but a superficial idea of the inner connection of economic relations obtaining within competition."[15] Thus the 'formula' (i.e., the idealizing assumption) which states that demand equates supply is adopted in order to eliminate the adventitious factor which is the fluctuations of demand and supply, and to concentrate on the essential factor, namely the value of the commodities, that is the time of socially necessary labour required to produce them. Such in general is the role of idealizations: idealizing assumptions are adopted to neutralize certain adventitious factors and thus to enable us to analyse relationships between the basic factors only, that is those which most strongly affect the quantities under consideration.

It is an open issue what criteria are used in science to single out those factors which are essential for a given factor, and next to order those essential factors by the effect they have upon the factor (magnitude) under consideration. An answer can be given only by empirical methodological research on the criteria of significance used by researchers in practice. (Because of linguistic considerations the terms *significance*, *significant* are more convenient to use that *essentiality*, *essential—L.N.*) From the theoretical point of view it is important that there *are some* criteria which, first, enable us to single out the set U of those factors which are significant for a given factor (i.e., the set of those factors which affect a given factor), and, secondly, enable us to order the set U. The second point means that we assume that the relation of being a more significant factor is *antisymmetric*: if A is more significant for Z than B is, then it is not the fact that B is more significant for Z than A is. It is also assumed that that relation is *transitive*: if A is more significant for Z than B is, and if B is more significant for Z than C is, then A is more significant for Z than C is. Hence if we consider a given set U of those factors which are possible determinants for a designated factor F, then the relation of being a more significant factor orders U. In U we can also single out factors which are equisignificant for F, namely such that one is not more significant for F than the other is. Let those factors which are equisignificant for F form sets $D_0, D_1, D_2, \ldots, D_k$. Factors A, B are in D_i if neither A is more significant for F than B is, nor B is more significant for F than A is. At the same time any factor in D_i is more significant for F than any factor in D_{i+1} is. Those factors which are in D_0 are thus the *most significant* of all in the initial set U: they are more significant than those in D_1, and next than those in D_2, etc. The sequence of sets $D_0, D_1, \ldots, D_k$ will be termed the *hierarchy of significance* for the initial set U.

Assume now that the set D_0 of those factors which are the most significant for F consists of $H_1, \ldots, H_n$ (the *main* factors). The set D_1 of less significant factors consists of $H_{n+1}, \ldots, H_m$. And so on, until we reach the set D_k of the least significant factors, which consists of $H_{w+1}, \ldots, H_z$. Thus the whole initial set U includes the main factors $H_1, \ldots, H_n$ and the *adventitious* factors $H_{n+1}, \ldots, H_m; \ldots; H_{w+1}, \ldots, H_z$.

If that hierarchy of significance of factors is fixed for U, then the role of making idealizing assumptions consists in disregarding in turn the least significant factors, i.e., those in D_k, then those in D_{k-1}, etc., until all factors except those which are in the most significant class, i.e., those in D_0, are eliminated from our considerations. Thus by making the idealizing assumption that $p_k(x) = d_k$ we disregard the factors $H_{w+1}, \ldots, H_z$. Our considerations then refer to the *first-order deep level*, which includes the factor F (as the dependent variable) and the factors in $D_0, D_1, \ldots, D_{k-1}$. It differs from the *surface level* (i.e., the sphere of facts) by the fact that the latter also includes those factors which are in D_k, that is those which are in the least significant class. By adopting the next idealizing assumption we eliminate those factors which are in D_{k-1}, and so on, until by making the assumption that $p_1(x) = \mathrm{d}_1$ we disregard the factors in D_1. At that stage our considerations refer to the *k-th-order deep level* (or, in other words, the *inner level*), i.e., the system which includes only the factor F and those factors which are the most significant for F, namely those in D_0, i.e., the factors $H_1, \ldots, H_n$. All the adventitious factors (those in $D_1, \ldots, D_k$) being now disregarded, we hypothetically establish the dependence of F on its most significant determinants, i.e., $H_1, \ldots, H_n$. If we assume that this dependence is described by the function f, then we thereby formulate Theorem (I.1), that is, an idealizational statement which has in its antecedent those idealizing assumptions which eliminate the disturbing factors, and in its consequent, a formula which describes the dependence of F on its most significant determinants $H_1, \ldots, H_n$.

It might also be said that Theorem (I.1) describes the *inner* (i.e., the deepest, k-th) level of the *significance structure* of F:

$$
\begin{array}{ll}
(k) & H_1, \ldots, H_n, \\
(k-1) & H_1, \ldots, H_n;\ H_{n+1}, \ldots, H_m, \\
 & \ldots\ldots\ldots\ldots\ldots\ldots\ldots \\
(0) & H_1, \ldots, H_n;\ H_{n+1}, \ldots, H_m;\ \ldots;\ H_{w+1}, \ldots, H_z.
\end{array}
$$

By adopting the idealizing assumptions which eliminate one by one the sets of adventitious factors we as it were proceed from the concrete to the abstract, that is, to the inner level, at which we have to do only with the main factors which affect the factor under consideration.

The approach described above enables us to comprehend the inner nature of regularities. A regularity is a dependence of a factor F on its most significant determinants, i.e., factors $H_1, \ldots, H_n$. The form of that dependence—if (I.1) is true—is determined by a function f. A regularity is thus a relation that holds between F and $H_1, \ldots, H_n$, that is, a relation that holds in the inner structure; it includes among its arguments only those factors which are the most significant for F. This, perhaps, may be treated as an interpretation of Lenin's well-known formulation that clearly refers to the Marxian statement on the inner nature of regularities: "A law is a *relation*. (...) A relation of *essences*, that is, a relation between essences."[16] "A law is an essential phenomenon. Hence *law* and *essence* are homogeneous concepts (of the same order), or, to put it strictly, of the same degree, and express a deeper understanding by man of phenomena, the world, etc."[17]

We shall refer to the law of value, as analysed above in a very simplified manner, in order to try to illustrate the concepts introduced above and to answer the question about the inner nature of regularities, to which that law refers. For the sake of simplification we assume that the law of value has the form of (I.2), and hence that it includes only one idealizing assumption. We can then say that the surface structure investigated by Marx includes the following factors: price P, value V, and fluctuations of demand and supply W. In this case the set D_1 has one element only, namely the factor W. By adopting the idealizing assumption which equates demand and supply we eliminate that factor W from our considerations. In this simplified example the inner structure includes two factors only: the price P and the value V (which is the only element of the set D_0 of the most significant determinants of P). As Marx wrote, "... political economy assumes that supply and demand coincide with one enother. Why? To be able to study phenomena in their fundamental relations, in the form corresponding to their conception, that is, to study them independently of the appearances caused by the movement of supply and demand."[18] The regularity referred to by the law of value (I.2) is the dependence of P upon V, represented, according to that law, by a linear function.

The set of concepts introduced above enables us to reconstruct the Marxian concept of the form of manifestation of a regularity. After having fixed, for a k-th-order deep structure (i.e., the inner structure), the idealizational law (I.1) we get from it its strict concretization (I.3). The latter refers to the deep structure of the $k-1$-th order which includes the following factors: F, its most significant determinants $H_1, \ldots, H_n$ and its less significant determinants in D_1, namely $H_{n+1}, \ldots$ $\ldots, H_m$. By (I.3), F depends on $H_1, \ldots, H_n, H_{n+1}, \ldots, H_m$ in the way described by the formula $f' = g(f, h)$. To put it in general terms, the consequent of (I.3) is a transformation of the consequent of (I.1) in accordance with certain additional conditions.[19] Now the dependence f' of F upon $H_1, \ldots, H_n, H_{n+1}, \ldots, H_m$ can be described as a *form of a first-order manifestation* of the regularity formulated in (I.1), that is, the dependence f of F upon $H_1, \ldots, H_n$. Thus, for instance, the dependence of prices P upon the value V and the fluctuations W of demand and supply, as described by (I.4), is a form of (first-order) manifestation of the dependence of prices upon value, as described in (I.2). Fluctuations of demand and supply occur here as the factor which modifies the basic regularity described in (I.2): "the market-value regulates the ratio of supply to demand, or the centre round which fluctuations of supply and demand cause market-prices to oscillate."[20] Hence the price of production, which assumes demand and supply, "is an utterly external and *prima facie* meaningless form of the value of commodities, a form as it appears in competition (...)."[21]

The form of manifestation of the second and higher orders may be defined in an analogous manner. They are described by the successive concretizations of (I.1), and they are dependences of F upon factors in D_0 and D_1 and D_2, D_0 and D_1 and D_2 and D_3, etc. The form of manifestation of the k-th order, that is that dependence of F upon the factors in $D_0, D_1, \ldots, D_k$ which occurs in the surface structure is a *tendency*. It can never be fixed with precision, because it is not possible effectively to take into account the effect of all those factors which disturb the functioning of the factor under consideration. The progress of science rather reveals new factors, which were not taken into account earlier.[22] This is why a strict concretization is usually not carried out to the end by introduction of corrections that would cover all idealizing assumptions; the concretization of an idealizational law is concluded by an approximate concretization. "Under capitalist production, the general law acts as the prevailing tendency only in a very complicated and approximate manner, as a never ascertainable average of ceaseless fluctuations."[23] The treatment of a tendency as the final form of the manifestation of regularities (i.e., the form of the manifestation of regularities in the surface structure, that is, in the sphere of what is empirically given) seems to correspond to the well-known formulations by the founders of the Marxist theory on the relationship between regularity and chance. For Marx, it was the sphere of competition which was the surface structure of economic facts. "But (...) the sphere of competition (...), considered in each individual case, is dominated by chance; (...) the inner law, which prevails in these accidents and regulates them, is only visible when these accidents are grouped together in large numbers (...)."[24] Engels states more or less the same, in general terms, when he says that whenever at the surface we see a play of chance, that chance is always governed by hidden inner laws, and the point is to discover them.[25] As can be seen, tendencies are in any case functions of relationships which are probabilistic in nature. The problem whether that opinion of the founders of the Marxist theory is to be explained by their epistemological or ontological ideas would require a separate study.

As can be seen from the above, the idealization procedure assumes a hierarchization as to significance of the set of those factors which are taken into account as possible determinants of some other factor. That hierarchization is made on the basis of statements in the form:

(II.1) *a factor A is more significant for a factor Z than a factor B is.*

Such statements may be termed *significance hypotheses.* The question arises, how such hypotheses are substantiated (in this paper we cannot treat that issue in detail). Now, substantiation of significance hypotheses may, for instance, be based on experiments, or, in more general terms, on the search for conditions in which one factor does operate while the other ceases to operate. This, perhaps, could explain the scientists' interest in 'strange' cases: they would be those affected by only one of the factors being compared; standard cases are those which are affected by both factors simultaneously. Such cases can be found not only today, but in the remote past as well. This might explain certain references in *Capital* to historical forms of economic activity.[26]

Note also that significance hypotheses are very often adopted on the basis of the philosophical system which a given researcher assumes: for instance, the statement that social stratification based on class division is deeper-reaching than that based on social status is substantiated in the Marxist theory by assumptions which are clearly philosophical in nature.[27] If this were so, the success of an idealizational theory based on specified significance hypotheses adopted on the basis of a philosophical doctrine would indirectly confirm that doctrine. This might be the way of finding some criteria of acceptance of philosophical doctrines and their relationships to specialized disciplines that would be more interesting than vague generalities about the heuristic significance of those doctrines for the development of specialized branches of science.

Laws as substantiated statements. We now proceed to discuss another property of regularities, namely to explain the meaning of the formulation that a regularity is a relationship between apparently incompatible facts. Let us examine those contexts in which Marx referred to the concept of apparent contradiction. In his commentary to one of the necessary consequences of the law of value he says: "This law clearly contradicts all experience based on appearance. (...) For the solution of this apparent contradiction, many intermediate terms are as yet wanted (...)."[28] *Intermediate terms* are for him a synonym of *concretization*. Hence the above formulation means that in reality the magnitude P, referred to in the law of value (I.2) does not equal $f(V)$; in general terms, the magnitude F, referred to in (I.1), does not in reality equal $f(H_1, \ldots, H_n)$. Nevertheless this disagreement with facts is merely apparent: when law (I.2) is concretized, an additional factor is taken into account (in the simplified case analysed here this applies to the fluctuation of demand and supply W), which reduces the actual deviations from (I.2). Should all the corrections be taken into consideration, such deviations would be reduced to an acceptable level. In general terms: the difference between the value of F and that of $f(H_1, \ldots, H_n)$, which can be established empirically, proves 'an apparent contradiction' between those magnitudes, because a concretization of (I.1), and hence the consideration of less significant factors, makes it possible to reduce the difference between the value of F and that of the magnitude on the right side of the consequent of some successive concretization of (I.1) (e.g., the value of $f'(H_1, \ldots, H_n, H_{n+1}, \ldots, H_m)$) to an acceptable level. It can thus be understood why Marx says that the task of science is just to explain how the law of value manifests itself; hence, "should anyone wish to explain in advance those facts which apparently contradict that law, he would have to formulate a science that would be prior to science."[29] In fact, when formulating an idealizational statement we never list all the determinants of the dependent factor, but we merely point to its dependence on the most significant factors. And if this is so, then observations will always show that the magnitude in question takes on (in the surface level that is, at the empirical level) values other than those which follow from that idealizational statement. That, however, does not refute that statement if it can be demonstrated that some of the successive concretizations is in agreement with experience.

That is not understood by vulgar economists, who believe to have made a great discovery if to the revealing of the inner connections they boastingly oppose the fact that the external manifestations are different from those connections. They in fact boast of clinging to appearances, which they take to be something final.[30]

To sum up, the formulation that there is an apparent contradiction between the determined and the determining factor means that an approximate concretization of an idealizational statement which describes those factors is false, but that the said idealizational statement can be strictly concretized so that the resulting statement is a sufficiently good approximation to empirical data (which means that its approximate concretization is true). A law is thus a statement which has been substantiated, its disagreement with empirical data being merely apparent. The type of substantiation is that which follows from the Marxian concept of substantiation and which differs from that which is now adopted in the main trends of the contemporary philosophy of science.[31]

Regularities as necessary connections. Consider now the contexts in which Marx used the term *necessary* as referred to *laws* (i.e., regularities). These contexts show that in Marx's formulations there is a clear connection between the concept of essence (nature) and that of necessity. "It therefore *follows* of itself from the *nature* of the capitalist process of accumulation, which is but one faced of the capitalist production process, that the increased mass of means of production that is to be converted into capital always finds a correspondingly increased, even excessive, exploitable worker population. As the process of production and accumulation advances therefore, the mass of available and appropriated surplus-labour, and hence the absolute mass of profit appropriated by the social capital, *must* grow."[32] (Italics in the text a mine— *L.N.*) Thus a regularity is necessary when it follows from the nature, the essence, of that of which it is a regularity. This applies precisely to regularities, that is, connections between the most significant factors, and not to forms of their manifestations. "The general and necessary tendencies[33] of capital must be distinguished from their forms of manifestation. It is not our intension to consider, here, the way in which the laws, immanent of capitalist production, manifest themselves in the movements of the individual mass of capital (....) (...) But this much is clear: a scientific analysis of competition is not possible, before we have a conception of the inner nature of capital, just as the apparent motions of the heavenly bodies are not intelligible to any but him, who is acquainted with their real motions, motions which are not directly perceptible by the senses."[34] That regularities are necessary, unlike their forms of manifestation, can be seen quite clearly from the following formulation. "The number of labourers employed by capital, (...)the mass of the surplus-value produced by it, and therefore the absolute mass of the profit produced by it, *can*, consequently, increase, (...). And this not only *can* be so. *Aside from temporary fluctuations* it *must* be so, on the basis of capitalist production."[35] (Italics in the text are mine— *L.N.*) Disturbances ('temporary fluctuations') are disregarded, and an idealizational statement is formulated, whose consequent refers to a regularity; that regularity is necessary as it follows from the nature of capitalist production. On the other hand, those regularities which originate merely from a uniform effect of adventitious factors are not necessary, but accidental. "There is no good reason why average conditions of competition, the balance between lender and borrower, should give the lender an interest rate of 3, 4, 5%, etc., or else a certain percentage of the gross profits, say 20% or 50%, on his capital. Whenever it is *competition* which determines anything, the determination is accidental, purely empirical, and only pedantry or fantasy would seek to represent this accident as a necessity."[36] (Italics in the text are mine— *L.N.*)

Consider now that underlies the concept of a regularity, for instance an economic one, which is 'a logical necessity' "proceeding from the nature of the capitalist mode of production."[37] If the interpretation of Marx's concept of essence, as adopted in this paper, is correct, then the numerous formulations of the above kind may be understood thus: a regularity is necessary because the relationship that describes it (i.e., the consequent of an idealizational statement) follows from the *essential body* of the knowledge on the basis of which that idealization is made. At the beginning of this Section we have shown the procedure of introducing idealizing assumptions in object-language formulations, that is, by indicating which factors are removed by the adoption of those assumptions. Statements which describe those factors are left out as a result of the adoption of idealizing assumptions, and form the *scope of abstraction*. The remaining statements, i.e., those which describe the most essential factors form the essential body of knowledge. Let N stand for the initial set of statements, and let the least subset which contains sentences contradicting the assumption that there exists an object fulfilling the idealizing assumption '$p_1(x) = d_1$' be denoted by N^{p_1}; it is removed from N. The remaining part of N will be denoted by N^{+p_1}. By adopting the next idealizing assumption we remove from N other statements thus singling out in N^{+p_1} the set of those statements which are in agreement with both idealizing assumptions, '$p_1(x) = d_1$' and '$p_2(x) = d_2$', namely the subset (N^{+p_1} *and* N^{+p_2}). By proceeding further in this manner we obtain the set of those statements which are in agreement with all the idealizing assumptions made, that is the set (N^{+p_1} *and* ... *and* N^{+p_k}), which we have termed the essential body of knowledge. The factors described by the statements in that set form the set of those factors which are the most significant for the factor under consideration (the dependent variable), and hence are elements of its inner structure. This intuitive formulation may now be worded with much more precision: the statement that a regularity f that holds between a factor F and factors $H_1, \ldots, H_n$ (which are for F the most significant ones) is *necessary* means the same as the statement that $F(x) = f(H_1(x), \ldots, H_n(x))$ follows logically from the set (N^{+p_1} *and* ... *and* N^{+p_k}).

Note that such a criterion ensures the implementation of that basic idea of the founders of the Marxist theory that necessity consists neither in exception-free sequence nor repeatability of events. On the contrary, it follows from that idea of theirs that necessary connections occur in the sphere of facts merely as tendencies. "But in theory it is assumed that the laws of capitalist production operate in their pure form. In reality there exists only approximation, but, this approximation is the greater, the more developed the capitalist mode of production and the less it is adulterated and amalgamated with survivals of former economic conditions."[38] Thus the interpretation of necessity as a constant sequence of events could only lead to the conclusion that there are no necessary relationships. According to Marx, they do exist, with the proviso that at the empirical level a necessary regularity manifests itself "as a tendency, i.e., as a law whose absolute action is checked, retarded, and weakened, by counteracting circumstances."[39]

According to the opinion of the founders of the Marxist theory, not only does not necessity consists in a 'constant sequence of events', but even the study of a constant sequence of events is not a criterion of the occurrence of a regularity. For if that which is necessary is only one of the factors which, taken together, shape that which can be observed (the others being the disturbing factors) then empirical observation alone can never sufficiently prove a necessity.[40] But we can verify idealizational laws by means of concretization. By proceeding from the main factors covered by the law under consideration and by gradually taking into account the disturbing factors we arrive at the explanation of the facts recorded by observation. In this manner we explain deviations from the initial idealizational statement: by pointing to those disturbing factors which account for the said deviations. But it is the direction in which we proceed that is essential: we have to start from the law that describes the operation of the main factors. "It is this law that explains the deviations, and not vice versa, the deviations that explain the law."[41] When we check whether a given connection is necessary in character we are, of course, bound by the principle of substantiation referred to earlier: deviations from a regularity do not refute a law if we can explain them by the intervention of disturbing factors.

The concept of law of science. It follows from the above that some characteristics quoted by Marx as properties of laws pertain to laws of science, i.e., certain statements, while others pertain to regularities. The latter characteristics can also be referred to laws of science by saying, for instance, that the statement that a law of science is inner in nature means the same as the statement that that formulation refers to an inner relationship (in the sense mentioned earlier). The same can be done with the property of necessity. The formulation that a regularity is a connection between apparently incompatible facts can be made to refer to statements in the manner discussed above; that will mean that the laws of science are statements which are substantiated in the Marxian sense of the word. Thus, it may be supposed that a law of science could be defined (presumably in agreement with Marx's opinion) in the following manner: a *law of science* is a strictly general idealizational statement, necessary in character, of an inner nature, and substantiated in the Marxian sense of the word. Factual statement would, accordingly, not be laws: this follows from the fact that the epistemological properties of laws of science, as conceived by Marx, are attributes of idealizational statements only. This sequel of the Marxian interpretation of the concept of law accounts for the fact that the extension of that term is narrower than that current in the contemporary terminological usage, according to which factual statements (at least those which are the final concretizations of idealizational statements) would also be termed laws. But this is why the Marxian concept of law reveals the theoretical heterogeneity of the present-day concept of law of science, which brings together such widely different types of statements as idealizational and factual ones (including the phenomenalistic statements of the type "every raven is black").[42] But according to the ideas advanced by the founders of the Marxist theory newly constructed concepts are not intended to reproduce received concepts, since that would mean a petrification of the vision of the world as embodied in those received concepts.[43]

III. CONCLUSION: THE MARXIAN MODEL OF EXPLANATION

As the present writer strove to demonstrate on another occasion, Karl Marx's methodological reflections include an outline of original ideas concerning the model of explanation and the construction of scientific theories. In Marx's works the schema of explanation is as follows:

$$T^k \dashv T^{k-1} \dashv \dots \dashv T^i \;\backsim\!\!\dashv\; T^0 \wedge P \rightarrow E,$$

where the index j shows the number of idealizing assumptions in the antecedent of T^j $(j = k, fk - 1, \dots, i, 0)$; $\dashv$ stands for the relation of strict concretization; $\backsim\!\!\dashv$ for the relation of approximation; P for the initial conditions of the factual statement T^0; and E for the explanandum. Now for the correctness of a given case of explanation it is necessary that T^k be a law in the sense analysed above.[44] This implies that a fact can be explained only if we point to its essence, as described by a law, and if we show how that essence manifests itself on the empirical level, which in turn is shown by the concretizations of the law involved. This is so because the schema of explanation is isomorphic with the significance structure of the magnitude under investigation. As can be seen, Marx's idea differs entirely from that which underlies both the positivist and the hypothetist model of explanation, namely that we understand something if we know that that something always occurs under given circumstances.

REFERENCES

[1] Marx, K. *Wages, Prices and Profits*, quoted after the Polish-language edition of the collected works of Marx and Engels, Vol. 16, Warszawa 1968, p. 141.

[2] Marx, K. *Capital*, Vol. 3, Moscow 1962, p. 186.

[3] A more comprehensive analysis of that concept and also of the concepts of idealizational law and concretization can be found in L. Nowak, *The Foundations of the Marxian Methodology of Sciences* (in Polish), Warszawa 1971, Chaps. II & III. See also L. Nowak, "Idealization: A Reconstruction of Marx's Ideas", *Poznań Studies in the Philosophy of the Sciences and the Humanities,* Vol. I, no 1, 1975.

[4] The index i may equal $1, 2, ..., k$. Note also that this schema covers both quantitative and qualitative laws (cf. the book quoted in footnote (3)).

[5] Marx, K. *Theories of Surplus Value*, quoted after the Polish-language edition, Part III, Warszawa 1966, p. 98.

[6] Engels, F. the review of Vol. 1 of *Capital*, written for *Demokratisches Wochenblatt*, quoted after the Polish-language edition of the collected works of Marx and Engels, *ed. cit.*, Vol. **16**, p. 259.

[7] Marx, K. *Theories of Surplus Value*, Part II, Moscow 1968, p. 174.

[8] Cf. the works quoted in footnote (3).

[9] Marx, K. *Capital*, *ed. cit.*, Vol. **3**, p. 220.

[10] Strictly speaking, what is involved here is not the function f, which is an ordinary numerical function, but the isomorphically corresponding function f, which holds between those objects which have the appropriate numerical measures assigned to them. This distinction is disregarded here for the sake of simplicity.

[11] Marx, K. *Capital*, *ed. cit.*, Vol. **3**, p. 307.

[12] Marx, K. *Capital*, *ed. cit.*, Vol. **3**, p. 797.

[13] *Ibid.*

[14] Marx, K. *Theories of Surplus Value*, Part II, *ed. cit.*, p. 165.

[15] Marx, K. *Capital*, Vol. **3**, *ed. cit.*, pp. 355–6.

[16] Quoted after the Polish-language edition of V. Lenin's *Philosophical Notebooks*, Warszawa 1956, p. 127.

[17] *Ut supra*, p. 126.

[18] K. Marx, *Capital*, Vol. **3**, *ed. cit.*, p. 186.
[19] For a more comprehensive discussion of the issue see the works quoted in footnote (3).
[20] Marx, K. *Capital*, Vol. **3**, *ed. cit.*, p. 178.
[21] *Ut supra*, p. 194.
[22] Nowakowa I. "Idealization and the Problem of Correspondence", *Poznań Studies in the Philosophy of the Sciences and the Humanities, Vol. I, no 1, 1975.*
[23] Marx, K. *Capital*, Vol. **3**, *ed. cit.*, p. 159.
[24] *Ut supra*, p. 807.
[25] Quoted after F. Engels, *Ludwig Feuerbach and the Decline of Classical German Philosophy* in the Polish-language edition of the selected works of Marx and Engels, Vol. 2, Warszawa 1949, p. 372.
[26] Kmita, J. 'Marxism in the Face of the Controversy: Realism Wersus Instrumentalism' (in Polish), *Studia Metodologiczne*, No. **8**, p. 34.
[27] Jasińska A. & L. Nowak, 'Methodological Foundations of Marxian Theory of Classes' in: D. H. DeGrood (Ed.), *East-West Dialogues*, Amsterdam, 1973.
[28] Marx, K. *Capital*, Vol. **1**, Moscow 1954, p. 307.
[29] Marx, K. Letter to Kugelmann of July 8, 1868, quoted after Marx's and Engels's Letters on *Capital*, in the Polish-langue edition, Warszawa 1957.
[30] *Ut supra*, p. 189.
[31] See Chap. XI of the book quoted in footnote (3).
[32] Marx, K. *Capital*, Vol. 3, ed. cit., p. 214.
[33] Marx sometimes used the terms *law* and *tendency* interchangeably, although—as can be seen from his formulations quoted in this paper—the prevailing meaning of the term *tendency* was that which he defined in the passage quoted in this paper on. p. 409.
[34] Marx, K. *Capital*, Vol. 1, *ed. cit.*, p. 316.
[35] Marx, K. *Capital*, Vol. 3, *ed. cit.*, p. 213.
[36] *Ut supra*, p. 356.
[37] *Ut supra*, p. 209.
[38] *Ut supra*, p. 172.
[39] *Ut supra*, p. 229.
[40] Engels, F. *The Dialectics of Nature*, quoted after the Polish-language edition, Warszawa 1953, p. 239.
[41] Marx, K. *Capital*, Vol. **3**, *ed. cit.*, p. 184.
[42] This applies also to the concept of law which the present writer used in the works quoted in footnote (3).
[43] L. Cf. Nowak, 'The Model of the Empirical Sciences as Seen by the Founders of the Marxist Theory' in: (eds.) M. Przełęcki, K. Szaniawski, R. Wójcicki, *25 Years of Logical Methodology in Poland,* Reidel (in print).
[44] See the works quoted in footnote (3).

K. SAMOCHWALOW

THE IMPOSSIBILITY THEOREM FOR UNIVERSAL THEORY OF PREDICTION*

INTRODUCTION

The problem of construction and justification of an empirical prediction theory (perhaps, it is the main aspect of the 'problem of induction') is investigated in the present paper. One fundamental obstacle arises here: the statement S "A prediction theory f_1 *will* predict *in the future* more successfully than a prediction theory f_2" is, generally speaking, meaningless because there is no method of verification or falsification of this statement. Neither a deduction from formerly established facts nor a distinct empirical observation is such a method. So we deal with the pseudoproblem here.

All the same, in spite of the opinion of a large number of authors, there is some hope to advance the solution of induction problem; there is (at least *prima facie*) the following possibility to justify an empirical prediction theory and (nevertheless) to escape the above pseudoproblem. We must take into account that in fact we wish to construct a prediction theory having features of two different kinds. The features of the first kind (the I-features) correspond to universal applicability, non-triviality, and coherence (quarantee against appearance of any alternative results of applications) of the desirable theory. The features of the second kind (the II-features) should provide truth of predictions accomplished with the help of the prediction theory. The problem of establishing these II-features includes obviously the same pseudoproblem as the statement S. Therefore this problem must be neglected as pseudoproblem too. But the I-features remain in force! It may well be that they are so strong that for any theory f, having the I-features, we can establish the following disjunction: the result of arbitrary application of the theory f is *either* the result of the same (i.e. under the same initial circumstances) application of certain fixed (independently of f) theory f^* *or* trivial result being constituent of initial information for considered application and being, therefore, uninteresting. If so, then cognizance of f solves the problem of justification of universal, non-trivial, and coherent prediction theory f because all interesting results of applications of *any such* f become *a priori* known at this rate. And no pseudoproblem arises here because these interesting results are determinated independently of any attempt to provide the truth of them.

The present paper is an attempt to investigate the possibility of justification of empirical prediction theory. It is natural that a considerable part of this investigation is devoted to elucidating a number of concepts from methodology of empirical sciences.

I. EMPIRICAL HYPOTHESIS

We proceed from the conviction that without essential restriction of generality a distinct act of empirical prediction may be considered as follows. Originally there is a protocol (a record) pr_0 of experiment over finite set D_0 of empirical objects. This protocol is considered as a mere registration of results of interaction of these objects with instruments used. Generally speaking, there also is an empirical hypothesis h_0 affirming that certain protocols can never be obtained, if h_0 is true, and if the experiments are made with the given instruments over any finite sets of objects (and not only the set D_0). The hypothesis h_0 may be regarded as a description of supposed characteristics of measuring instruments, and the protocol pr_0 may be regarded as a record of results of measurements of elements of D_0 carried out with these instruments. The hypothesis h_0 is supposed to be such that the protocol pr_0, corresponding to the experiment made over the set D_0, is admissible relative to this hypothesis (i.e. the hypothesis h_0 conforms to the protocol pr_0). Otherwise we must state that the hypothesis is refuted by this experiment and it must be revised as a wrong initial information. This is precisely the case when our initial assumptions concerning characterristics of measuring instruments are wrong.

A distinct act of prediction, say $\langle h_0, pr_0, D_0\rangle \to h_1$ is the one that starts from the initial hypothesis h_0 and using information involved in the protocol pr_0, concerning elements of D_0, leads to a new hypothesis h_1 such that:

(i) h_1 is in a sense more (or at least not less) informative than h_0;
(ii) pr_0 is admissible relative to h_1.

The act of prediction is considered as successful (or true) till a new set (of empirical objects) is found such that the protocol of experiment of the kind mentioned above over this set is not admissible relative to h_1 and is admissible relative to h_0. It is an obvious, *trivial* act of prediction, i.e. any act of the type $\langle h_0, pr_0, D_0 \rangle \to h_0$, is always successful in this sense, being completely uninteresting.

So far, an individual act of prediction has been concerned. As for the *theory of prediction*, it is natural to consider it as a function f of the type:

$$f(\langle h_0, pr_0, D_0 \rangle) = h_1$$

Of course, certain restrictions that arise from our intention to impart certain desirable features to prediction theory should be imposed upon this function f.

Let us introduce or specify a number of concepts that will enable us to express our ideas definitely enough from now on.

First of all let us specify the concept of empirical hypothesis. It is clear that any such hypothesis is empirically meaningful only if there is a realizable in principle method of its experimental examination. Otherwise the acceptability or non-acceptability of the hypothesis is not determinated by any observations and the hypothesis proves to be empirically meaningless. Therefore, we identify an *empirical hypothesis* h with ordered ternary $\langle v, I^v, T^v \rangle$,

$$h = \langle v, I^v, T^v \rangle,$$

where:

1) v is a finite non-empty set of symbols of the type $P_i^{m_i}$ e.g.: $v = \{P_1^{m_1}, \ldots, P_k^{m_k}\}$. Let us call the set v a *vocabulary* (of hypothesis h), and the symbol $P_i^{m_1}$ a *predicate of degree* m_i.

2) I^v is a certain set too. It has the same cardinal number as v. Each element of I^v, say $p_i^{m_i}$, is a concrete way of carrying out observations that secures, given an arbitrary non-empty finite set, say D, of some

sort of objects, observation of 3-valued m_i-ary relation $P_i^D(x_1,\dots,x_{m_i})$ defined on this set D.[1] Each element $P_i^{m_i}$ of v is considered as the name of the corresponding element $\mathbf{P}_i^{m_i}$ of I^v. In other words, to fix a concrete set I^v means to provide the existence of a concrete (although, generally speaking, unknown) one-valued function O_{I^v} that assigns some model (with 3-valued relations) $M^{v,D}$, having the signature v and the domain D, to every finite non-empty set D of arbitrary empirical objects:

$$O_{I^v}(D) = M^{v,D}.$$

It is supposed that every (i.e. for every D) originally unknown value $O_{I^v}(D)$ of the function O_{I^v} can be established *a posteriori*—by observing the objects of D through the use of elements $p_i^{m_i}$ of I^v. The purpose of these elements consists wholly in securing our ability to carry out observations of this kind.

Let us call the way of carying out observations $p_i^{m_i}$ the *intensional procedure* and the set I^v the *intensional basis* (in vocabulary v).[2]

The concept of instruments intended for making experiments examining an empirical hypothesis is closely connected with the concept of intensional basis of this hypothesis: these instruments are considered as means to fix an intensional basis. Having chosen some sort of instruments, i.e. having chosen some sort of things and having made arrangements about the use of these things for realization of some sort of observations, we thereby get: (1) a certain (although, generally speaking, unknown) function O_{I^v} of the kind mentioned above[3], and, thus, (2) a fixed intensional basis I^v, obligatory for any empirically meaningful hypothesis concerning *these* things and *this* arrangement (i.e. concerning the possible observations *of these* things when using them according to *this* arrangement).

3) T^v is a certain algorithm. Let us call T^v a *test algorithm* (for a hypothesis *h*) *in a vocabulary v*. A few concepts must be specified in order to describe it. Let us call a symbol expression of the form $P_i^{m_i}a_1 \dots \dots a_{m_i}$, or $\bar{P}_i^{m_i}a_1 \dots a_{m_i}$, $\tilde{P}_i^{m_i}a_1 \dots a_{m_i}$, where $P_i^{m_i} \in v$ and $a_1, \dots, a_{m_i}$ are the symbols (not obligatorily all different) of some fixed enumerable alphabet α, an *elementary sentence* (*in the vocabulary v*).

Let β be an arbitrary finite (non-empty) subset of the set α. Let an arbitrary non-empty finite set pr^v of elementary sentences in the vocabulary v be a *protocol* (*in the vocabulary v*) iff for some β this set satisfies the following conditions:

a) for each $P_i^{m_i} \in v$ and for any $a_1, \ldots, a_{m_i} \in \beta$, either $P_i^{m_i} a_1, \ldots, a_{m_i} \in pr^v$, or $\overline{P}_i^{m_i} a_1, \ldots, a_{m_i} \in pr^v$, or $\tilde{P}_i^{m_i} a_1, \ldots, a_{m_i} \in pr^v$;

b) for each $P_i^{m_i} \in v$ and for any $a_1, \ldots, a_{m_i} \in \beta$, if $P_i^{m_i} a_1 \ldots a_{m_i} \in pr^v$, then $\overline{P}_i^{m_i} a_1, \ldots, a_{m_i} \notin pr^v$, $\tilde{P}_i^{m_i} a_1, \ldots, a_{m_i} \notin pr^v$, and if $\overline{P}_i^{m_i} a_1, \ldots, a_{m_i} \in pr^v$, then $P_i^{m_i} a_1, \ldots, a_{m_i} \notin pr^v$, $P_i^{m_i} a_1, \ldots, a_{m_i} \notin pr^v$, and if $\tilde{P}_i^{m_i} a_1, \ldots, a_{m_i} \in pr^v$, then $P_i^{m_i} a_1, \ldots, a_{m_i} \notin pr^v$, $\overline{P}_i^{m_i} a_1, \ldots, a_{m_i} \notin pr^v$.

Let pr^v be an arbitrary protocol and let β be precisely that subset of alphabet α relative to which the protocol pr^v satisfies the conditions a) and *b*) given above. Then we shall often write $B(pr^v)$ instead of β and call the set $B(pr^v)$ a *basis of the protocol* pr^v.

Let us call the cardinal number $\overline{\overline{B}}(pr^v)$ of the set $B(pr^v)$ a *power of the protocol* pr^v.

Let protocols pr_1^v and pr_2^v be called *isomorphic* (symbolically, $pr_1^v \simeq pr_2^v$) iff they can be made identical through one-to-one renaming of the elements of their bases.

Let us return to the definition of a test algorithm. *Test algorithm* (*in the vocabulary* v) is an arbitrary algorithm T^v that satisfies the following conditions:

(i) T^v is applicable to any protocol in the vocabulary v, and for every such protocol pr^v this algorithm delivers one of the two possible values 0 and 1: either $T^v(pr^v) = 1$, or $T^v(pr^v) = 0$;

(ii) for any two protocols pr_1^v and pr_2^v in the vocabulary v if $pr_1^v \simeq pr_2^v$, then $T^v(pr_1^v) = T(pr_2^v)$;

(iii) for every natural number $n \geqslant 1$ there exists a protocol pr^v in the vocabulary v, such that $\overline{\overline{B}}(pr^v) = n$ and $T^v(pr^v) = 1$.

The meaning of the hypothesis $h = \langle v, I^v, T^v \rangle$ is defined by certain convention concerning the way of experimental examination of h. This way consists in the following.

First of all an arbitrary non-empty set D of some sort of objects is fixed. Let D be the set $\{\mathbf{a}_1, \ldots, \mathbf{a}_n\}$. Elements of the set D are one-one denoted by arbitrary symbols of α. Let β be the set $\{a_1, \ldots, a_n\}$ of all these names. Then 3-valued relations $P_1^D(x_1, \ldots, x_{m_1}), \ldots, P_k^D(x_1, \ldots, x_{m_k})$ are observed with the help of intensional procedures $\mathbf{P}_1^{m_1}, \ldots, \mathbf{P}_k^{m_k}$ of I^v on the set D; i.e. for the given D the value $O_{I^v}(D)$ of the function O_{I^v} is found. Let the model $M^{v,D}$ be this value: $O_{I^v}(D) = M^{v,D}$. The diagram of the model $M^{v,D}$ is recorded in the form of the corresponding protocol pr^v having the vocabulary v and the basis $B(pr^v) = \beta$. That is, for each $P_i^D(x_1, \ldots, x_{m_i})$, $i = 1, \ldots, k$, and for any m_i-tuple $\langle \mathbf{a}_{j1}, \ldots$ $\ldots, \mathbf{a}_{jm_i}\rangle$ of objects of D, an elementary sentence $P_i^{m_i}a_{j1} \ldots a_{jm_i}$ (or $\overline{P}_i^{m_i}a_{j1} \ldots a_{jm_i}$, or $\tilde{P}_i^{m_i}a_{j1} \ldots a_{jm_i}$) is written iff $P_i^D(\mathbf{a}_{j1}, \ldots, \mathbf{a}_{jm_i}) = \text{true}$ (or $P_i^D(\mathbf{a}_{j1}, \ldots, \mathbf{a}_{jm_i}) = \text{false}$, or $P_i^D(\mathbf{a}_{j1}, \ldots, \mathbf{a}_{jm_i}) = \text{meaningless}$, respectively). Then the algorithm T^v is applied to the obtained protocol pr^v (it is always possible because of (i)). If $T^v(pr^v) = 0$, then the hypothesis h is considered to be refuted by the given (over the given set D) experiment. If $T^v(pr^v) = 1$, then the hypothesis h is considered to be confirmed (but not proved) by the given experiment.

The fact that T^v is an algorithm, i.e. an effective function, defined on the set of all protocols in the vocabulary v, guarantees the finite (hencefore, empirical) feasibility of any (for every D) such distinct act of examination of the hypothesis.

By virtue of (ii) the fact of refutation (or confirmation) of the hypothesis does not depend on the way of designation of objects of D. And by virtue of (iii) the fact of refutation cannot be established a priori. Thus, the conditions (ii) and (iii) exclude nonintersubjective and/or a priori false hypotheses from the consideration.

Moreover, our reasonings conform to the "principle of falsification": any empirical hypothesis h can be refuted by a single experiment, but it can never be proved once and for all.

Let I^v be an arbitrary fixed intensional basis $\{\mathbf{P}_1^{m_1}, \ldots, \mathbf{P}_k^{m_k}\}$ in the vocabulary $v = \{P_1^{m_1}, \ldots, P_k^{m_k}\}$. Let w be an arbitrary vocabulary $\{R_1^{s_1}, \ldots, R_L^{s_L}\}$. Let F be an algorithm, such that:

a) F is applicable to every protocol pr^v in the vocabulary v;

b) the value $F(pr^v)$ of F for the given protocol pr^v is a protocol pr^w in the vocabulary w;

c) for every pr^v, $B(pr^v) = B(F(pr^v))$;

d) for every pr_1^v and for every pr_2^v, if $pr_1^v \simeq pr_2^v$ then $F(pr_1^v) \not\simeq F(pr_2^v)$;

e) for every pr_1^v and for every pr_2^v if $pr_1^v \not\simeq pr_2^v$, then $F(pr_1^v) \not\simeq F(pr_2^v)$.[4]

If some algorithm, say F_v^w, meets the requirements a), b), c), d), e) for some v and some w, then we shall often write: $F_v^w \in \mathscr{S}$. Class $\mathscr{S}$ is non-empty (see [1]).

For a fixed I^v and a fixed $F_v^w \in \mathcal{S}$, the definite intensional basis $F_v^w I^v$ in the vocabulary w, say $F_v^w I^v = \{r_1^{s_1}, \dots, r_L^{s_L}\}$, is fixed by the couple $\langle I^v, F_v^w \rangle$ at the cost of the following convention. We say that an arbitrary protocol pr^w in the vocabulary w registers results of observation of an arbitrary set D *by procedures* $r_1^{s_1}, \dots, r_L^{s_L}$ iff $pr^w = F_v^w(pr^v)$ and pr^v is a protocol (in the vocabulary v) registering results of observation of the same set D by procedures $p_1^{m_1}, \dots, p_R^{m_k}$. Again, the fact that F_v^w is an algorithm meeting the requirements a), b), c), provides the empirical existence of an intensional basis $F_v^w I^v$, given that the intensional basis I^v exists empirically. The condition d) exludes dependence of the results of observation of any set D by procedures of $F_v^w I^v$ on the way of designation of elements of D. The purpose of the condition e) will be clear below.

Let us call any such intensional basis $F_v^w I^v$ a *non-creative F_v^w-modification of the basis I^v.*

Let $h_1 = \langle v, I^v, T^v \rangle$ and $h_2 = \langle w, I^w, T^w \rangle$ be arbitrary empirical hypotheses. Let us call the hypothesis h_2 a *non-creative F_v^w-modification of the hypothesis* h_1 iff there exists an algorithm $F_v^w \in \mathcal{S}$, such that:

(i) I^w is a non-creative F_v^w-modification of I^v (in symbols, $I^w = F_v^w I^v$);

(ii) for every protocol pr^w in the vocabulary w, the test algorithm T^w meets the following requirement:

$$(1.1) \qquad T^w(pr^w) = \begin{cases} 1 & \textit{if there exists a protocol } pr^v \textit{ in the vocabulary } v \\ & \textit{such that } pr^w \approx F_v^w(pr^v) \textit{ and } T^v(pr^v) = 1; \\ 0 & \textit{otherwise} \end{cases}$$

Let us write $h_2 = F_v^w h_1$ iff h_2 is a non-creative F_v^w-modification of h_1, and let us write $T^w = F_v^w T^v$ iff the equality (1.1) holds. It is clear that for any $h = \langle v, I^v, T^v \rangle$ and for any $F_v^w \in \mathcal{S}$ there exists (one and only one) non-creative F_v^w-modification $F_v^w h$ of h, namely:

$$F_v^w h = \langle w, F_v^w I^v, F_v^w T^v \rangle.$$

On the other hand, if $h_2 = F_v^w h_1$, then the hypotheses h_1 and h_2 are refuted or confirmed simultaneously; i.e. any experiment examining h_1 over an arbitrary set D refutes or confirms the hypothesis h_1 iff (here we appeal in particular to condition *e*) for F_v^w) the corresponding examining experiment over the same set D refutes or, respectively, confirms the hypothesis h_2.

For these reasons we consider that the following methodological principle MP takes place:

Any hypothesis h_1 (having the vocabulary v, for example) is acceptable iff every (i.e. for every $F_v^w \in \mathcal{S}$) non-creative F_v^w-modification of this h_1 is acceptable, too.

II. POSTULATES FOR THE PREDICTION THEORY

Requirements imposed on the function f are justified, as we have already said, by our intension to regard this function as a prediction theory, having definite desirable features. Let us give an exposition of the corresponding postulates for f.

Let us call $\langle T^v, pr^v \rangle$ an *admissible pair*, iff T^v is a test algorithm and pr^v is a protocol, such that $T^v(pr^v) = 1$. Let π be the set of all possible admissible pairs, and let τ be the set of all possible test algorithms.

We want prediction theory to be universally applicable in empirical researches. In exact terms it means the following requirement RI.

The function f must be such that for arbitrary pr_0^v, D_0 and $h_0 = \langle v, I^v, T_0^v \rangle$ if $T_0^v(pr_0^v) = 1$ and if pr_0^v is a record of results of observation of D_0 with the help of I^v, then

$$(1.2) \qquad f(\langle h_0, pr_0^v, D_0 \rangle) = \langle v, I^v, A_f(\langle T_0^v, pr_0^v \rangle) \rangle,$$

where A_f is a certain one-valued mapping from π into τ, defined (independently of v, D_0, and I^v) on the whole π so that if $A_f(\langle T_0^v, pr_0^v \rangle) = T_1^{v_1}$, then $v = v_1$.

Then we want the result of every application of the theory to be conformed to the initial evidence upon which any such application is rested. That is to say, we consider that the function A_f must satisfy the condition R2:

for any $\langle T_0^v, pr_0^v \rangle \in \pi$ and for any $T_1^v \in \tau$, if $A_f(\langle T_0^v, pr_0^v \rangle) = T_1^v$, then $T_1^v(pr_0^v) = 1$.

The following requirement, R3, expresses non-triviality of supposed prediction theory.

(i) *For any $\langle T_0^v, pr_0^v \rangle \in \pi$, $T_1^v \in \tau$, and pr^v, if $A_f(\langle T_0^v, pr_0^v \rangle) = T_1^v$ and $T_0^v(pr^v) = 0$, then $T_1^v(pr^v) = 0$*

(ii) *There exist such* $\langle T_0^v, pr_0^v\rangle \in \pi$ *and such* pr^v *that for every* $T_1^v \in \tau$, *if* $A_f(\langle T_0^v, pr_0^v\rangle) = T_1^v$, *then* $T_0^v(pr^v) = 1$ *and* $T_1^v(pr^v) = 0$.

The following requirement R4 expresses a certain aspect of coherence of prediction theory:

for arbitrary $\langle T^v, pr^v\rangle \in \pi$, $F_v^w \in \mathscr{S}$,

$$(2.2) \qquad F_v^w A_f(\langle T^v, pr^v\rangle) = A_f(\langle F_v^w T^v, F_v^w(pr)\rangle).$$

If the function A_f does not satisfy R4, then there exist such admissible pair $\langle T_0^v, pr_0^v\rangle$ and such algorithm $^*F_v^w \in \mathscr{S}$ that

$$(3.2) \qquad {}^*F_v^w A_f(\langle T_0^v, pr_0^v\rangle) \neq A_f(\langle {}^*F_v^w T_0^v, {}^*F_v^w(pr_0^v)\rangle).$$

What does this inequality mean? It may well be that in some sort of empirical research we shall have the ternary $\langle\langle v, I^v, T_0^v\rangle, pr_0^v, D_0\rangle$ as initial information for an application of the theory f. In accord with R1 the result of such application is a hypothesis h_1, such that

$$(4.2) \qquad h_1 = f(\langle\langle v, I^v, T_0^v\rangle, pr_0^v, D_0\rangle) = \langle v, I^v, A_f(\langle T_0^v, pr_0^v\rangle)\rangle.$$

By virtue of MP and of R1, accepting the ternary $\langle\langle v, I^v, T_0^v\rangle, pr_0^v, D_0\rangle$ as an appropriate initial information for an application of the theory f, we must accept the ternary $\langle w, {}^*F_v^w I^v, {}^*F_v^w T_0^v\rangle, {}^*F_v^w(pr_0^v), D_0\rangle$ as an appropriate initial information for an application of f, too. As a consequence of this latter application we shall have a hypothesis h_1^* such that

$$(5.2) \qquad h_1^* = \langle w, {}^*F_v^w I^v, A_f(\langle {}^*F_v^w T_0^v, {}^*F_v^w(pr_0^v)\rangle)\rangle.$$

Let us consider the non-creative $^*F_v^w$-modification of the hypothesis h_1 i.e., the hypothesis h_2 such that

$$(6.2) \qquad h_2 = {}^*F_v^w h_1 = \langle w, {}^*F_v^w I^v, {}^*F_v^w A_f(\langle T_0^v, pr_0^v\rangle)\rangle.$$

Again in accord with MP the hypotheses h_1 and h_2 are acceptable simultaneously. On the other hand, the hypotheses h_1 and h_1^* are acceptable simultaneously too because they are the results of applications of the theory f to the simultaneously acceptable initial data $\langle\langle v, I^v, T_0^v\rangle, pr_0^v, D_0\rangle$ and $\langle\langle w, {}^*F_v^w I^v, {}^*F_v^w T_0^v\rangle, {}^*F_v^w(pr_0^v), D_0\rangle$. So the hypotheses h_1^* and h_2 are acceptable simultaneously, too.

But this situation is fraught with certain alternatives, given that inequality (3.2) holds. Obviously, at this rate there exists such protocol pr_1^w in the vocabulary w that

$$(7.2) \qquad T_1^w(pr_1^w) \neq T_2^w(pr_1^w),$$

where $T_1^w = A_f(\langle {}^*F_v^w T_0^v, {}^*F_v^w(pr_0^v)\rangle)$ and $T_2^w = {}^*F_v^w A_f(\langle T_0^v, pr_0^v\rangle)$ Let us suppose that

$$(7.2') \qquad T_1^w(pr_1^w) = 0 \quad \textit{and} \quad (7.2'') \qquad T_2^w(pr_1^w) = 1$$

Let n be the power of the protocol pr_1^w: $n = \overline{\overline{B}}(pr_1^w)$. Let D be an arbitrary set (of arbitrary objects) having n as its cardinal number. By virtue of our conventions and by suposition (7.2'), accepting the hypothesis h_1^*, we assert in particular that *it is not possible* that results of observation of the set D with the help of the intensional basis ${}^*F_v^w I^v$ are described by the protocol which is isomorphic with the protocol pr_1^w. But by virtue of the same conventions and by supposition (7.2''), accepting the hypothesis h_2, we assert in particular that *it is possible* that the same results are described by the protocol which is isomorphic with the protocol pr_1^w. Thus, alternative is available.

In the same fashion we get a similar alternative again if we suppose that the equalities $T_1^w(pr_1^w) = 1$ and $T_2^w(pr_1^w) = 0$ take place instead of (7.2') and (7.2''), respectively.

Hencefore there is no guarantee against appearance of alternatives when applying the theory f, if the inequality (7.2), and thereby the inequality (3.2), take place. But it means in turn that the requirement R4 is one of the necessary conditions for coherence of the theory f.

One more necessary condition for coherence of the theory is the one expressed by the requirement R5 below.

For any vocabulary v, let v' be an arbitrary vocabulary such that

(i) if $v = \{P_1^{m_1}, \ldots, P_k^{m_k}\}$, then $v' = \{Q_1^{m_1}, \ldots, Q_k^{m_k}\}$ and

(ii) $v \cap v' = \emptyset$

For any protocol pr^v in the vocabulary v, let ${}^*pr^{v'}$ be a protocol in the vocabulary v' which is obtained from the protocol pr^v by means of substitution of every occurence of the symbol $P_i^{m_i}$ in the protocol pr^v by occurence of the corresponding symbol $Q_i^{m_i}$.

For any test algorithm T^v (in the vocabulary v) let ${}_*T^{v'}$ be a test algorithm (in the vocabulary v'), such that for any pr^v and for any $pr^{v'}$,

$T^v(pr^v) = {}^*T^{v'}(pr^{v'})$ iff $pr^{v'} \approx {}_*pr^{v'}$.

It is clear that the following requirement R5 is a necessary condition for coherence of the theory f.

R5: *For any* $\langle T^v, pr^v\rangle \in \pi$, $T_1^v \in \tau$, *and* v', *if* $A_f(\langle T^v, pr^v\rangle) = T_1^v$, *then* $A_f(\langle {}^*T^{v'}, {}_*pr^{v'}\rangle) = {}^*T_1^{v'}$.

III. IMPOSSIBILITY OF UNIVERSAL, NON-TRIVIAL, COHERENT, AND USEFUL PREDICTION THEORY

In the previous section we have formulated five requirements R1-R5 imposed on the function f. These requirements in no way appeal to attempts to guarantee truth of predictions accomplished with the help of f, and, hencefore, they are not bound up with the pseudoproblem mentioned in Introduction. On the other hand, below we intend to show that these requirements are strong enough so that any prediction theory, say f, that meets them, in every concrete instance of its application yields either a trivial result or a result which can also be obtained by applying (under the same circumstances) certain in advance fixed theory f^*.

Let us consider a function A, defined on the whole class π by the equality

$$(1.3) \qquad A(\langle T_0^v, pr_0^v\rangle) = |T_0^v pr_0^v|,$$

where $\langle T_0^v, pr_0^v\rangle$ is an arbitrary element of π, and $|T_0^v pr_0^v|$ is a function defined for every protocol pr^v in the vocabulary v as follows:

$$|T_0^v pr_0^v|(pr^v) = \begin{cases} 0 & \textit{if } \overline{\overline{B}}(pr^v) = \overline{\overline{B}}(pr_0^v), \textit{ and protocols } pr^v, \\ & pr_0^v \textit{ are not isomorphic;} \\ T_0^v(pr^v) & \textit{otherwise.} \end{cases}$$

Theorem. For every pair $\langle T_0^v, pr_0^v\rangle \in \pi$ and for every function A_f that meets requirements R1-R5, either

$$A_f(\langle T_0^v, pr_0^v\rangle) = A(\langle T_0^v, pr_0^v\rangle) = |T_0^v pr_0^v|$$

or

$$A_f(\langle T_0^v, pr_0^v\rangle) = T_0^v.$$

Proof. (See [1])

Now let us consider a function f^* defined by the condition: for arbitrary $h_0 = \langle v, I^v, T_0^v \rangle$, pr_0^v, D_0, if $T_0^v(pr_0^v) = 1$ and if pr_0^v is a record of results of observation of objects of D_0 with the help of I^v, then

$$f^*(\langle h_0, pr_0^v, D_0 \rangle) = \langle v, I^v, A(\langle T_0^v, pr_0^v \rangle) \rangle, \tag{2.3}$$

where the function A satisfies the equality (1.3).[5] In accord with Theorem, this function f^* has the following feature. For above arbitrary h_0, pr_0^v, D_0, and for an arbitrary prediction theory f, that meets requirements R1-R5, either

$$f(\langle h_0, pr_0^v, D_0 \rangle) = f^*(\langle h_0, pr_0^v, D_0 \rangle) \tag{3.3}$$

or

$$f(\langle h_0, pr_0^v, D_0 \rangle) = h_0 \tag{4.3}$$

The equality (4.3) corresponds to the trivial result of application of the prediction theory f (see p. 421). Thus, we have the following:

Basic Conclusion (BC). Any universal, non-trivial, and coherent prediction theory in every concrete instance of its application yields either a trivial result or a result that can be obtained by applying the function f^*, defined by (2.3).

Of course, we have not proved that the class $\mathscr{F}$ of universal, non-trivial, and coherent prediction theories is non-empty. (We have not even proved that the function f^* by itself satisfies necessary conditions R1–R5 for membership in $\mathscr{F}$). But, as a matter of fact, such proof is not promising. More specifically, let us suppose the class $\mathscr{F}$ to be non-empty. Then, by virtue of (BC), it does not contain any theory, which will be capable of realizing some non-trivial prediction that is inaccessible for f^*. On the other hand, the class of predictions corresponding to f^* is so narrow that the function f^* seems to be practically useless in empirical researches. At any rate, one can re-discover none of the well-known empirical laws by using only the function f^*, if one claims that:

a) any well-known empirical law is some empirical hypothesis $h = \langle v, I^v, T^v \rangle$ such that for some natural number n_0 and for every natural number $n \geqslant n_0$ there exists such protocol pr^v that $\overline{\overline{B}}(pr^v) = n$ and $T^v(pr^v) = 0$.[6]

b) discovery of an empirical law is, in the last analysis, an individual act of prediction $\langle h_0, pr_0, D_0 \rangle \rightarrow h$ such that the test algorithm T_0^v

(for the hypothesis h_0) meets the following condition: $T_0^v(pr^v) = 1$ for every pr^v.[7] With these reservations we pass from (BC) to

Affirmation (A). There exists no prediction theory that is universal, non-trivial, coherent, and useful simultaneously.

IV. CONCLUSION

If we take (A) for granted then, thereby, we submit for consideration the following (in particular) dilemma. Either any attempts to guess (it is already prohibited to say 'to justify') a universal, non-trivial, and useful prediction theory are *a priori* hopeless or these attempts should be conducted at the cost of violation of coherence of the expected prediction theory. The latter looks more doubtful (but more interesting) than the former. At any rate we must keep in sight that if some day and somehow one guesses a non-trivial, universal, and useful prediction theory, then while applied, this theory will not be guaranteed against appearance of alternatives (of kind mentioned above).

This viewpoint is not new. In fact a near standpoint is, for instance, that of H. Smokler and M. D. Rohr when they postulate non-invariantness of induction relative to translations from one language into another equivalent language (see [3]).

BIBLIOGRAPHY

[1] Самохвалов К. Ф., 'О теории эмпирических предсказаний', *Вычислительные системы*, 1973, вып. 55, 3—35, Новосибирск.

[2] Tichy, P., 'Intension in terms of Turing machines', *Studia Logica*, 1969, t. XXIV.

[3] Smokler, H., Rohr, M. D., 'Confirmation and Translation', in: *Philosophical Logic* (ed. I. W. Davis, D. I. Hockney, and W. K. Wilson), D. Reidel Publish Company, Dordrecht, 1969, pp. 172–180.

REFERENCES

* This paper is a version of [1].

[1] By 3-valued m-ary relation $P^X(x_1, ..., x_m)$ defined on the set X we mean a function of m arguments having the set X and the set {true, false, meaningless} as its domain and its range, respectively.

[2] The term has been borrowed (with some modification of its meaning) from [2].

[3] Of course, the existence of such function is provided for an *arbitrary* choice of things and for an *arbitrary* arrangement about the use of them, at the cost of the

use of 3-valued relations (with supplementary value 'meaningless') instead of usual 2-valued ones.

[4] '$x \not\cong y$' is abbreviation for 'x, y are not isomorphic'.

[5] Let us note that, obviously, for every $\langle T_v^0, pr_0^v \rangle \in \pi$, $|T_0^v pr_0^v|$ (i.e. $A(\langle T_0^v, pr_0^v \rangle)$) is a test algorithm in the vocabulary v by virtue of the definition of $|T_0^v pr_v^0|$.

[6] It is resonable by virtue of the factual history of natural sciences: in fact, all well--known laws admit such re-formulation.

[7] Otherwise the initial hypothesis h_0 by itself should be previously discovered, and so on.

HÅKAN TÖRNEBOHM

SCIENTIFIC KNOWLEDGE-FORMATION

ABSTRACT. Two kinds of scientific knowledge-formation are distinguished:
I. Piecemeal knowledge-formation, and
II. Synthesizing knowledge-formation (in physics = theory formation).

I. PIECEMEAL KNOWLEDGE-FORMATION

Schematic description.

Piecemeal knowledge-formation is a string of operations.

1. A question is raised.
2. One or more tentative answers are proposed. Hypotheses are framed.
3. A testing plan is designed.
4. A testing plan is implemented. A body of evidence is collected as a result.
5. An hypothesis *h* is confronted with a body of evidence. A verdict is passed on *h* in the light of the evidence *e*:
   "*h* is (strongly, moderately, weakly) supported by *e*" or
   "*h* is (strongly, moderately, weakly) undermined by *e*".
6. A decision is made either
   (a) to reject *h* altogether or
   (b) to amend *h* by replacing *h* by another hypothesis *h′* such that all evidence which is favorable for *h* is also favorable for *h′* and no evidence which is unfavorable for *h* undermines *h′*, or
   (c) to gather more evidence, or
   (d) to accept *h*.
7. An hypothesis is promoted, if it is accepted. The tester puts forth arguments on its behalf before a court of 'umpires' = recognized authorities in the field.
8. If the 'umpires' are convinced by the arguments, then they confirm the hypothesis. Otherwise the tester has to iterate earlier operations.

Queries concerning piecemeal knowledge-formation.

Under what conditions does an hypothesis deserve to be confirmed?
How should a testing plan be designed?
What is involved in confrontations: hypotheses versus evidence?
What is the rationale of various decisions adopted after a verdict?
In particular, how should a tester act when an hypothesis faces negative evidence?
What long run trends are to be expected if piecemeal knowledge-formation is not combined with synthesizing?

Instrument.

In order to deal with these questions a formalism *KF* (for knowledge-formation) is used.

Outline of KF.

KF contains

1. A theory of probability on propositions (used as an auxiliary to other parts of *KF*)
   (Functor: *P*)
2. A theory of content
   (Functor: *C*)
3. A theory of matching
   (Functor *M*)

and

4. A theory of partial truth
   (Functor *T*)

The probability- and content theory are linked to each other by means of these postulates:

$$C(p) = -\log P(p)$$

$$C(p/g) = -\log P(p/q)$$

The functor *M* in the matching theory is linked to the functor *C* in the content theory by this postulate:

$$M(p/q) = \frac{C(p) - C(p/q)}{C(p)}$$

The theory of graded truth is based on a notion of truth-core and the notion of matching. A *truth-core* of a proposition p is a proposition p^* such that

(1) p entails p^*
(2) p^* is true, and
(3) p^* entails every true proposition which is entailed by p.

The functor T for degree of truth is linked to the functor M for matching and the notion of truth-core by this postulate:

$$T(p) = M(p/p^*)$$

(Only non-empty propositions have a T-value!)
In other words: The degree of truth of a proposition is the degree to which it is covered by its truth-core.

The most important theorems in *KF* from the point of view of applications are these:

(1) *A distribution theorem for matching.*

$$M(p/q \wedge r) = M(p/q) + \frac{C(r)}{C(p)}[1 - M(r/q)] - \frac{C(r/p \wedge q)}{C(p)} \qquad \text{(D)}$$

(2) *A theorem about the relation of degree of truth to two-valued truth.*

(a) $T(p) = 1$ *if and only if* p is true.
(b) *If* $T(p) < 1$, *then* p *is false.*

(3) *A theorem about the relation between degree of truth and matching.*

$$T(p) \geqslant M(p/q^*).$$

The degree of truth of p is at least as large as the degree to which p is covered by the truth-core of another proposition q (no matter which one).

Answers to queries.

Using *KF* we arrive at these answers to the queries raised above concerning piecemeal knowledge-formation.

Criterion on knowledgehood.

An hypothesis h descerves to be confirmed to the extent that $T(h)$ is large. There are good grounds to hold that $T(h)$ is large if a body of evidence e has been assembled such that both $T(e)$ and $M(h/e)$ are large.

Relation between evidence and hypotheses.

The relation between a body of evidence e and an hypothesis h on which a verdict is passed is conceived to be as follows:

e is favorable for h if the content of e overlaps with that of h.
e is unfavorable for h if e is favorable for the negation of h.

$M(h/e)$ has such properties that it may be regarded as a *degree of positive evidence* when it is positive and as a *degree of unfavorable evidence* when it is negative. When $M\ (h/e) > 0$ then $M(h/e)$ measures the degree to which (the content of) e covers (the content of) h.

Testing plans.

Two kinds of testing plans are employed:

(1) *Elimination plans* are designed when a researcher believes that an hypothesis h has a low T-value.

(2) *Confirmative plans* are designed when a researcher believes that an hypothesis h has a high T-value.

The distribution theorem (D) is important in a study of confirmation plans and their implementations.

A body e of evidence ought to satisfy these requirements according to (D):

1. e is composed of a body of already confirmed hypotheses k and information of data type d. Thus $e = k \wedge d$.

2. k covers h to a large extent.

3. d adds as little as possible to the contents of h and k. *Ideal situation*: d is entailed by $h \wedge k$.

4. d has a large content, so that $\dfrac{C(d)}{C(h)}$ is large.

5. d is independent as much as possible of k.

If e satisfies these conditions, then $M(h/e)$ is large according to (D): e satisfies a requirement of relevance. e should also satisfy a requirement of truth: $T(e)$ ought to be large. If e satisfies these two requirements, then $T(h)$ is large according to the theorem $T(h) \geqslant M(h/e^*)$ and the findings that $M(h/e)$ is large, and that $T(e)$ is large.

Negative evidence.

(a) h has first been confronted with a body of favorable evidence e with a large degree of truth.

(b) h has later on been confronted with negative evidence e' which also has a high degree of relevance and truth.

How should a researcher assess such a situation?

$T(h)$ is large because of (a). h is false because of (b).

If a researcher holds the opinion that an hypothesis deserves to be confirmed, only if it is known to be perfectly true, then he should reject h. If, however, he adopts the more reasonable criterion of knowledgehood, according to which an hypothesis h deserves to be confirmed *to the extent* that it has a high degree of truth and there are good grounds to hold that this is the case, then he should not reject h. He should instead try to amend h by modifying h into an hypothesis h' which is supported both by e and by e'. Such an operation may hopefully have the effect that $T(h') > T(h)$. If it has, then it is a refinement. Negative evidence in the context of confirmation plans provides an opportunity for refining hypotheses before they are confirmed.

Long run trends in piecemeal knowledge-formation which is not combined with synthesizing.

Hypotheses which have been confirmed may be used as evidence in the process of testing later hypotheses. An old confirmed hypothesis k will not be challenged when it is confronted with a new hypothesis h in the process of testing h.

Result: Once confirmed hypotheses will remain confirmed.

Knowledge will grow in bulk but the mapping qualities of confirmed hypotheses (= their degree of truth) show no tendency of being raised.

Piecemeal knowledge-formation in isolation from synthesizing knowledge-formation lacks a machinery for revising once confirmed hypotheses.

II. SYNTHESIZING KNOWLEDGE-FORMATION

Schematic description.

General and special theories are formed in cooperation with piecemeal knowledge-formation.

A general theory is composed of a formalism F and a world-picture

W: $G = (F, W)$. A development of a general theory is a sequence $\Rightarrow (F, W) \Rightarrow (F', W') \Rightarrow (F'', W'')$ ——— of such pairs.
W maps features of the world which are found in a number of domains of the world X_1, X_2 ——— X_n ——— .

Special theories in the making map these territories: S_1 maps X_1, S_2 maps X_2 etc. A family of special theories S_1, S_2 ——— are served by a general theory G which maps common features of X_1, X_2 ——— etc. by means of its world-picture part W as follows:

(a) G supplies premises to arguments which are construed in the process of building special theories.

(b) The formalism in G is employed as a soft-ware instrument in the evolution of special theories.

Special theories serve a general theory as follows: World-picture assumptions in a general theory G are linked to the level of observation by way of arguments which are formed in the evolution of special theories served by G. There are two kinds of premises in arguments within a special theory S_i that is served by a general theory G.

(1) There are premisses which come from G (G-propositions).

(2) There are premisses which refer to specific features of X_i (S-propositions).

The conclusions are S-propositions. If the conclusion of an argument in a special theory S_i describes a feature of X_i which is also described by an already confirmed hypothesis, then the argument is a tentative solution of an explanatory problem. It serves also to assimilate a piece of knowledge into a theory. Arguments which have these functions may be called *assimilative arguments.* If the conclusion of an argument in S_i describes a feature of X_i which has not earlier been described by a confirmed hypothesis, then we have an *anticipatory argument.* Such argument are successors of assimilative arguments.

We may sketch a typical sequence in the evolution of a special theory as follows:

(1) A problem of explanation is raised when an hypothesis p has been confirmed.

(2) An assimilative (or explanatory) argument is construed, the conclusion p' of which describes the same feature of the world as p does.

(3) Untested hypotheses among the premisses in the assimilative argument need to be tested. Therefore problems of testing arise.
(4) Anticipatory arguments are framed as a step towards a solution of problems of testing.
(5) Conclusions of anticipatory arguments serve as virtual evidence which becomes genuine if
(6) They are confirmed.

Queries concerning synthesizing.

It is noted that a conclusion p' of an assimilative argument may be and often is incompatible with an hypothesis p to be assimilated:

$$M(p'/p) < 0.$$

How should a researcher assess such a situation? How should he act? It happens frequently that builders of special theories must revise hypotheses which they frame in their attempts to assimilate empirical findings. How should such revisions be carried out? What losses and gains are involved in opting for various possible moves?

How are world-picture assumptions affected by the evolution of special theories and by revisions which are enforced upon builders of special theories when their anticipations fail to come out true?

Tentative answers.

Let p be a confirmed hypothesis which describes a feature f in a domain X_i. Let p' be the conclusion of an assimilative argument, constructed in an attempts to explain f.

p' differs from p both in form and content. The couple of propositions (p, p') raises this question: Which one of them has the highest degree of truth? In order to answer this question a researcher must design a *discriminative* test. His task is to gather a body of evidence e' which is positive for one of them and negative for the other one. If it turns out that e' favors p', then p will be disconfirmed and p' will be confirmed in its place. Here we have a revision after a completed confirmation. One confirmed hypothesis is replaced by another one with a higher degree of truth.

A discriminative test may have another outcome than a confirmation of one of the two propositions p and p'. It may occur that a third hypothesis p'' will be confirmed which is judged to have a higher degree of truth than both p and p'.

If that should happen the set of premisses from which p' was deduced will have to be revised.

Here a tactical problem arises: Which one(s) of the premises ought to be revised?

Revision of a G-proposition will have the effect that a number of arguments in which it has been employed need to be revised also.

If many arguments in several special theories are affected, then the revision program will be so extensive that it may constitute a serious obstacle for the formation of new knowledge by way of anticipatory arguments. It is therefore to be expected that researchers will be less inclined to revise G-propositions at a late stage in the formation of a family of special theories served by a general theory G.

It is therefore to be expected that world-pictures belonging to general theories tend to become immunized against revision in 'normal science'. Such a trend can be justified as follows: World-picture assumptions are covered by confirmed conclusions to an increasing degree the more often they are employed as premisses in assimilative and anticipatory arguments. It is therefore desirable that they are not replaced by others too light-heartedly.

SUMMARY

Piecemeal knowledge-formation lacks a machinery for revising confirmed hypotheses. Synthesizing knowledge-formation is equipped with such a machinery. In fact frequent revisions on different levels is a characteristic feature of synthesizing research. These revisions are accompanied by discriminative testing in which confirmed hypotheses compete with hypotheses formed in attempts to assimilate knowledge into theories.

Synthesizing knowledge-formation has these main characteristics: Bodies of hypotheses are organized into special theories. The components of a special theory are subject to perpetual revisions leading to refinements of the mapping quality of the theory (its degree of truth).

General theories are bult up which serve and are served by special theories.

This paper is an abstract from the following recent works by its author:

'An essay on knowledge-formation' (Zeitschrift)ur allgemeine Wissenschaftsheorie, **viii**, 1975.

'On piecemeal knowledge-formation', in R. J. Bogdan (Ed.), *Local induction*, Copyright by D. Reidel Publishing Company, Dordrecht-Holland.

EDWARD C. ULIASSI AND ROGER J. CHACON

THE METHODOLOGY OF BEHAVIORAL THEORY CONSTRUCTION: NOMOLOGICAL-DEDUCTIVE AND AXIOMATIC ASPECTS OF FORMALIZED THEORY

The pervasiveness of mathematical formulations of empirical behavioral theory notwithstanding, consideration by behavioral scientists of the methodological foundations of nomological-deductive and axiomatic theory in the context of actual prototypes has been notably absent (cf. McPhee). Even when characterized as 'deductive' or 'axiomatic', theory in these sciences has rarely been more than partially and implicitly formal; and even when cast in formal language, such formalization has not been rigorously carried out. Furthermore, such theoretical formulations have rarely been coextensive in generality with the interactional boundaries of the behavioral systems to which they are meant to apply, heuristically adducing instead, as descriptive-explanatory generalizations about human behavior, postulates of limited scope. Such postulates, neither general nor elementary, consequently provide little basis for nomological theory capable of systematizing either the transmission or the discovery of knowledge in the behavioral sciences generally or in the domains of application of the specialized theories.

This is particularly the case with reference to studies in subject areas involving empirical dissimilation, pervasive in these sciences: limited accessibility to evidence not amenable to public communication, the absence of historical or ethnographic data because of institutional or natural constraints, secrecy (as with reference to political behavior), institutionalized dissimilation (as in historical reconstructions in the absence of incriminating or other recorded evidence), cultural restrictions of time and space (as in the study of alien cultures), institutional restrictions interfering with direct access to data (as in the ethnographic study of covert behavior resistant to direct study by those outside a particular culture), or in the study of phenomenologically private experience; or for other reasons likely to reveal systematic bias in evidence and inductive generalizations, all of which make behavioral phenomena recalcitrant to direct empirical investigation. Although such recalcitrance is endemic to the behavioral sciences, in which formal inference is abjured on 'positive' grounds by most empiricists, it is nevertheless similar to the inaccessibility revealed by the data of other sciences of nature (e.g., physics), in which such formal methods typically provide inference strategies in the construction of positive theory.

The resulting absence of behavioral theory articulating natural systems with analytic systems in the manner of the advanced natural sciences has generated such recent but atypical methodological criticism as Homans' *The Nature of Social Science*, programmatically advocating theoretical reconstruction through what are putatively described as 'deductive' and 'axiomatic' formulations nomologically founded upon such general and elementary 'laws' as those of behavioral psychology, or represented by behavioral restatements of utility theory, etc.; laws generally familiar or truistic (such as operant conditioning) because of their widely experienced character as empirical generalizations. Such methodological proposals have in turn been criticized: by some, for their epistemological assumption that nomothetic schemata can be superimposed upon unique historical or other contextual configurations for which idiographic descriptions or other forms of humanistic inquiry are regarded as more appropriate (v. Hula); and, by others, for using the terms and rhetoric of deductive and axiomatic formalization as their spurious justification, without adequate reference to their epistemological and logical requirements in the empirical sciences (v. Diesing).

Invariably, however, the context for such criticism is a small body of published 'deductive' and 'axiomatic' theory (cf. McPhee; Rudner; Brodbeck) whose primitive or embryonic character largely justifies such criticism on methodological grounds, while providing little of the expository context for a methodological assessment of the utility of formal methods *sui generis* in the behavioral sciences. The resultingly inadequate 'ontological reach' of such behavioral theories is nevertheless—and paradoxically—ascribed frequently to the alleged inappropriateness of such nomological and formal methods themselves, considered as epistemological strategies for the behavioral sciences (v. Krimerman; Brodbeck). The limited 'ontological reach' of such behavior theories as the basis for the systematization of empirical knowledge in these sciences, we affirm to the contrary, has different roots, specifically the *nonnomological* and *informal* character generally revealed by such theories. Behavioral theories fail to approximate formal theory in the advanced natural sciences particularly for the following reasons:

(1) the absence of 'external' reference to behavioral regularities, expressed as laws, and coextensive with the intended behavioral domains of application of such theories; i.e., their absence of unassailable heuristic generality, with reference to these domains, in contrast to laws with equivalent heuristic functions in such natural sciences as celestial mechanics; and

(2) the absence of attention, in such literary or inadequately-formalized behavior theories, to the consistency and completeness of the structure of 'internal' or syntactical relations implicit in such theories.

Some of the more empirically-oriented behavioral scientists have tended instead to study *ad hoc*, and generally on the basis of empirical generalizations founded upon informal taxonomies and inexplicit logical structures, overtly-evidenced relations limited in their spatio-temporal reference, and thus easily disconfirmed; and devoid of any systematization of formal relations, such as an explicit deduction or the use of a model provide. This is the case despite the fact that such implicit formal relations are contained—even if inadvertently suppressed—within the ordinary-language, mathematical, or other notational devices in which they are typically expressed. Empiricist doubts and positive disdain for such putatively-formal behavioral theory have been exaggerated not only by the limited empirical generality but also, and relatedly, by the *ad hoc* and unrealistic character of such heuristic postulates as have typically been advanced (as in Riker's game-theoretic formulation of 'positive' political theory, or within the limiting parameters of such heuristic models as *n*-person zero-sum games, such constructions as the rationality assumption of traditional utility theory, etc.). Thus, despite the admonition of such programmatic methodologists as Homans, recent theory in the behavioral sciences has continued to be inadequately general, and imperfectly attentive to the formal structures which provide their implicit syntax. Therefore, the putative 'laws,' 'deductive' formulations, and 'axiomatic' models of such theories are suspect, failing to reveal the explanatory-descriptive power of formalized theory founded upon unassailable laws of exceptional generality.

Our objective, then, is to construct, in contrast to extant formulations, an illustrative prototype of formal behavioral theory which is nomological (i.e., its postulates embody biogenetic and psychological covering laws at a very high level of generality), uses symbolic logic (*PCI*), and may lend itself to axiomatization. The presentation of the elements of a prototype behavioral theory reveals the covert epistemological, methodological, and logical assumptions often suppressed or masked by statements of theory in the vernacular. Such formalization escapes the theoretical limitations and the limited generality of typological constructions correlating through pseudo-laws arbitrarily-defined predicates of their subject populations without the rigor imposed by a formalization. It also permits significant re-presentations of reality which often appear, due to their unfamiliarity, not only unorthodox but also seemingly 'unrealistic' ways of discussing human behavior; yet which escape the 'misplaced concreteness' typical of informal idiography, historical or ethnographic typologies, empirical generalizations, and other *ad hoc* generalizations set forth in the absence of such formalization. Formal statement in unfamiliar terms also reduces the hypostatization of theoretical constructs (as when analytic systems are regarded as real), hypostatization which is common in the behavioraf sciences, particularly when theories (such as Easton's theory of the 'political system' or Parsons' theory of the 'social system') are stated in natural languages with reference to abstracted relations some o which are familiar in ordinary experience.

To illustrate these points, we take what appears to be a behavioral law of remarkable generality, resultantly familiar in content, and consider two different but related aspects of formalization, which we shall call the 'illative-directive' and the 'axiomatic' aspects, respectively.

(1) The 'illative-directive' aspect includes:

(a) the illative or validational phase, involving the use of logical methods (predicate calculus with identity) to test the validity of arguments; and

(b) the directive or exploratory phase, consisting of the

generation, through such inferences, of what are in effect 'hypotheses-schemata.' It may be the case that, as the following simple deductions illustrate, these hypotheses merely make explicit what already is implicitly or latently known, or more-or-less directly experienced. Nevertheless, we believe that the conclusions of these arguments are hypotheses, subject to empirical test, which in principle may eventually lead us to lesser known phenomena.

(2) The 'axiomatic' aspect of formalization involves the use of the covering law as the basis for an interpretation of an axiomatization of some calculus.

We proceed to some examples. We take first the illative-directive phase of formal theory. The heuristic function of this phase is well-known to natural scientists; we wish now to illustrate its use for the behavioral scientist.

Embracing at the outset utilitarian principles broadly conceived, we state the following well-known psychological generalization: *people act in their own self-interest*, i.e., a person's behavior continues, retains, or reinforces whatever he finds rewarding. In symbols we can express this fact as

$$(1) \qquad \forall x(Px \rightarrow \forall y(Iyx \rightarrow Rxy))$$

We also note that whatever increases the achievement of valued objects for a given person is in that person's self-interest. In translation, using obvious symbolization, we obtain

$$(2) \qquad \forall x(Px \rightarrow \forall y(Gyx \rightarrow Iyx))$$

Now, suppose that Px. Using predicate calculus on (1) and (2) we can get

$$(3) \qquad \forall x(Px \rightarrow \forall y(Gyx \rightarrow Rxy))$$

Verbally expressed, (3) says that a person's behavior continues, retains, or reinforces whatever increases his achievement of valued objects. This conclusion, which we shall call the 'principle of operant conditioning,' may serve as one of the basic premisses for many important conclusions in the behavioral sciences.

As an example, we shall derive the 'principle of reciprocity' in political behavior. In order to proceed, we need three additional premises. First, from (3) we may deduce that

$$(4) \qquad \forall x\forall y((Px \wedge Py) \rightarrow (Gyx \rightarrow Rxy))$$

In other words, given any individuals x and y, not necessarily different, if individual y increases the achievement of valued objects for individual x, then x's behavior continues, retains, or reinforces y's efforts. Second, we give an explicit definition of the term 'support.'

'Support' $=_{Df}$ 'an increase in the achievement of valued objects.' To paraphrase, if x supports y, then x increases y's achievement of valued objects.

(5) $\forall x \forall y((Px \wedge Py) \rightarrow (Sxy \rightarrow Gxy))$

Third, we also note that if x's behavior continues, retains, or reinforces y's efforts, then in effect x is supporting y. Using obvious abbreviations, we have

(6) $\forall x \forall y((Px \wedge Py) \rightarrow (Rxy \rightarrow Sxy))$

If we assume $Pw \wedge Pz$, it follows by *PCI*, from premises (4), (5), and (6) that

(7) $\forall x \forall y(Px \wedge Py) \rightarrow (Syx \rightarrow Sxy))$

This is the 'principle of reciprocal support.' It states that given any individuals x and y, if y supports x, then x supports y.

We turn now to the 'principle of reciprocal opposition.' First we need an explicit definition of 'opposition.'

'Opposition' $=_{Df}$ 'a decrease in the achievement of valued objects.' In other words, if x *opposes* y, then x decreases y's achievement of valued objects. In symbols this becomes

(8) $\forall x \forall y((Px \wedge Py) \rightarrow (Oxy \rightarrow Lxy))$

Moreover, proposition (4) can be recast, *mutatis mutandis*, as

(9) $\forall x \forall y((Px \wedge Py) \rightarrow (Lxy \rightarrow Eyx))$

In English (9) says that a person's behavior leads to the cessation, abandonment, or extinction of whatever decreases his achievement of valued objects. Note that when x's behavior decreases y's achievement of valued objects, x is in effect opposing y. This may be formalized as

(10) $\forall x \forall y((Px \wedge Py) \rightarrow (Lxy \rightarrow Oxy))$

Now, from premises (8), (9), and (10), by a proof similar to the one used to derive the 'principle of reciprocal support,' we deduce the 'principle of reciprocal opposition,' which affirms that, for given individuals x and y, if y opposes x, then x opposes y. Symbolized this becomes

$$(11) \qquad \forall x \forall y((Px \wedge Py) \rightarrow (Oyx \rightarrow Oxy))$$

Finally, propositions (7) and (11), by conjunction, express the 'principle of reciprocity.'

$$(12) \qquad \forall x \forall y((Px \wedge Py) \rightarrow (Syx \rightarrow Sxy)) \wedge \forall x \forall y((Px \wedge Py) \rightarrow \rightarrow (Oyx \rightarrow Oxy))$$

Such a theoretical strategy can by extension describe complex behavioral systems. Moving from a level of analysis describing laws governing the behavior of individuals (such as the utilitarian principle above) to derived laws (such as reciprocity) governing their transactional relations, propositions relating to the patterned interactivity or institutionalized behavior of specified individuals can be successively achieved, in order to create typologically successive approximations to the characteristics of actual individuals in homologous situations. In this way, statements about systematically-interrelated complexes of behavior can be deduced (e.g., those in a constitutional or 'political system' defined by the 'authoritative allocation of values,' as by Easton), revealing, from laws about the elementary behavior of individuals, deductions about patterned behavior of realistic ideal types.

Note that in such illative-directive uses of formalized theory, the initial behavioral principles—like such principles as *entropy*—are assumed, unexplained, and may appear descriptively 'uninformative', due to their universal acceptance, generality of reference, and extensive confirmation; yet, *ipso facto*, they are useful as the basis for further deductions.

The second aspect of formalization (i.e., the *axiomatic aspect*) calls for the axiomatization of the theory within some calculus. The interpretation of such an axiomatic system or calculus should be compatible with the covering law and the propositions of the theory should be satisfiable with respect to the model. Several axiomatizations within first-order function calculus or set theory are available; hence, our task in this more ambitious undertaking will be merely to indicate that such axiomatization is possible. The calculus proposed to express the theory is one of the sets of axioms for the propositional calculus, augmented by variables and quantification rules (e.g., Łukasiewicz, 1930; Rosser, 1953; etc.).

Łukasiewicz's system consists of three axioms:

Axiom 1. *CpCqp*

Axiom 2. *CCpCqrCCpqCpr*

Axiom 3. *CCNpNqCqp*

Disregarding quantifiers for the moment, our covering law, $\forall x(Px \rightarrow \rightarrow \forall y(Iyx \rightarrow Rxy))$, has the following form: $P \rightarrow (Q \rightarrow R)$, which fits the antecedent of Łukasiewicz's second axiom. Hence, it is possible that the covering law is compatible with a *wff* in this system.

As was noted above, the axioms would have to be reinforced by means of variables and additional postulates or rules regarding the use of quantifiers. Should this maneuver prove unfeasible (cf. Church on the independence of primitive functors), it is possible to use some standard extended propositional calculus (e.g., Russell, 1906; Tarski and Łukasiewicz, 1930). We are not developing this aspect of formalization here; at this initial stage, we are merely indicating the possibility of its being carried out. Nevertheless, we expect to carry it out at a later date.

The intended interpretation permits deductively-quided empirical investigation, including more rigorously controlled data searches when traditional empirical methods may reveal significantly unrepresentative data for the previously-indicated reasons, all of which make behavioral phenomena recalcitrant to direct empirical investigation (v. p. 1); it further serves to validate isolated empirical generalizations to assure that unrepresentative data do not interfere with our stronger theoretical conclusions, such as those derived by a network of consistent nomological statements.

Since the covering laws of a general theory must be principles of well-established generality, they are likely to be 'trans-disciplinary' and, in a constitutive sense, 'elementary.' These fundamental behavioral laws are affirmed by Homans to be the most general principles of behavioral psychology (such as operant conditioning, or related principles of political economy within the utilitarian tradition) and are regarded as necessary to establish a systematic 'social science' along nomological-deductive and axiomatic lines. Such theory is regarded as fundamentally deterministic even when permitting the 'illusion of choice' (Homans). Since it is a macro-behavioral and generic theory, it may be formulated compatibly with more specific theories of subsystems, including, for example, positive political theory (cf. Riker), exchange theory, decision theory, etc. This may be illustrated by construing, within the covering law, 'utilitymaximizing' choice and purposeful 'rational behavior', frequently postulated in exchange and utility theory, *not* as the initial postulates but as epiphenomenal manifestations of the deterministic principle of self-interest. It is evident that in this framework, an implicitly 'contractarian' and 'teleological' tradition, vestigially remaining and posed as 'positive' theory (cf. Riker), is replaced with a deterministic schema founded upon evolutionary-historical necessity and operant behavioral theory. In such theory, rational choice is largely excluded in favor of conditional necessity, with 'rational man' heuristically replaced with 'operant man'; and *homo economicus* with the more comprehensive, if Aristotelian-sounding, *homo politicus.*

The generality of reference of behavioral theory may be extended for exploratory purposes by the 'surplus meaning' contained in the terms of the *interpretation*, i.e., by the connotation or denotation of such terms as 'self-interest,' not exhausted by nor coterminous with any single index in all instances. Defining 'interest' as 'achievement of valued objects of activity' and regarding the 'political system' as an entity for the 'authoritative allocation of values' in any society (as by Easton), for example, the theory generates deductions about individuals, collectivities, and other typological approximations to 'real' individuals, collectivities, etc. Such theory may quasi-idiographically describe successively an entire conceptually-isolated political or other behavioral system, generating valid inferences of empirical relevance, i.e., serving as hypotheses-schemata for controlled or directed empirical study. The heuristic advantages of behavioral theory in its exploratory form, afforded by the semantic openness of some of its constructs, is compatible with its testability, through 'reduction definitions' or other indexing; thus, it has the 'virtues of its defects.' It is evident that such a strategy of theory construction reveals a nomothetic emphasis which is absent in the American behavioral sciences, despite extensive discussions in their methodological literature regarding the value of formal theory.

Without affirming the epistemological priority of formalized theory construction—and, indeed, in the context of a contrary thesis recognizing idiographic, phenomenological, biographic, humanistic and other modes of inquiry useful in identifying the parameters for such theory — emergent theoretical constructions capable of systematizing the largely inarticulated findings of these sciences may be expected. The use of such formal theory, paradoxically, permits nomothetic reinterpretation of apparent anomalies of political or other behavior often used to justify their 'historicist' study. The approach suggested in this paper also tends to free the behavioral sciences from ideological biases that are likely to becloud substantive discussions in these disciplines.

The preceding remarks propose a subtler 'logical' empiricism to complement the grosser 'positive' empiricism prevalent in the American behavioral sciences, and a theoretical strategy to complement the more direct empiricism and vacuous formalism endemic in these sciences today. Formalized theory should be distinguished from the 'formalizations' which tend to 'mathematicize' and thereby legitimate empiricist emphases without producing nomological theory which is both illative-directive and lends itself to axiomatization.

BIBLIOGRAPHY

Brodbeck, May (ed.). *Readings in the Philosophy of the Social Sciences*. New York: Macmillan, 1968.

Braithwaite, R. B. 'Models in the Empirical Sciences', in *Proceedings of the* 1960 *International Congress for Logic, Methodology, and Philosophy of Science*. Edited by E. Nagel, P. Suppes, and A. Tarski. Stanford: Stanford University Press, 1962.

Church, Alonzo. *Introduction to Mathematical Logic*. Vol. I. Princeton: Princeton University Press, 1956.

Diesing, Paul. *Patterns of Discovery in the Social Sciences*. Chicago: Aldine-Atherton, 1971.

Easton, David. *A Systems Analysis of Political Life*. New York: Wiley, 1965.

Eulau, Heinz. *Macro-Micro Political Analysis: Accents on Inquiry*. Chicago: Aldine, 1969.

Homans, George. *The Nature of Social Science*. New York: Harcourt, Brace, and World, 1967.

Hula, Erich: 'Comment' on Hans Jonas 'The Practical Uses of Theory'. In Maurice Natanson (ed.), *Philosophy of the Social Sciences: A Reader*. New York: Random House, 1963.

Krimerman, Leonard (ed.). *The Nature and Scope of Social Science: A Critical Anthology*. New York: Appleton-Century-Crofts, 1969.

McPhee, William N. (Institute for Research in the Behavioral Sciences, University of Colorado, Boulder, Colorado (U.S.A.)). *Formal Theory in the Behavioral Sciences*. (In preparation).

Prior, A. N. *Formal Logic*. 2d ed. Oxford: Clarendon Press, 1962.

Riker, William, and Ordeshook, Peter. *An Introduction to Positive Political Theory*. Englewood Cliffs: Prentice-Hall, 1973.

Rosser, J. Barkley. *Logic for Mathematicians*. New York: McGraw-Hill, 1953.

Rudner, Richard. *The Philosophy of Social Science*. Englewood Cliffs: Prentice-Hall, 1966.

Suppes, P. 'Models of Data', in *Proceedings of the* 1960 *International Congress for Logic, Methodology, and Philosophy of Science*. Edited by E. Nagel, P. Suppes, and A. Tarski. Stanford: Stanford University Press, 1962.

JAN M. ŻYTKOW

INTERTHEORY RELATIONS ON THE FORMAL AND SEMANTICAL LEVEL

I

I.1. In section I scientific theory is treated in a formal way i.e. as a certain subset of the set of well formed formulae, closed in respect to consequence operation.

Such a theory is presented in the well known standard manner which can be shortly formed as follows

$$\mathscr{L} = \langle V, WFF \rangle$$
$$T = \langle \mathscr{L}, Cn, Ax \rangle.$$

Language ($\mathscr{L}$) is described by its *vocabulary* (V) consisting of a set of primitive terms and by an inductive definition of a set of *well formed formulae* (WFF).[1] Some of primitive terms are interpreted in a standard way as terms of logic and mathematics, others—some predicates and names of functions[2]—have special meaning connected with properties of systems described by the theory.

Let Cn be a set of logical and mathematical laws and rules of inference. $Cn(X)$, where X is a subset of WFF, is a set of these and only these formulae which can be inferred from X by means of Cn.

Besides strict logical inference other type of inference is also used in science. E.g. if

(1) $\quad B_0(x) \gg B_1(x)$

than the formula

(2) $\quad A(x) = B_0(x)$

can be approximately inferred from the formula

(3) $\quad A(x) = B_0(x) + B_1(x).$

Perhaps the exact formal status of expresions of the type (1) is not clear but in fact they are used in science. So we include them to WFF.

By Cn_{app} let us name Cn enlarged by the set of rules of approximative inference.[3]

Now, by pointing out certain finite subset Ax of WFF we can present scientific theory in an axiomatic way as a set $Cn(Ax)$ or $Cn_{app}(Ax)$. It is standard understanding of the term "theory" in mathematical logic.

On the other hand set $Cn(Ax)$ contains basic laws of natural systems of a certain kind. But there are other laws and sentences e.g. particular laws describing special functions of forces or equations of fetters in mechanics. Each of these sentences is valid in not as wide class of systems as the set of $Cn(Ax)$.

In our formal picture we shall call this set $\{\alpha_i\}$, $i \in I$. Each of these sentences can use new primitive terms, not used in the set $Cn(Ax)$—e.g. some physical or other constants. So we have to enlarge V and WFF to make it possible to form these sentences. These enlarged sets we will indicate V^i, WFF^i, V^I, WFF^I, respectively.

One can add definitions of new terms to the theory. It demands an earlier enlarging of the vocabulary and the set of WFF. We assume that definitions considered in this paper are equivalences and therefore are translative and non-creative.

We will consider relations between two theories indicated by indices 1 and 2. Similarly all structures connected with these theories get the same indices. We do not, however, specify Cn_1 and Cn_2 since in each case they can be constructed in one standard way.

I.2. *Equivalence of theories*

a) Simple case—languages of both theories are the same

Definition 1. Two formalisms $T_1 = \langle \mathscr{L}_1, Ax_1 \rangle$ and $T_2 = \langle \mathscr{L}_1, Ax_2 \rangle$ are equivalent iff $Cn(Ax_1) = Cn(Ax_2)$.

b) General case. Vocabularies V_1 and V_2 are assumed to have at least the same mathematical and logical terms. For the remaining—specific predicates and function symbols—there may even happen to be no common ones.

Let $D_2(V_1 - V_2)$ be a set of sentences defining terms belonging to $V_1 - V_2$ in a language $\mathscr{L}_2$. Let us correspondingly call $D_1(V_2 - V_1)$ a set of definitions of terms from $V_2 - V_1$ in the language $\mathscr{L}_1$.

Definition 2. Two formalisms $T_1 = \langle \mathscr{L}_1, Ax_1 \rangle$ and $T_2 = \langle \mathscr{L}_2, Ax_2 \rangle$ are equivalent iff there exists such a language $\mathscr{L}$ with a vocabulary $V = V_1 \cup V_2$ and such sets of definitions $D_1(V_2 - V_1)$, $D_2(V_1 - V_2)$ that $Cn(Ax_1 \cup D_1(V_2 - V_1)) = Cn(Ax_2 \cup D_2(V_1 - V_2))$.
E.g. equivalent in this sense are 1° newtonian and lagrangean formalisms of classical mechanics, 2° mechanics and theory of electric circuits (this formal equivalence is employed in analogue computers).

Taking into account sentences of the type α_i one can formulate further conditions of equivalence.

Definition 3. If $\langle \mathscr{L}_1, Ax_1 \rangle$ is equivalent to $\langle \mathscr{L}_2, Ax_2 \rangle$, then the sentence $\alpha_{1,i} \in WFF_1$ is equivalent to $\alpha_{2,i} \in WFF_2$ iff $Cn(Ax_1 \cup D_1 (V_2^i - V_1^i) \cup \alpha_{1,i}) = Cn(Ax_2 \cup D_2(V_1^i - V_2^i) \cup \alpha_{2,i})$

I.3. *Reduction of theories*

a) Simple case—$V_1 \subset V_2$.

Definition 4. Formalism $T_1 = \langle \mathscr{L}_1, Ax_1 \rangle$ reduces to $T_2 = \langle \mathscr{L}_2, Ax_2 \rangle$ iff there exists a set $B \subset WFF_2$ including perhaps statements of the kind (1) that $Cn(Ax_1) \subset Cn_{app}(Ax_2 \cup B)$ $Cn(Ax_1) \cap Cn_{app}(B) = \emptyset$.

If we want to find in $\mathscr{L}_1$ the rules corresponding to the rules $\alpha_{2,i}$ we have, moreover, to specify a set of conditions

$$\alpha_{1,i} \in Cn_{app}(Ax_2 \cup B \cup \alpha_{2,i}) - Cn_{app}(Ax_2 \cup B)$$

b) General case—$V_1 - V_2 \neq \emptyset$.

Definition 5. Formalism $T_1 = \langle \mathscr{L}_1, Ax_1 \rangle$ reduces to $T_2 = \langle \mathscr{L}_2, Ax_2 \rangle$ iff there exists a set $B \subset WFF_2$ and a set of definitions $D_2(V_1 - V_2)$ that $Cn(Ax_1) \subset Cn_{app}(Ax_2 \cup B \cup D_2(V_1 - V_2)$ $Cn(Ax_1) \cap Cn_{app}(B) = \emptyset$.

In the case of the rules $\alpha_{2,i}$ one can easily found corresponding definitions.

It happens that $Cn(Ax_1) = Cn_{app}(Ax_2 \cup B) \cap WFF_1$ but more often $Cn_{app}(Ax_2 \cup B)$ contains laws in the language $\mathscr{L}_1$ not known before.

Intertheory relations, from the viewpoint of scientific practice, seem to be insufficiently presented on a formal level. It is necessary to add further relations—relations between sets of systems described by theories in question.

II

II.1. In section 2 we turn to the interpretation of the formalism. It was formerly generally accepted in methodology that scientific theory describes nature as a whole, as a one system (of course not all properties of it). It seems to be more convenient in methodology and widely accepted in science to assume that scientific theory describes many different natural systems consisting of a finite, generally not too large number of bodies or parts. Common to all these systems is the set of properties, functions and relations taken into account in description.

Empirical interpretation of the language consists in attributing to each specific term $f_1, \ldots, f_n$ and to each kind of variables $x_1, \ldots, x_m$ certain procedure or a set of procedures.

In the case of observational terms $f_1, \ldots, f_k$ ($k \leqslant n$) it consists in pointing out a set of instruments and measurements $\Phi_1, \ldots, \Phi_k$ sufficient to find out values of the functions or denotations of the predicates.

In the case of theoretical terms $f_{k+1}, \ldots, f_n$ the corresponding procedures $\Phi_{k+1}, \ldots, \Phi_n$ consist in suitable measurements performed by means of procedures $\Phi_1, \ldots, \Phi_k$ and then in employment of the basic laws of the theory.

Variables can be of different kinds in different theories. They can relate to time instants, space points and particles or parts forming natural systems. Suitable operational definitions $\mathscr{X}_1, \ldots, \mathscr{X}_m$ have to be introduced.

So we get an ordered set

$$\mathscr{X}_1, \ldots, \mathscr{X}_m, \Phi_1, \ldots, \Phi_k, \Phi_{k+1}, \ldots, \Phi_n$$

as an interpretation of the language. Applying these operational definitions to a given natural system s one gets a system

$$\mathfrak{M}_s^* = \langle S_1^*, \ldots, S_m^*, F_1^*, \ldots, F_n^* \rangle$$

very similar to Wójcicki's ([1], [2]) *operational system.* Finite sets S_i^* ($i = 1, \ldots, m$) are formed from elements distinguished by their operational definitions e.g. set of time instants is formed of distinguishable indications of a given clock during an experiment. Arguments of the functions and predicates are elements of corresponding cartesian products of S_i^* (or rather of their subsets for we do not require all possible observations that can be made). Values of the functions are intervals of real numbers (instead of real numbers). Predicates are vague: for a given predicate f_i there are elements belonging neither to the denotation of f_i nor to the denotation of $\neg f_i$. In this way we take into account so called experimental error. This error should be possibly small of course.

The system $\mathfrak{M}_s^*$ can not be recognized as an intended model of the language. In order to obtain an intended model and make it possible to introduce the notion of truth in its standard meaning and to check up the theory it is necessary:

1° to enlarge certain universes S_i^* to coherent subsets of real numbers in order to enable feasibility of differentiations,

2° to replace $F_1^*, \dots, F_n^*$ by proper functions (with real numbers as their values) and exact predicates. This should be done in the limits of experimental error,

3° to extend functions and predicates to enlarged universes.

It is necessary to impose several further conditions upon these extensions to avoid an excess of possibilities.

Performance of these steps leads to the set $\mathbf{M}_s$ of *intended models* $\mathfrak{M}_s = \langle S_1, \dots, S_m, F_1, \dots, F_n \rangle$ of the language.

In respect to $\mathfrak{M}_s$ the notion of truth can be defined in a standard way. True sentences from *WFF* form the set $Ver(\mathfrak{M}_s)$. Now, in respect to $\mathfrak{M}_s^*$ the notion of *approximate* or *relative truth* can be defined. Let $\alpha \in WFF$ and $X \subset WFF$.

Definition 6. $\alpha \in Rel(\mathfrak{M}_s^*) \equiv \exists \mathfrak{M}_s \in \mathbf{M}_s(\alpha \in Ver(\mathfrak{M}_s))$.

Definition 7. $X \in Rel(\mathfrak{M}_s^*) \equiv \exists \mathfrak{M}_s \in \mathbf{M}_s(X \subset Ver(\mathfrak{M}_s))$.

Let us notice that it is possible that each sentence belonging to a certain set Y is relatively true but set Y as a whole is not.

We would like to say that theory *describes* s if it is relatively true in $\mathfrak{M}_s^*$. But the larger is a set of sentences the more difficult it is to fulfil the conditions of relative truth and the more valuable is the verification.

Let us admit that the vocabulary contains a set of singular names of a sufficient number of types. Atomic sentences based on the observational terms f_i $(i = 1, \dots, k)$ form a set At.

Definition 8. Set $X \subset WFF$ is *empirically complete* iff

$$\forall \beta \in At(\beta \notin Cn(X) \rightarrow Cn(X \cup \{\beta\}) = WFF).$$

Definition 9. Natural system s is an *examined application of* T if there exists an empirically complete[4] set of sentences X such that $Cn(Ax) \subset X \subset \mathrm{Rel}(\mathfrak{M}_s^*)$. Set $\mathscr{E}_T$ consists of the natural systems being examined applications of T.

Given $\mathscr{X}_1, \ldots, \mathscr{X}_m, \Phi_1, \ldots, \Phi_n$ and $\mathscr{E}_T$, one can distinguish in nature a set $\mathscr{P}_T$ of *possible applications* of the formalism. In respect to $\mathscr{P}_T - \mathscr{E}_T$ the theory is a program which should but do not have to be realized. $\mathscr{T} = \langle T, \mathscr{P}_T, \mathscr{E}_T \rangle$ forms an *interpreted theory*.

The other interpretation of a theory consists in aiming at a system $\langle T, \mathscr{X}_1, \ldots, \mathscr{X}_m, \Phi_1, \ldots, \Phi_n \rangle$ (or shortly $\langle T, \mathscr{X}, \Phi \rangle$)

In the case of particular laws one can define $\mathscr{E}_{\alpha_i}$ and $\mathscr{P}_{\alpha_i}$ as subsets of $\mathscr{E}_T$ and $\mathscr{P}_T$. $\mathscr{E}_{\alpha_i}$ and $\mathscr{P}_{\alpha_i}$ consist of systems, in description of which laws α_i were or should be used. $A_i = \langle \alpha_i, \mathscr{P}_{\alpha_i}, \mathscr{E}_{\alpha_i} \rangle$ forms an *interpreted particular law*.

One can easily notice that certain relations hold, namely $\mathscr{P}_{\alpha_i} \subset \mathscr{P}_T$, $\mathscr{E}_{\alpha_i} \subset \mathscr{E}_T$, for each sentence or set of sentences X $\mathscr{E}_X \subset \mathscr{P}_X$.

Given two interpretations of the theory

$$\langle T, \mathscr{P}_T, \mathscr{E}_T \rangle \quad \text{and} \quad \langle T, \mathscr{X}, \Phi \rangle$$

one can say more about intertheory relations. Some definitions are proposed in the paper but they seem not to be final. They can be combined to obtain more satisfactory conditions. It seems, however, that one and the best solution does not exist. Such relations as equivalence and reduction are connected with many different intuitions and no definition can fulfil all of them.

II.2. *Equivalence of interpreted theories.*

Definition 10. Theories $\mathscr{T}_1 = \langle T_1, \mathscr{P}_{T_1}, \mathscr{E}_{T_1} \rangle$ and $\mathscr{T}_2 = \langle T_2, \mathscr{P}_{T_2}, \mathscr{E}_{T_2} \rangle$ are *equivalent formulations of the same theory* iff

1° formalisms T_1 and T_2 are equivalent

2° $\mathscr{P}_{T_1} = \mathscr{P}_{T_2}$ and $\mathscr{E}_{T_1} = \mathscr{E}_{T_2}$.

As on the formal level, we can formulate conditions of equivalence of particular laws.

Definition 11. Particular laws (or sets of laws) $A_1 = \langle \alpha_1, \mathscr{P}_{\alpha_1}, \mathscr{E}_{\alpha_1} \rangle$ and $A_2 = \langle \alpha_2, \mathscr{P}_{\alpha_2}, \mathscr{E}_{\alpha_2} \rangle$ of theories $\mathscr{T}_1$ and $\mathscr{T}_2$, respectively, are *equivalent* iff

1° theories $\mathscr{T}_1$ and $\mathscr{T}_2$ are equivalent, α_1 is equivalent to α_2 (def. 3).

2° $\mathscr{P}_{\alpha_1} = \mathscr{P}_{\alpha_2}$ and $\mathscr{E}_{\alpha_1} = \mathscr{E}_{\alpha_2}$.

It should be possible to establish an equivalence relation in a set of procedures (operational definitions) by means of which the terms are interdepreted (p. 455). Two such procedures, roughly, are equivalent if they lead to the same results when applied to the same objects. Formal characteristic of this relation will not be done in the paper. Let us however assume that this relation is properly introduced.

Definition 12. Theories $\mathcal{T}_1 = \langle T_1, \mathfrak{X}\emptyset_1\rangle$ and $\mathcal{T}_2 = \langle T_2, \mathfrak{X}\emptyset_2\rangle$ are *equivalent* iff

1° formalisms T_1 and T_2 are equivalent

2° corresponding terms have equivalent operational definitions.

If there is no possibility to establish equivalence relation between formalisms we can sometimes speak of an empirical equivalence.

Definition 13. Theories $\mathcal{T}_1 = \langle T_1, \mathfrak{X}\emptyset_1, \mathcal{E}_{T_1}\rangle$ and $\mathcal{T}_2 = \langle T_2, \mathfrak{X}\emptyset_2, \mathcal{E}_{T_2}\rangle$ are *empirically equivalent in respect to some system s* iff

1° $s \in \mathcal{E}_{T_1} \cap \mathcal{E}_{T_2}$

2° sets of arguments and observational terms have equivalent operational interpretations.

Definition 14. Theories $\mathcal{T}_1 = \langle T_1, \mathfrak{X}\emptyset_1, \mathcal{E}_{T_1}\rangle$ and $\mathcal{T}_2 = \langle T_2, \mathfrak{X}\emptyset_2, \mathcal{E}_{T_2}\rangle$ are *empirically equivalent* iff

1° for every $s \in \mathcal{E}_{T_1} \cap \mathcal{E}_{T_2}$ theories $\mathcal{T}_1$ and $\mathcal{T}_2$ are empirically equivalent

2° $\mathcal{E}_{T_1} = \mathcal{E}_{T_2}$

II.3. *Reduction of interpreted theories.*

Definition 15. Theory $\mathcal{T}_1 = \langle T_1, \mathcal{P}_{T_1}, \mathcal{E}_{T_1}\rangle$ *reduces* to theory $\mathcal{T}_2 = \langle T_2, \mathcal{P}_{T_2}, \mathcal{E}_{T_2}\rangle$ iff there exists a set $B \subset WFF_2$ such that

1° formalism T_1 reduces to $T_2 \cup \mathrm{B}$

2° $\mathcal{P}_{T_1} = \mathcal{P}_{T_2} \cap \mathcal{P}_B$, $\mathcal{E}_{T_1} = \mathcal{E}_{T_2} \cap \mathcal{E}_B$.

Definition 16. Particular law (or set of laws) $A_1 = \langle \alpha_1, \mathcal{P}_{\alpha_1}, \mathcal{E}_{\alpha_1}\rangle$ of $\mathcal{T}_1$ *reduces* to $A_2 = \langle \alpha_2, \mathcal{P}_{\alpha_2}, \mathcal{E}_{\alpha_2}\rangle$ of $\mathcal{T}_2$ iff there exists a set $B \subset WFF_2$ such that

1° $\mathcal{T}_1$ reduces to $\mathcal{T}_2$ by means of B

2° $\alpha_1 \in Cn_{app}(Ax_2 \cup B \cup \alpha_2) - Cn_{app}(Ax_2 \cup B)$

3° $\mathcal{P}_{\alpha_1} = \mathcal{P}_{\alpha_2} \cap \mathcal{P}_B$ and $\mathcal{E}_{\alpha_1} = \mathcal{E}_{\alpha_2} \cap \mathcal{E}_B$.

BIBLIOGRAPHY

[1] Wójcicki, R., 'Semantyczne pojęcie prawdy w metodologii nauk empirycznych' Semantic Notion of Truth in Methodology of Empirical Sciences), *Studia Filozoficzne*, **3** (1969), English translation will appear in *Dialectic and Humanism*.

[2] Wójcicki, R., 'Set theoretic representation of empirical phenomena', *Reports of the seminar on formal methodology of empirical sciences*, Wrocław, November 1973.

REFERENCES

[1] Beside a standard inductive definition there are some further conditions imposed upon the set *WFF*. They are connected with empirical meaning of formulae i.e. with their independence of choice of a measurement scale. These conditions are, however not considered in this paper.

[2] We shall assume that the functions take real numbers as their values.

[3] Since we present scientific theory only in a very sketchy way we will not discuss the problems connected with the concept of Cn_{app}—e.g. a problem of *avoiding contradiction* in the set $Cn_{app}(X)$.

[4] However it is not always necessary to verify a theory using empirically complete sets of sentences. We can do it by means of weaker sets.